ENCYCLOPEDIA OF PHYSICS

EDITED BY

S. FLÜGGE

VOLUME IV

PRINCIPLES OF ELECTRODYNAMICS AND RELATIVITY

WITH 15 FIGURES

SPRINGER-VERLAG
BERLIN · GÖTTINGEN · HEIDELBERG
1962

HANDBUCH DER PHYSIK

HERAUSGEGEBEN VON

S. FLÜGGE

BAND IV

PRINZIPIEN DER ELEKTRODYNAMIK UND RELATIVITÄTSTHEORIE

MIT 15 FIGUREN

SPRINGER-VERLAG
BERLIN · GÖTTINGEN · HEIDELBERG
1962

ISBN-13: 978-3-642-45975-7 e-ISBN-13: 978-3-642-45973-3
DOI: 10.1007/978-3-642-45973-3

Softcover reprint of the hardcover 1st edition 1962

Contents.

Classical Electrodynamics.

By

MELBA PHILLIPS.

With 14 Figures.

Introduction.

Early observations of electric and magnetic phenomena were purely qualitative. The affinity of lodestone for iron and that of freshly rubbed amber for small objects were noted at least as early as the Golden Age of Greece, but no systematic study was recorded. Much later investigation of magnetism was stimulated by practical use of the mariners' compass, a device introduced in the west probably during the eleventh century. The epistle of PETER PEREGRINE[1] (Peter the Pilgrim) on the magnet, written in 1269, revealed considerable clarity in the concepts of polarity and magnetic fields, and WILLIAM GILBERT[2] in *De Magnete* (1600) noted the impossibility of isolating magnetic poles. GILBERT also emphasized the ubiquity of *electrics* (materials on which a charge could be produced by friction) and gave the first known description of an electrical measuring instrument, a primitive electroscope, but electrical repulsion was overlooked until its discovery by NICOLO CABEO[3] (1629).

The fundamental concepts of electricity began to emerge clearly only in the 18th century with the distinction between conductors and insulators (STEPHEN GRAY[4] 1729) and the observation that there are two kinds of charge which attract each other but are self-repelling (CHARLES FRANCOIS DE CISTERNAY DUFAY[5] 1734). Charge was designated as positive or negative by BENJAMIN FRANKLIN[6] (1747), who thus implied the principle of conservation of charge. FRANKLIN also demonstrated the absence of electrostatic forces inside a charged conductor, from which JOSEPH PRIESTLEY[7] inferred the inverse square law of interaction by analogy to the similar result in gravitation (1767). The force relation was put to direct test by CHARLES-AUGUSTIN COULOMB[8] in 1785, and with his experiment the quantitative science of electricity began. [It should be noted, however, that the law of force between magnetic poles was first discovered by JOHN MICHELL[9] (1750), and that the inverse square law for electric charges was known to HENRY

[1] Epistola Petri Peregrini de Maricourt de magnete, 1269. Printed in Augsburg (Germany) 1558.

[2] GULIELMI GILBERTI de Magnete, magneticisque Corporibus, et de magno Magnete Tellure Physiologia nova. London 1600.

[3] N. CABEO: Philosophia magnetica 1629.

[4] ST. GRAY: Phil. Trans. Roy. Soc. Lond. **37**, 18, 227, 285, 397 (1731).

[5] C. DU FAY: Mém. Acad. Sci. 1733, 1734, 1737. — Phil. Trans. Roy. Soc. Lond. **38** (1734).

[6] B. FRANKLIN: New Experiments and Observations on Electricity.

[7] J. PRIESTLEY: The History and Present State of Electricity, with Original Experiments. London 1767.

[8] CH. A. COULOMB: Mém. Acad. Sci. 1785. (Later memoirs in 1786, 1787, 1788, 1789.)

[9] Memoirs of JOHN MICHELL, edit. by Sir A. GEIKIE. Cambridge, 1918.

CAVENDISH[1] before 1785, probably on theoretical grounds. CAVENDISH also anticipated FARADAY in the discovery of specific inductive capacity, but he did not publish these results.]

Controlled production of current electricity followed the observation by LUIGI GALVANI[2] that a frog's leg twitches when placed in contact with two dissimilar metals (1791). ALESSANDRO VOLTA[3] demonstrated that the effect did not depend on animal tissue, and invented the voltaic battery in 1799, thus making possible experiments with appreciable steady currents which culminated in HANS CHRISTIAN OERSTED'S[4] discovery (1819) of the magnetic effects of a current. Quantitative investigation followed rapidly the announcement of this discovery: JEAN-BAPTISTE BIOT and FÉLIX SAVART[5] showed (1820) that the strength of the effect is inversely proportional to the distance from a long straight wire carrying a current, and ANDRÉ-MARIE AMPÈRE[6] not only achieved (1820 to 1824) the generalization known as AMPÈRE's law, but also arrived at the hypothesis that all magnetism may be considered an electrical phenomenon. After the discovery of electromagnetic induction by MICHAEL FARADAY[7] (1831) and JOSEPH HENRY[8] (1830), continued efforts were made, notably by WILHELM EDUARD WEBER[9], to describe all electromagnetic phenomena in terms of forces between charges, some of which are velocity dependent. The concept of field, "stresses" in an all pervading medium which transmit electric and magnetic actions from one charge to another, is largely due to FARADAY. It was also FARADAY who discovered the effects of dielectrics on electromagnetic phenomena and the magnetic properties diamagnetism and paramagnetism, and who conceived the propagation of "ray vibrations". FARADAY's basic ideas were extended and developed in the mathematical treatment of JAMES CLERK MAXWELL[10], of which the electromagnetic theory of light was the most important consequence (1865). With the experimental verification of electric waves by HEINRICH RUDOLPH HERTZ[11] (1887) and his solution of the wave equation the classical theory of electromagnetic fields in continuous media became logically complete.

MAXWELL'S theory, while satisfactory for empty space, failed to account for such optical properties of dielectrics as dispersion. This failure was in large measure eliminated by the electron theory, of which H. A. LORENTZ[12] was the chief architect, even prior to the experimental discovery of the electron by J. J. THOMSON[13] in 1897. Electron theory remains a field theory, but the properties of matter and the origins of all fields are traced to elementary molecular charges. The chief difficulty with electromagnetic theory as it existed at the beginning of the present century lay in its dynamic consequences. The theory predicted that electric and magnetic effects are propagated only with a finite velocity, and on

[1] The Electrical Researches of Hon. Henry Cavendish, edit. by J. CLERK MAXWELL. 1879.

[2] ALOYSII GALVANI de Viribus Electricitatis in Motu Musculari Commentarius. Bononiae 1791.

[3] A. VOLTA: Phil. Trans. Roy. Soc. Lond. **83**, 10, 27 (1793). — Phil. Mag. **4**, 59, 163, 306 (1799).

[4] H. CH. OERSTED: Experimenta circa effectum conflictus electrici in acum magneticam. Copenhagen 1820; also J. Chem. Phys. **29**, 275 (1820) and Ann. Philosophy **16**, 273 (1820).

[5] J.-B. BIOT, and F. SAVART: Ann. Chim. Phys. **15**, 222 (1820). — J. Phys. **91** (1820).

[6] A.-M. AMPÈRE: Mém. Acad. Sci. **6**, 175 (1825).

[7] Experimental Researches in Electricity, MICHAEL FARADAY.

[8] J. HENRY: Amer. J. Sci. **22**, 408 (1832).

[9] W.E. WEBER: Leipziger Abh. **1846**, 209. — Pogg. Ann. Phys. u. Chem. **73**, 193 (1848).

[10] Phil. Trans. **155**, 459 (1865). Also Scientific Papers of J. CLERK MAXWELL.

[11] WIEDEMANN'S Ann. Phys. u. Chem. **31**, 421 (1887); **36**, 1 (1889). Also H. HERTZ: Untersuchungen über die Ausbreitung der elektrischen Kraft (1892) or Electric Waves (1894, 1900).

[12] A. LORENTZ: The Theory of Electrons (1909); also Collected Papers.

[13] J. J. THOMSON: Phil. Mag. (5) **44**, 298 (1897).

the basis of Galilean kinematics it should therefore be possible to differentiate between one inertial system and another as to absolute motion—something which is not possible within Newtonian mechanics itself. Efforts to detect such distinctions failed. Moreover, as EINSTEIN[1] pointed out, observable phenomena such as induced electromotive forces depend on relative velocities only, whereas the field descriptions in MAXWELL's theory involve apparently absolute velocities. It was the consideration of these phenomena that led in 1905 to the special theory of relativity, which preserved electromagnetic field formalism intact but amended mechanics so as to be compatible with it. Electron theory, in which fields are traced to their material sources, was both simplified and clarified by the theory of relativity, and we shall not here consider in detail those difficulties in electrodynamics which disappeared with the advent of special relativity. Moreover, we shall confine our attention in this article to the electrodynamics of media at rest, since an unambiguous treatment of electromagnetic effects in moving media is possible only by explicit application of relativity theory, and the subject is treated elsewhere in this volume.

Detailed successes of electron theory, especially in accounting for characteristic radiation, for the behavior of electrons in metals, and for magnetism, were possible only after the introduction of the quantum theory. Nevertheless classical electron theory provides the framework for the atomistic interpretation of electromagnetic properties of matter. The fundamental difficulties of understanding the electron itself remain; the existence of the electron (or any other charged or magnetic particle) as a stable entity is inexplicable on the basis of electrodynamics alone, and the problem has as yet been only by-passed in quantum theory. It was a major achievement of classical electron theory to exhibit the roots of these fundamental questions.

We should note that although it is possible to formulate electromagnetic theory consistently in terms of delayed action at a distance[2] the field formulation is used exclusively here.

The rationalized mks (meter-kilogram-second) system of units, comprising the practical units of engineering, is now in very common use, but the Gaussian system in which the velocity of light occurs explicitly retains its popularity in theoretical physics. Gaussian units are used here, but a table of conversion factors is included as an appendix[3].

I. The basic laws of electrodynamics: MAXWELL's equations.

1. COULOMB's law and the electrostatic field in vacuum. According to the empirical law of COULOMB two point charges, q_1 and q_2, act on each other with a force which is directly proportional to the product of the charges and inversely proportional to the distance between them. If $\boldsymbol{r}$ is the vector distance from q_1 to q_2, the force $\boldsymbol{F}_2$ on q_2 in vacuo is given by

$$\boldsymbol{F}_2 = K\frac{q_1 q_2}{r^3}\boldsymbol{r}, \tag{1.1}$$

where K depends on the units employed. That like charges repel and unlike charges attract each other is implicit in Eq. (1.1).

---

[1] A. EINSTEIN: Ann. d. Physik **17**, 891 (1905).

[2] See, e.g., J. A. WHEELER and R. P. FEYNMAN: Rev. Mod. Phys. **21**, 425 (1949). This theory is examined and extended in Dr. BERGMANN's accompanying article on special relativity.

[3] The question of units is discussed at some length in the general references [*15*] and [*19*]. The most extensive work of the development of electromagnetic theory is "A History of the Theories of Aether and Electricity" by Sir EDMUND WHITTAKER, Ref. [*23*]. See also L. ROSENFELD, Nuovo Cimento (10) **4**, Supplemento, p. 1630 (1957).

The electrostatic unit of charge, and thence the electrostatic system of units (esu) is based on cgs mechanical units with K set equal to unity; in the Gaussian system all charges and electric fields are measured in esu. The electric field or electric intensity $\boldsymbol{E}$ at any point in space is defined such that the force on a point test charge q is given by

$$\boldsymbol{F} = q\,\boldsymbol{E} \tag{1.2}$$

in the limit as q goes to zero. As an operational definition Eq. (1.2) can be applied only in the absence of a ponderable medium, and its validity is in practice further limited by the atomicity of charge. Nevertheless Eq. (1.2) is conceptually satisfactory for the definition of the macroscopic electric field intensity, although it must be further specified that the velocity of the test charge is vanishingly small to eliminate the possibility of magnetic forces.

It follows from the definition of electrostatic field intensity and from Eq. (1.1) that the field produced by a point charge q in empty space is, in electrostatic units,

$$\boldsymbol{E} = \frac{q\,\boldsymbol{r}}{r^3} = -\operatorname{grad}\frac{q}{r}. \tag{1.3}$$

The righthand expression is added as a mathematical identity: the field of a point charge may be written as the gradient of a scalar function. The electric field obeys the principle of superposition, and therefore the vacuum field due to a spatial distribution of charge of density ϱ is given by

$$\boldsymbol{E} = \frac{\varrho\,\boldsymbol{r}}{r^3}\,dV = -\int \varrho \operatorname{grad}\frac{1}{r}\,dV = -\operatorname{grad}\int\frac{\varrho\,dV}{r}. \tag{1.4}$$

The volume integral is to be taken over all sources, while of course the gradient operator refers to the field point alone.

Eq. (1.4) implies that the electrostatic field is irrotational, since it can be expressed as the gradient of a scalar[1]. Thus, for a static field produced by charges,

$$\operatorname{curl}\boldsymbol{E} = 0. \tag{1.5}$$

The electrostatic potential ϕ is defined (except for an arbitrary constant) by the equation

$$\boldsymbol{E} = -\operatorname{grad}\phi, \tag{1.6}$$

a definition which expresses the electrostatic vector field $\boldsymbol{E}$ in terms of a scalar field ϕ at every point in space.

COULOMB's law may be put into the very useful form known as GAUSS's flux theorem[2] by integrating Eq. (1.3) over a surface enclosing the charge,

$$\int \boldsymbol{E}\cdot d\boldsymbol{S} = \int\frac{q\,\boldsymbol{r}\cdot d\boldsymbol{S}}{r^3} = 4\pi q = 4\pi\int\varrho\,dV, \tag{1.7}$$

where substitution for the charge q of the integral of the charge density over the volume inclosed by the surface is again justified by the principle of superposition. But the surface integral on the left is equal to the integral of the divergence of $\boldsymbol{E}$ over this same volume, and therefore, in the absence of material media,

$$\operatorname{div}\boldsymbol{E} = 4\pi\varrho. \tag{1.8}$$

The consequences of this equation for the potential are traced in Chap. II.

---

[1] The utility for electrical problems of a scalar potential, first introduced in the theory of gravitation, was pointed out by SIMEON DENIS POISSON, Bull. Soc. Philomathique **3**, 388 (1813). See also Mém. Institut, Pt. **1**, 1, Pt. **2**, 163 (1811). The word *potential* was introduced by GEORGE GREEN in 1828.

[2] CARL F. GAUSS (1840), Werke, **5**, 197–242. Engl. trans. in TAYLOR, Scientif Memoirs **3**, 153–196 (1843).

The numerical factor 4π, which appears when the field of a point charge is integrated over a sphere that incloses the charge, will be carried into all field equations involving sources. In what are called rationalized units this factor is suppressed by inserting 4π in the denominator of the expression for the Coulomb force, Eq. (1.1). (It is understood that for consistent rationalization 4π would also be introduced in the expression for magnetic forces.) The electrostatic system of units is unrationalized; a rationalized variant of the Gaussian system, called the Heaviside-Lorentz system, has never achieved widespread use.

2. The displacement vector and the polarization field. The extension of electrostatics to include the presence of material media necessitates the introduction of partial fields in addition to the total field defined by Eq. (1.2). This necessity can be exemplified in a simple way by considering the subject historically. So far as electrostatic phenomena are concerned, conductors may be defined as substances which have no internal fields. Therefore the potential is constant throughout a conductor, and there are no internal charges. Any surface distribution of charge on a conductor corresponds to an external field at right angles to the conducting surface. The value of the external field in vacuo at the surface may be written in terms of a surface charge density τ by applying GAUSS's theorem to a small volume enclosing an element of the conductor surface to obtain

$$E_n = 4\pi\,\tau. \tag{2.1}$$

Between a pair of parallel plates, whose dimensions are large compared with their separation so that edge effects may be neglected, a uniform field may be set up corresponding to positive and negative surface charge density τ on the two plates. Then the line integral of $\boldsymbol{E}$ from one plate to the other is

$$\int_1^2 \boldsymbol{E}\cdot d\boldsymbol{s} = \phi_2 - \phi_1 = 4\pi\,\tau\,d, \tag{2.2}$$

where d is the plate separation. The quotient of the total charge on the positive plate by the difference of potential between the plates is called the capacity of the configuration of conductors, equal to $A/4\pi\,d$ in this example if A is the area of each plate. FARADAY found that this quotient is affected by the presence of insulating material between the plates. If the potential difference is maintained unchanged while the insulating material is introduced to fill the intervening space, the original field $\boldsymbol{E}$ must retain its previous value, since $\int \boldsymbol{E}\cdot d\boldsymbol{s} = \phi_2 - \phi_1$ follows from the definition of potential. The charge on the plates is changed, however, and the only free movable charges involved here are those on the plates. The field intensity is therefore not wholly determined by the free ("true") charges. It is convenient to introduce a second field vector which is related to the "true" movable charges in much the same way that $\boldsymbol{E}$ is related to the charges in vacuo. This vector field is called the *electric displacement* or the displacement vector, and is denoted by $\boldsymbol{D}$. It satisfies the condition everywhere that

$$\operatorname{div}\boldsymbol{D} = 4\pi\,\varrho, \tag{2.3}$$

where ϱ is interpreted as the true movable charge density. In empty space $\boldsymbol{D}=\boldsymbol{E}$ in electrostatic units, although $\boldsymbol{D}$ is not dimensionally equal to $\boldsymbol{E}$ in other systems. For isotropic insulators

$$\boldsymbol{D} = \varepsilon\,\boldsymbol{E}, \tag{2.4}$$

where ε is the *dielectric constant* or *specific inductive capacity* of the medium. The factor relating $\boldsymbol{D}$ to $\boldsymbol{E}$ in other systems of units is called the *permittivity* of the medium.

The difference between $\boldsymbol{D}$ and $\boldsymbol{E}$ is due to the *polarization* of the medium:

$$\boldsymbol{D} = \boldsymbol{E} + 4\pi \boldsymbol{P}. \tag{2.5}$$

The polarization $\boldsymbol{P}$ depends on the inaccessible sources associated with the medium, and can in fact be interpreted as a description of those sources. The relation between $\boldsymbol{P}$ and its sources will be given in Sect. 9, but we may note here that the tracing of $\boldsymbol{P}$ to its (inaccessible) sources is responsible for the coefficient 4π that appears in Eq. (2.5), since unrationalized units are employed. The additive relation (2.5) is entirely general, whereas Eq. (2.4) applies only to a limited class of media.

Eq. (2.3) is one of the set of fundamental relations in macroscopic electrodynamics known as MAXWELL's equations, that one whose experimental basis is COULOMB's law of electrostatic interaction.

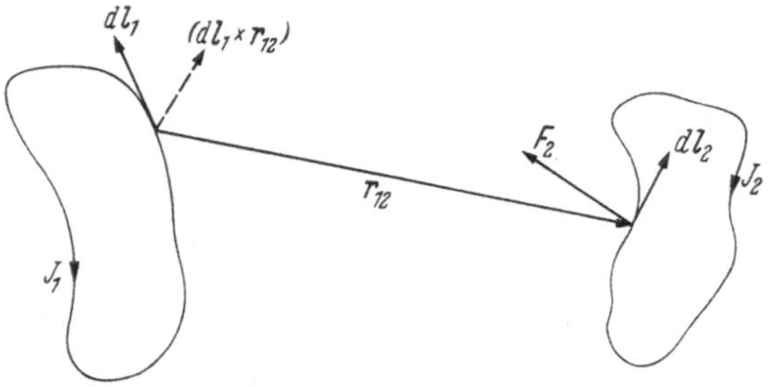

Fig. 1. Illustrating AMPÈRE's law.

3. AMPÈRE's law and the magnetic induction field in vacuo. The magnetic effect of a current discovered by OERSTED is equivalent to the interaction of currents, and the integral which yields the force on one current circuit produced by another as observed experimentally by AMPÈRE is now known as AMPÈRE's law. The existence of steady currents, such as may be produced by voltaic batteries, is assumed. In a linear circuit the same current flows through every cross section of the conductor carrying the current. With reference to Fig. 1, let $\boldsymbol{r}_{12}$ be the vector distance from line element $d\boldsymbol{l}_1$ of a circuit carrying current J_1 to $d\boldsymbol{l}_2$ of a circuit in which the current is J_2. Then the force on the second circuit produced by the first may be written

$$\boldsymbol{F}_2 = K J_1 J_2 \oint\oint \frac{d\boldsymbol{l}_2 \times (d\boldsymbol{l}_1 \times \boldsymbol{r}_{12})}{r_{12}^3}, \tag{3.1}$$

where K is a constant depending on the units employed. The electromagnetic system of units is based on cgs mechanical units with K equal to unity; the unit of current thus defined is called the *abampere*. The set of units based on this definition is called the electromagnetic system (emu); in Gaussian units magnetic quantities are measured in emu, although current itself is sometimes measured in esu because of its close relation to charge. In this article current will be expressed in abamperes.

Eq. (3.1) is written as action at a distance in empty space, but this form of AMPÈRE's law can be expressed immediately in terms of a field produced by the first circuit:

$$\boldsymbol{F}_2 = J_2 \oint d\boldsymbol{l}_2 \times \boldsymbol{B}, \tag{3.2}$$

where, with $\boldsymbol{r}$ the vector from the source point to a field point,

$$\boldsymbol{B} = J_1 \oint \frac{d\boldsymbol{l}_1 \times \boldsymbol{r}}{r^3}. \tag{3.3}$$

The quantity $\boldsymbol{B}$ is called the *magnetic induction field*; its defining equation, in terms of the force on a linear current element, is (3.2) above. Note that $\boldsymbol{B}$ so defined does not exert a force on an element of current that satisfies NEWTON's third law. The total force given by the integral (3.1) or (3.2) and the analogous force produced on the entire circuit 1 by the current J_2 in circuit 2 do satisfy the action-reaction principle, however.

Both the defining equation for $\boldsymbol{B}$ and the vacuum expression for $\boldsymbol{B}$ in terms of the current producing it can be readily extended from linear to volume currents. If $\boldsymbol{j}$ is the current density (in abamperes/cm²), Eq. (3.2) may be replaced by the volume integral

$$\boldsymbol{F} = \int (\boldsymbol{j} \times \boldsymbol{B})\, dV. \tag{3.4}$$

Similarly, the magnetic induction field produced by a current distribution of density $\boldsymbol{j}$ in vacuum is given by the generalization of Eq. (3.3):

$$\boldsymbol{B} = \int \frac{\boldsymbol{j} \times \boldsymbol{r}}{r^3}\, dV. \tag{3.5}$$

The form of Eq. (3.5) leads to differential relations of great importance. Formally,

$$\boldsymbol{B} = -\int \boldsymbol{j} \times \operatorname{grad}\left(\frac{1}{r}\right) dV = \operatorname{curl} \int \frac{\boldsymbol{j}}{r}\, dV, \tag{3.6}$$

since the vector differential operator acts at the field point and the integration is taken over the source distribution. It follows that $\boldsymbol{B}$ is a solenoidal vector, since the divergence of the curl of any vector vanishes:

$$\operatorname{div} \boldsymbol{B} = 0. \tag{3.7}$$

The vector from which $\boldsymbol{B}$ may be derived by taking the curl is called the vector potential[1], and may be identified in Eq. (3.6):

$$\boldsymbol{B} = \operatorname{curl} \boldsymbol{A} \tag{3.8}$$

where

$$\boldsymbol{A} = \int \frac{\boldsymbol{j}}{r}\, dV. \tag{3.9}$$

If the vector potential is defined quite generally as the vector whose curl is the magnetic induction $\boldsymbol{B}$, $\boldsymbol{A}$ may differ from Eq. (3.9) by some function with vanishing curl. The divergence of $\boldsymbol{A}$ is not here defined, but the condition for steady current is that $\operatorname{div} \boldsymbol{j} = 0$; if the currents are not only steady but also finite in space, so that the surface integral of $\boldsymbol{j} \cdot d\boldsymbol{S}/r$ can be made to vanish, it follows formally that the divergence of Eq. (3.9) is zero. The divergence of $\boldsymbol{A}$ is not assumed to vanish for nonstationary currents.

The curl of $\boldsymbol{B}$ may be evaluated in terms of its current sources. By a vector identity the curl of Eq. (3.6) becomes

$$\left.\begin{aligned} \operatorname{curl} \boldsymbol{B} &= \operatorname{curl}\operatorname{curl} \int \frac{\boldsymbol{j}}{r}\, dV \\ &= -\int \boldsymbol{j}\, \nabla^2\left(\frac{1}{r}\right) dV + \operatorname{grad}\operatorname{div} \int \frac{\boldsymbol{j}}{r}\, dV. \end{aligned}\right\} \tag{3.10}$$

The last term vanishes under the assumption that $\operatorname{div} \boldsymbol{A} = 0$. The Laplacian of $1/r$ vanishes everywhere except at the origin, and may be written

$$\nabla^2\left(\frac{1}{r}\right) = -4\pi\,\delta(\boldsymbol{r}), \tag{3.11}$$

[1] This analysis of AMPÈRE's law, with the introduction of the vector potential, is traced to FRANZ NEUMANN: Berl. Abh. 1845 and 1848, reprinted as Nos. 10 and 36 of OSTWALD's Klassiker.

where $\delta(\boldsymbol{r})$ is the Dirac δ-function, defined by its integral properties

$$\int \delta(\boldsymbol{r})\, dV = 1, \qquad \int f(r)\, \delta(\boldsymbol{r})\, dV = f(0), \tag{3.12}$$

if $f(r)$ is continuous at $r=0$. Therefore

$$\operatorname{curl} \boldsymbol{B} = 4\pi \boldsymbol{j} \tag{3.13}$$

in vacuo. Eq. (3.13) is the differential equivalent of what is called AMPÈRE'S circuital law,

$$\oint \boldsymbol{B} \cdot d\boldsymbol{s} = 4\pi J, \tag{3.14}$$

where J is the total current encircled by the line integral. This is seen by applying STOKES' theorem to the vector $\boldsymbol{B}$.

4. Magnetic field intensity and magnetization. Historically the subject of magnetism began with the study of ferromagnetic materials, and FARADAY'S investigation of para- and diamagnetism made it necessary to take into account the magnetic properties of all material media. In modern electromagnetic theory all magnetic effects are traced to current distributions, including the inaccessible currents originally postulated by AMPÈRE. On this basis fields are divided into those directly related to "true" currents and those which depend entirely on the inaccessible currents in a medium. The magnetic induction $\boldsymbol{B}$ is retained as the field produced by the sum-total of *all* currents in accord with Eq. (3.5), and thus remains solenoidal—at least for steady currents. The field whose curl is given by the true current density only is called (for historical reasons) the magnetic field intensity, and is denoted by $\boldsymbol{H}$:

$$\operatorname{curl} \boldsymbol{H} = 4\pi \boldsymbol{j}_{\text{true}}. \tag{4.1}$$

In empty space $\boldsymbol{B}$ and $\boldsymbol{H}$ are identical in electromagnetic units, and differ only by a constant factor in other systems of units. In the absence of true currents $\boldsymbol{H}$ is irrotational. Unlike $\boldsymbol{B}$, it need not be solenoidal, since $\boldsymbol{H}$ may differ from the induction field by a vector with non-vanishing divergence. The magnetic field intensity is therefore not wholly determined by Eq. (4.1) in the presence of magnetic media.

The difference between $\boldsymbol{B}$ and $\boldsymbol{H}$ in a medium is (in unrationalized units) $4\pi\boldsymbol{M}$, where $\boldsymbol{M}$ is called the *magnetization*[1] of the medium:

$$\boldsymbol{B} = \boldsymbol{H} + 4\pi \boldsymbol{M}. \tag{4.2}$$

Like the polarization in electrostatics, $\boldsymbol{M}$ plays the dual role of partial field and description of the inaccessible sources, in this case the magnetization currents. These relations are elaborated in Chap. III.

5. The displacement current. It is an axiom of electrodynamics in accord with experience that charge is conserved. This axiom is stated as an equation of continuity which relates the divergence of current density to the time rate of change of charge density. In this article charge is written in esu, as defined from COULOMB'S law, and current is in emu as defined from AMPÈRE'S law. From the emu of current a unit quantity of charge can be obtained. It is clear from an examination of the formulas that the relation between these two units of charge has the dimensions of a velocity, and WILHELM WEBER and RUDOLPH KOHLRAUSCH showed that the emu of charge is 3×10^{10} cm/sec times the electrostatic charge unit[2]. The equality of this conversion factor and the velocity of light was an

[1] The magnetization was introduced by POISSON, along with the magnetic scalar potential: Mém. Acad. Sci. 5, 247 (1824).

[2] R. KOHLRAUSCH and W. WEBER: Pogg. Ann. Phys. u. Chem. 99, 10 (1856). See also OSTWALD'S Klassiker, No. 142.

important influence on the work of MAXWELL, and it became customary to designate the velocity constant as c. In Gaussian units the continuity equation contains this factor explicitly:

$$\operatorname{div} \boldsymbol{j} + \frac{1}{c} \frac{\partial \varrho}{\partial t} = 0. \tag{5.1}$$

The continuity of charge brings into sharp focus a difficulty that arises in computing magnetic fields due to currents that vary with time. According to the mathematical theory of vector fields $\boldsymbol{B}$ is determined uniquely by Eqs. (3.7) and (3.13)—a vector field is determined if its divergence and curl are given at all points of space, provided that the integral of the source and circulation densities over all space is finite. But if the total current density is to satisfy Eq. (3.13) it must be solenoidal, since div curl $\boldsymbol{B} \equiv 0$. This conclusion is incompatible with Eq. (5.1) for nonstationary currents. The paradox was resolved by MAXWELL by maintaining zero divergence for the induction field and redefining the current density so as to make it solenoidal under all circumstances. This is accomplished as follows: by Eq. (2.3) the accessible charge density is a measure of the divergence of the electric displacement $\boldsymbol{D}$, and therefore the continuity equation yields

$$\operatorname{div}\left(\boldsymbol{j} + \frac{1}{4\pi c} \frac{\partial \boldsymbol{D}}{\partial t}\right) = 0, \tag{5.2}$$

where $\boldsymbol{j}$ is interpreted as the accessible current density corresponding to the accessible charge density of Eq. (2.3). The term in $\partial \boldsymbol{D}/\partial t$, called the displacement current, is therefore added to the true current density, and thus AMPÈRE's law is amended to make it compatible with the continuity equation for nonsteady currents. Eq. (3.13) is replaced, with the help of Eq. (4.2), by

$$\operatorname{curl} \boldsymbol{B} = \operatorname{curl} \boldsymbol{H} + 4\pi \operatorname{curl} \boldsymbol{M} = \frac{1}{c} \frac{\partial \boldsymbol{D}}{\partial t} + 4\pi (\boldsymbol{j} + \operatorname{curl} \boldsymbol{M}). \tag{5.3}$$

The symbol $\boldsymbol{j}$ without subscript will be used henceforth in this article to represent the true accessible current density.

Physically the displacement current comprises two dissimilar terms,

$$\frac{1}{4\pi} \frac{\partial \boldsymbol{D}}{\partial t} = \frac{1}{4\pi} \frac{\partial \boldsymbol{E}}{\partial t} + \frac{\partial \boldsymbol{P}}{\partial t}, \tag{5.4}$$

the vacuum displacement current and the polarization current. The latter can be traced to the motion of inaccessible charges in a dielectric, but the first term, originally attributed to the "polarization of vacuum", has no significance as the motion of actual charges. Its introduction has been amply justified by the consequences in electromagnetic theory.

Eqs. (3.7) and (5.3), or the equivalent of (5.3),

$$\operatorname{curl} \boldsymbol{H} = \frac{1}{c} \frac{\partial \boldsymbol{D}}{\partial t} + 4\pi \boldsymbol{j}, \tag{5.5}$$

are, in addition to (2.3), members of the set of four fundamental relations called MAXWELL's equations.

6. Electromotive force and FARADAY's law of induction. The fields and charges introduced in electrostatics are of course related to the currents responsible for magnetic fields. An electric field within a conductor produces a flow of charge, and for many homogeneous conductors it is related to the current density by OHM's law,

$$c \boldsymbol{j} = \sigma \boldsymbol{E}, \tag{6.1}$$

where σ (measured in reciprocal seconds in esu) is the conductivity of the conducting material. But a current cannot be maintained by an electrostatic field alone; energy is expended continuously in a steady current at a rate given by $c\boldsymbol{j}\cdot\boldsymbol{E}$ ergs/sec/cm³, whereas the electrostatic field is conservative. An electromotive force is required, i.e., a force per unit charge must be supplied whose line integral about an electric circuit does not vanish. The internal microscopic field of a voltaic battery, for example, cannot be irrotational, although it may supply a conservative field externally.

FARADAY discovered experimentally that an electromotive force is produced by a varying magnetic field. One of the original experiments exhibiting this "induced" electromotive force is shown in Fig. 2. The first member of a double helix is connected to a voltaic battery, the second to a galvanometer. If the switch in circuit 1 is opened or closed a momentary current is observed in circuit 2. The quantitative expression for the electromotive force corresponding to this secondary current is

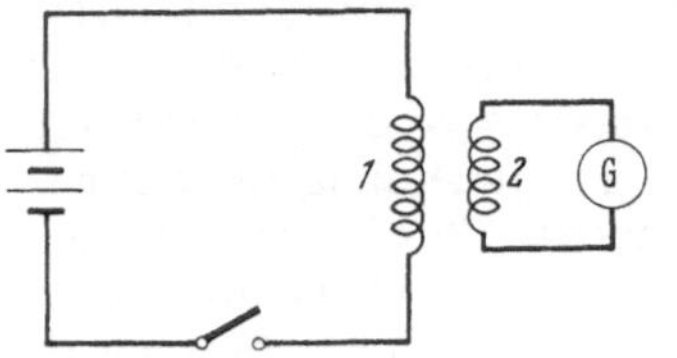

Fig. 2. Electromagnetic induction experiment.

$$\mathcal{E} = \oint \boldsymbol{E}\cdot d\boldsymbol{s} = -\frac{1}{c}\frac{d\Phi}{dt}, \tag{6.2}$$

where Φ is the flux of the magnetic induction field (in emu) produced by the first circuit which passes through the area bounded by the secondary circuit:

$$\Phi = \int \boldsymbol{B}\cdot d\boldsymbol{S}. \tag{6.3}$$

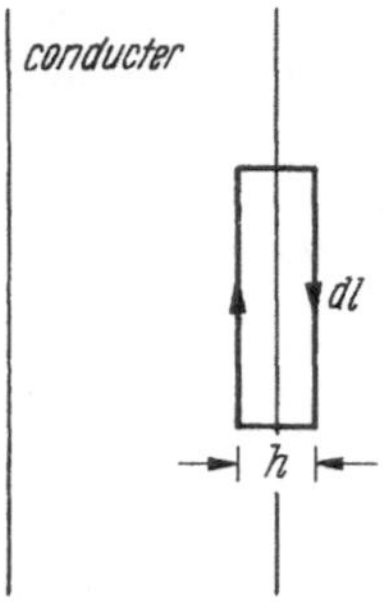

Fig. 3. Loop to show boundary conditions on the tangential component of $\boldsymbol{E}$.

The minus sign in Eq. (6.2) reflects the principle, first stated by FRIEDRICH EMIL LENZ[1], that an induced current is always such as to oppose the change of field that produces it. Eq. (6.2) is valid for all methods of changing the magnetic fields through a circuit, including self induction (first discovered by JOSEPH HENRY) due to changing currents set up by an impressed electromotive force in a single circuit.

The law of induction was generalized by MAXWELL so as to be independent of the presence of a conducting circuit. The line integral of $\boldsymbol{E}$ about an elementary loop one side of which is inside the conducting wire and the other outside (Fig. 3) vanishes as the width of the loop goes to zero, not only for a conservative field, but quite generally if the validity of Eq. (6.2) for the loop is granted. Since the tangential field is thus continuous across the boundary of the conductor, it seems plausible that this field would exist in the absence of the conductor. Then by STOKES' theorem,

$$\oint \boldsymbol{E}\cdot d\boldsymbol{s} = \int \operatorname{curl}\boldsymbol{E}\cdot d\boldsymbol{S} = -\frac{1}{c}\frac{d}{dt}\int \boldsymbol{B}\cdot d\boldsymbol{S} = -\frac{1}{c}\int \frac{\partial \boldsymbol{B}}{\partial t}\cdot d\boldsymbol{S}, \tag{6.4}$$

where the surface of integration is bounded by the line integral. Eq. (6.4) implies the equality of the integrands,

$$\operatorname{curl}\boldsymbol{E} = -\frac{1}{c}\frac{\partial \boldsymbol{B}}{\partial t} \tag{6.5}$$

Eq. (6.5) completes the set of field equations whose physical content consists of the fundamental empirical laws of electromagnetic phenomena.

[1] F. E. LENZ: Pogg. Ann. Phys. u. Chem. **31**, 483 (1834).

In addition to these fundamental laws of electromagnetism there is the important phenomenological relation, discovered by GEORG SIMON OHM[1], which relates the current density within a medium to the electric field intensity. According to OHM's law these quantities are directly proportional,

$$c\boldsymbol{j} = \sigma \boldsymbol{E}, \tag{6.6}$$

where σ is the electrical conductivity characteristic of the medium. For metallic conductors the linear relation (6.6) is valid over a large range of current densities, although for unisotropic crystals it must be replaced by a tensor equation. For linear circuits carrying steady currents the equivalent relation is usually written

$$V = RJ, \tag{6.7}$$

where R is said to be the resistance of a conductor such that it carries a total current J when a difference of potential V is applied across the ends of the conductor.

II. Electrostatics.

7. The electrostatic field in vacuo. Since curl $\boldsymbol{E}=0$ in electrostatics the electric field intensity may be derived from the scalar potential defined by Eq. (1.6). Eq. (1.8), valid in empty space, then leads to a differential equation for the potential,

$$\nabla^2 \phi = -4\pi \varrho, \tag{7.1}$$

which is known as POISSON's equation[2]. A particular solution of POISSON's equation in vacuum is the integral over all space,

$$\phi(\boldsymbol{x}) = \int \frac{\varrho(\boldsymbol{\xi})}{r} dV' \tag{7.2}$$

where $\boldsymbol{\xi}$ represents the space coordinates ξ_i of the charge distribution of density $\varrho(\boldsymbol{\xi})$, $\boldsymbol{x}$ represents the three coordinates x_i of the field point for which the potential is so determined, and r is the distance from source to field point. If $\boldsymbol{R}$ is the radius vector from the origin to the field point, $\boldsymbol{r}=\boldsymbol{R}-\boldsymbol{\xi}$, and the integral is to be taken over the source coordinates.

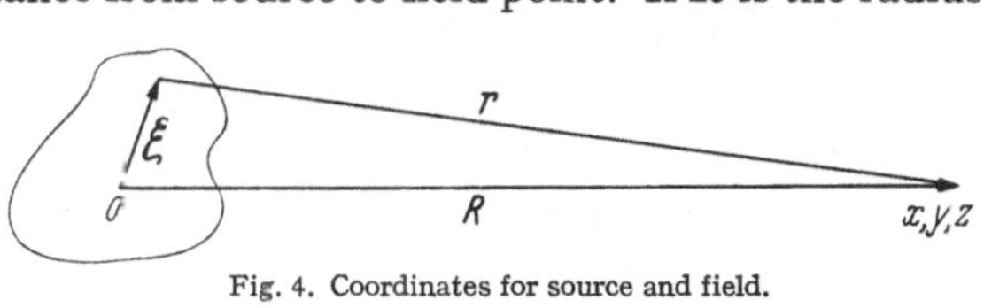

Fig. 4. Coordinates for source and field.

Eq. (7.2) reduces to the familiar Coulomb expression, $\phi=q/r$, for a point charge. Equally useful expressions may be obtained for the potential at large distances from other simple charge configurations. Let the origin of coordinates be within a finite charge distribution whose maximum spatial dimension is a (Fig. 4). A TAYLOR's expansion may then be made in terms of ξ_i/R for a field point $R(x_i) > a$ and substituted in Eq. (7.2):

$$\left.\begin{aligned}\phi = \Big\{\frac{1}{R}\int \varrho\, dV' + \left[\frac{\partial}{\partial \xi_i}\left(\frac{1}{r}\right)\right]_{\xi=0} \int \xi_i \varrho\, dV' + \\ + \frac{1}{2!}\left[\frac{\partial^2}{\partial \xi_i \partial \xi_j}\left(\frac{1}{r}\right)\right]_{\xi=0} \int \xi_i \xi_j \varrho\, dV' + \cdots\Big\}.\end{aligned}\right\} \tag{7.3}$$

[1] G. S. OHM: Pogg. Ann. Phys. u. Chem. **7**, 117 (1826). — Die galvanische Kette. 1827.

[2] LAPLACE showed that the gravitational scalar potential of LAGRANGE satisfies $\nabla^2\phi = 0$ in free space: Mém. Acad. Sci. (1782, published in 1785), p. 113. POISSON generalized the equation to points within the source, Eq. (7.1), in the same paper which exhibited its usefulness for electrical problems: Bull. Soc. Philomathique **3**, 388 (1813).

Here ξ_i, ξ_j, etc. represent the coordinate components describing the distribution of charge, and summation is implied for indices repeated in the same term. The integrals over V' are moments of the charge distribution, the n-th moment corresponding to the $(n+1)$-th term of the series. The zero moment, $p^{(0)} = \int \varrho \, dV' = q$, is just the total charge, and the first term of (7.3) is the Coulomb potential of this charge. In the second term the expression

$$p_i^{(1)} = \int \xi_i \varrho \, dV' \tag{7.4}$$

is the i-th component of the *dipole moment*; in general the 2^n multipole corresponds to the n-th term of the series. If two point charges $+q$ and $-q$ are separated by the vector distance $\boldsymbol{l}$ (directed from $-q$ to $+q$), then

$$\boldsymbol{p} = q \, \boldsymbol{l} \tag{7.5}$$

is the elementary expression for the moment of a simple dipole. The six components of the *quadrupole moment*,

$$p_{ij}^{(2)} = \int \xi_i \xi_j \varrho \, dV' \tag{7.6}$$

are those of a symmetric tensor of second rank. Octupole and higher moments are similarly tensors of higher rank.

The potential due to an arbitrary finite charge distribution, Eq. (7.3), is formally equivalent to a sum of partial potentials,

$$\Phi = \sum_{n=0}^{\infty} \Phi_n, \tag{7.7}$$

where

$$\Phi_n = \frac{p_{ij\ldots n}^{(n)}}{n!} \, \frac{\partial^n}{\partial \xi_i \, \partial \xi_j \ldots} \left(\frac{1}{r}\right) = \frac{(-1)^n p_{ij\ldots n}^{(n)}}{n!} \, \frac{\partial^{(n)}}{\partial x_i \, \partial x_j \ldots} \left(\frac{1}{r}\right). \tag{7.8}$$

The partial potentials of Eq. (7.8) may be interpreted as representing charge singularities of order 2^n at the coordinate origin, each of which may be obtained by differentiation from the preceding term. (This is of course the reason for the name multipole, and for the classification of multipoles noted above.) Thus

$$\Phi_1 = \frac{\partial \Phi_0}{\partial \xi_i} \, d\xi_i = q \, d\xi_i \frac{\partial}{\partial \xi_i} \left(\frac{1}{r}\right) = - p_i^{(1)} \frac{\partial}{\partial x_i} \left(\frac{1}{r}\right) = - \boldsymbol{p}^{(1)} \cdot \operatorname{grad} \left(\frac{1}{R}\right) = \frac{p_i^{(1)} x_i}{R^3}, \tag{7.9}$$

in which (for $R \gg a$) R may be written interchangeably with r when the derivatives are expressed in terms of the field coordinates x_i. The same formula is obtained by adding to the potential of a point charge $-q$ at the origin that of q displaced by $d\boldsymbol{\xi}$ from the origin, and permitting $d\boldsymbol{\xi}$ to vanish while maintaining $q \, d\boldsymbol{\xi}$ constant. Similarly,

$$\Phi_2 = \frac{\partial \Phi_1}{\partial \xi_j} \, d\xi_j = p_i^{(1)} \, d\xi_j \frac{\partial^2}{\partial x_i \, \partial x_j} \left(\frac{1}{R}\right) \tag{7.10}$$

represents the potential due to two dipoles, $\boldsymbol{p}^{(1)}$ displaced by $d\boldsymbol{\xi}$ from $-\boldsymbol{p}^{(1)}$. If the expression $2(p_i \, d\xi_j)$ is identified as the ij-th component of a quadrupole moment, Eq. (7.10) may be identified with Eq. (7.8) for $n=2$. In general the 2^n multipole moment of a singular cluster of point charges generated by superposing two equal and opposite 2^{n-1} poles is given by

$$p^{(n)} = n \, p^{(n-1)} \cdot d\boldsymbol{\xi}, \tag{7.11}$$

where $d\boldsymbol{\xi}$ is the displacement of $p^{(n-1)}$ from $-p^{(n-1)}$, in the limit that $p^{(n-1)}$ approaches infinity and $d\boldsymbol{\xi}$ approaches zero in such a way that their product remains finite. Outside a sphere which wholly contains an arbitrary distribution of finite charge density ϱ, the potential of the charge distribution is identical with that of a set of multipoles located at the origin of the sphere. The relative order of magnitude of successive multipole potentials is a/R, so that at large distances from the source the field will be essentially given by the first non-vanishing term of (7.7).

The 2^n multipole moment is a tensor of rank n, symmetric in all the indices. The components of the quadrupole and higher multipoles are not entirely independent, however, but satisfy the relation that contraction over any pair of indices gives zero. Thus the field of a quadrupole depends on only five independent quantities, not six. This follows from the fact that $1/R$ satisfies the Laplace equation for $r \neq 0$. That is,

$$\delta_{ij} \frac{\partial^2}{\partial x_i \partial x_j} \left(\frac{1}{R}\right) = 0, \tag{7.12}$$

where δ_{ij}, the Kronecker delta symbol, is equal to unity if $i=j$ and is zero otherwise. The quadrupole potential can therefore be written as

$$\phi_2 = \frac{1}{2} \int \left(\xi_i \xi_j - \frac{1}{3} \delta_{ij} \xi^2\right) \varrho \, dV' \frac{\partial^2}{\partial x_i \partial x_j} \left(\frac{1}{R}\right), \tag{7.13}$$

and the quadrupole moment may be explicitly defined as

$$Q_{ij} = \int (\xi_i \xi_j - \tfrac{1}{3} \xi^2) \, \varrho(\boldsymbol{\xi}) \, dV'. \tag{7.14}$$

Here ξ is the magnitude of the coordinate vector $\boldsymbol{\xi}$. For a system of point charges q_s, whose coordinate vectors are $\boldsymbol{\xi}_s$, the quadrupole moment is, similarly,

$$Q_{ij} = \sum_s q_s (\xi_i \xi_j - \delta_{ij} \xi^2)_s. \tag{7.15}$$

In general $2n+1$ parameters are needed to describe a multipole of order 2^n, including its orientation in space.

8. GREEN's formula and the uniqueness of the electrostatic potential. The principles and methods of determining static fields are to be found elsewhere in this Encyclopedia (Vol. XVI), and will not be discussed here. We must note, however, that as a solution of (7.1) the potential (7.2) is unique only if the total charge is finite, i.e., LAPLACE's equation (with $\varrho = 0$ everywhere) has solutions which correspond to charges at infinity. For the uniqueness of (7.2) the charge must be confined to a finite region, so that the potential itself may be set equal to zero at infinity. This may be seen as a limiting case of GREEN's formula, which expresses the potential within a finite region in terms of the charge distribution within a boundary and the conditions on the bounding surface or surfaces.

GREEN's formula is derived from GREEN's theorem[1], which states that

$$\int (\psi \nabla^2 \phi - \phi \nabla^2 \psi) \, dV = \int (\psi \operatorname{grad} \phi - \phi \operatorname{grad} \psi) \cdot d\mathbf{S} \tag{8.1}$$

where ϕ and ψ are arbitrary scalar fields. Let ϕ be the electrostatic potential and let $\psi = 1/r$, where r is the distance from a fixed point within the region to the variable integration point. Then substitution in (8.1) yields

$$\int \frac{\nabla^2 \phi}{r} dV - \int \phi \nabla^2 \left(\frac{1}{r}\right) dV = \int \left[\frac{\operatorname{grad} \phi}{r} - \phi \operatorname{grad} \left(\frac{1}{r}\right)\right] \cdot d\mathbf{S}. \tag{8.2}$$

[1] "An essay on the application of mathematical analysis to the theories of electricity and magnetism" (Nottingham, 1828); reprinted in The Mathematical Papers of the late GEORGE GREEN.

But $\nabla^2\phi = -4\pi\varrho$, and $\nabla^2(1/r) = -4\pi\,\delta(\boldsymbol{r})$, where $\delta(\boldsymbol{r})$ is the Dirac delta function defined by the integral properties (3.12). Therefore the potential at the fixed point is given by GREEN's formula,

$$\phi = \int \frac{\varrho\, dV}{r} + \frac{1}{4\pi}\int\left[\frac{1}{r}\,\frac{\partial\phi}{\partial n} - \phi\,\frac{\partial}{\partial n}\left(\frac{1}{r}\right)\right] dS, \tag{8.3}$$

where $\partial/\partial n$ is written for the positive normal component of the gradient at the bounding surface. It is clear that, if the region is extended to include all space, ϕ must be such that the surface integral vanishes in order that (8.3) be identical with (7.2). This is equivalent to setting the total charge (not just the charge density) at infinity equal to zero.

For a finite region the first term of (8.3) represents the direct contribution to the potential of the charge distribution inside the region, while the surface integral takes the place of the integral over the remainder of space. The surface integrals permit physical interpretation as charge configuration on the boundary. The normal derivative of ϕ is E_n, which by (2.1) equals $4\pi\tau$ at the surface of a conductor. The first surface integral is thus equivalent to a distribution of charge on a conductor bounding the region, with the surface charge density given by $\tau = (1/4\pi)\,\partial\phi/\partial n$. The last term of (8.3) has the form of (7.9) if the dipole is a layer of moment normal to the surface. In other words, a surface dipole density of moment per unit area given by $\phi/4\pi$ and directed toward the outward normal would produce the potential represented by the last term of (8.3). These results may be stated by saying that the charges outside a given region may be replaced by equivalent charge and dipole layers on the boundary of the region without affecting the potential at interior points. In practice the formula is useful for solving potential problems involving regions with conducting boundaries, on which the surface charge may be that induced by the presence of charges within the bounding surface.

9. Dielectric media (see Vol. XVII of this Encyclopedia). Within a region of uniform dielectric the formal considerations of the preceding section would apply to the electric displacement $\boldsymbol{D}$, but such a treatment is neither conceptually useful nor of practical value in the treatment of physical problems. The historical justification for the introduction of more than one electric field concept involved discontinuities in electrical properties from one kind of matter to another. The boundary conditions on the field intensity $\boldsymbol{E}$ and the displacement vector $\boldsymbol{D}$ at any discontinuity may be derived from the field equations (1.5) and (2.3), or, more precisely, from the integral forms of these equations which express the conservative nature of the electrostatic field and GAUSS's flux theorem respectively.

Consider the interface between two media of different dielectric properties. The line integral of $\boldsymbol{E}$ about an elementary loop shown in Fig. 5, in which the length of path normal to the boundary is vanishingly small, must be zero, and therefore the tangential component of the field is continuous across the interface. This condition may be written vectorially in terms of $\boldsymbol{n}$, a unit normal to the boundary surface between media 1 and 2:

$$\boldsymbol{n}\times(\boldsymbol{E}_1 - \boldsymbol{E}_2) = 0. \tag{9.1}$$

Actually, as noted in Sect. 6 where the special example of a conducting boundary was considered in relation to the generalization of electromagnetic induction, this condition is satisfied even by non-conservative fields.

The flux theorem may be applied to an infinitesimal cylindrical volume of vanishingly small altitude (Fig. 6) which includes the flat area dS of the interface. The lateral area of the cylinder is negligibly small, and the theorem reduces to

$$\boldsymbol{n}\cdot(\boldsymbol{D}_1-\boldsymbol{D}_2)=4\pi\,\tau, \tag{9.2}$$

where τ is the true charge per unit area on the surface. If there are no true charges, as would normally be the case for insulating media, the normal component of $\boldsymbol{D}$ is continuous across the interface.

The contribution to the field due to a dielectric medium possessing no true charges may be formally expressed in terms of the partial field $4\pi\boldsymbol{P}$ of Eq. (2.5). We may write

$$\operatorname{div}\boldsymbol{E}=\operatorname{div}\boldsymbol{D}-4\pi\operatorname{div}\boldsymbol{P}=4\pi(\varrho-\operatorname{div}\boldsymbol{P})=4\pi(\varrho+\varrho_P), \tag{9.3}$$

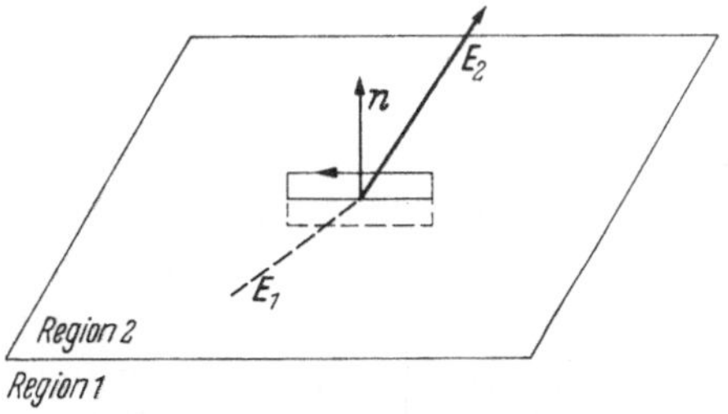

Fig. 5. To show boundary condition on $\boldsymbol{E}$.

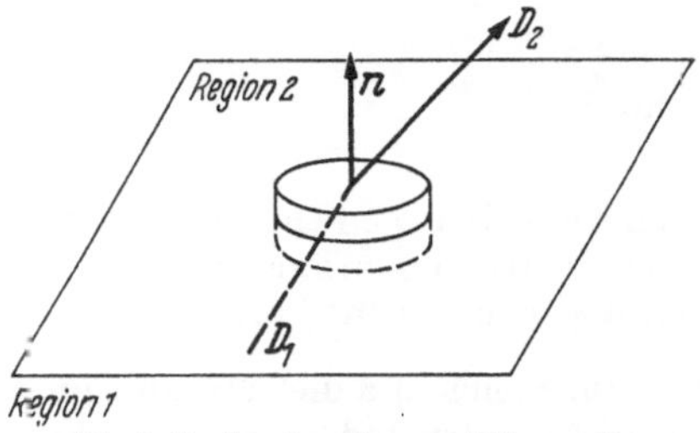

Fig. 6. To show boundary condition on $\boldsymbol{D}$.

and thus $4\pi\boldsymbol{P}$ is that part of the field whose flux arises from an inaccessible charge density ϱ_P equal to $-\operatorname{div}\boldsymbol{P}$. But it is more significant physically, and more useful in the consideration of discontinuous and non-uniform dielectrics, to define $\boldsymbol{P}$ as a source of field intensity. Consider an elementary volume dipole moment $\boldsymbol{P}\,dV$, where $\boldsymbol{P}$ is the electric dipole moment per unit volume. Then the (vacuum) potential due to a distribution of polarization density $\boldsymbol{P}$ is, by Eq. (7.9),

$$\phi=\int P_i\frac{\partial}{\partial\xi_i}\left(\frac{1}{r}\right)dV'. \tag{9.4}$$

Eq. (9.4) may be transformed to

$$\phi=\int\frac{\partial}{\partial\xi_i}\left(\frac{P_i}{r}\right)dV'-\int\frac{1}{r}\,\frac{\partial P_i}{\partial\xi_i}\,dV=\int\frac{\boldsymbol{P}\cdot d\boldsymbol{S}}{r}+\int\frac{\varrho_P}{r}\,dV', \tag{9.5}$$

where $-\operatorname{div}\boldsymbol{P}$ (taken at the source point) has again been identified with the inaccessible charge density designated as ϱ_P. The surface integral over the boundary of the polarized region represents an inaccessible polarization surface charge of density equal to the component of $\boldsymbol{P}$ normal to the surface. Polarization is sometimes defined by the mechanism of separating to some extent bound positive and negative charges; $\boldsymbol{P}$ is then equal in magnitude to the charge crossing the unit area whose normal is the direction of this separation. These definitions of $\boldsymbol{P}$ are entirely equivalent.

Even with electrostatic fields, for which $\operatorname{curl}\boldsymbol{E}=0$, there is no requirement that $\boldsymbol{P}$ or $\boldsymbol{D}$ be irrotational, but only that $\operatorname{curl}\boldsymbol{D}=4\pi\operatorname{curl}\boldsymbol{P}$. Thus $\boldsymbol{D}$ at any point in space may be derived formally from a scalar potential (whose source is the density of free charges) and a vector potential whose vortex density is $\operatorname{curl}\boldsymbol{P}$. This alternative is not often followed, however, since the polarization sources for the field $\boldsymbol{E}$ have more immediate physical significance than the circulation density of the displacement vector $\boldsymbol{D}$ or the polarization $\boldsymbol{P}$.

Thus far the polarization has been considered independent of $\boldsymbol{E}$, and the relations are equally valid for permanently polarized bodies (electrets) and those in which the polarization is itself produced by an impressed electric field. For most insulators there is no permanent polarization, and if the insulator is isotropic,

$$\boldsymbol{P} = \chi \boldsymbol{E}, \tag{9.6}$$

where χ, the *electric susceptibility*, is a positive coefficient. The susceptibility is related to the dielectric constant of Eq. (2.4),

$$\varepsilon = 1 + 4\pi\chi. \tag{9.7}$$

For non-isotropic media $\boldsymbol{P}$ may not be parallel to $\boldsymbol{E}$, and yet each component be proportional to $\boldsymbol{E}$. The relations thus remain linear, but both the susceptibility and the dielectric constant are symmetric tensors,

$$P_i = \chi_{ij} E_j \tag{9.8}$$

and

$$D_i = \varepsilon_{ij} E_j. \tag{9.9}$$

Relations between the fields in a medium, such as (9.6), (9.8) and (9.9), are called constitutive equations, and owe their existence to the observed properties of a limited class of media.

10. Fields in a dielectric medium. In Maxwell theory all properties of material media are treated as continuous. The phenomenological definition of the field intensity as the force per unit (infinitesimal) charge fails in a ponderable dielectric, and it is customary to imagine cavities in which such forces would be in principle observable. In a needle shaped cavity parallel to the field end effects may be neglected and the entire field is tangential to the boundaries; the electric field intensity $\boldsymbol{E}$ would then be the same in the cavity as in the dielectric, and would determine the force on a test charge. Within a disk shaped cavity whose major surfaces are perpendicular to the field the edge effects may be neglected; since the normal component of the displacement $\boldsymbol{D}$ is continuous, and since within the cavity $\boldsymbol{E} = \boldsymbol{D}$, the force on a test charge would give a measure of $\boldsymbol{D}$ in the medium. In general the force on a test charge within a cavity depends on the geometry of the cavity and its orientation with respect to the field.

The justification for maintaining any concepts of continuous field in matter which is in fact discontinuous depends on the validity of averaging atomic fields. From the point of view of electromagnetic theory each neutral or ionized atom or molecule is characterized by a distribution of charge and current which is ordinarily confined to the spatial region occupied by the molecular system and which exhibits more or less irregular time variation. On a microscopic scale the fields are vacuum fields. Thus, in the medium,

$$\operatorname{div} \boldsymbol{e} = 4\pi \varrho_a \tag{10.1}$$

where $\boldsymbol{e}$ and ϱ_a are the microscopic field and charge density, respectively. Macroscopic fields are averages of the corresponding microscopic quantities taken over spatial regions and time intervals which are large compared with atomic dimensions and periods. It is assumed that these space-time intervals are physically infinitesimal with respect to the macroscopic scale, so that average values of microscopic quantities over such regions can be treated as continuous space-time functions. The irregular fluctuations of the microscopic fields and sources are thus smoothed out by the process of averaging. This argument could come into question only if the macroscopic field varies appreciably within a distance corresponding to atomic dimensions, or within a time corresponding to atomic periods; it is obviously rigorous for static fields.

The space-time average of the microscopic electric field may be identified by considering the fundamental properties of averages. If the average of a function $f(x', y', z', t')$ over a physically infinitesimal interval V, T is defined by

$$\bar{f}(x, y, z, t) = \frac{1}{VT} \int\limits_{V} dV' \int\limits_{T} dt' f(x', y', z', t'), \tag{10.2}$$

it can be readily shown that

$$\frac{\partial \bar{f}}{\partial x} = \overline{\frac{\partial f}{\partial x}}. \tag{10.3}$$

The proof of Eq. (10.3) follows from writing x' as x plus a new variable of integration and similar substitutions for the other variables. Therefore

$$\operatorname{div} \bar{\boldsymbol{e}} = 4\pi \overline{\varrho_a}. \tag{10.4}$$

But the average of the atomic charge density is (to a very good approximation) the true and polarization charge densities on a macroscopic scale. Therefore, by comparison with (9.3), the space time average of the microscopic field is identified with the macroscopic field $\boldsymbol{E}$. From a phenomenological point of view, $\boldsymbol{E}$ is the average field on a fast moving charge traversing the medium. (See Ref. [*18*] for a discussion of higher order terms in the average charge density, which correspond to higher order partial fields.)

The field experienced by a molecule composing the dielectric depends not only on $\boldsymbol{E}$ but also on the volume polarization $\boldsymbol{P}$. It is useful to compute this field in order to relate the macroscopic properties of a dielectric to molecular properties. Consider an isotropic dielectric in which the molecules have no dipole moment in the absence of an impressed field. It is assumed that in the presence of a field the polarization $\boldsymbol{P}$ is entirely due to molecular dipoles induced by the field. Let a particular molecule be surrounded by a sphere whose radius b is small compared to the dimensions of the dielectric but large enough that the sphere contains many molecules. The total field at the center of a perfectly random distribution of molecular dipoles within the sphere is zero, and the effect of the dielectric can be computed from the surface polarization charge τ_P on the spherical surface. The magnitude of this effect, which will be in the same direction as $\boldsymbol{P}$ and the electric intensity $\boldsymbol{E}$, may be called the internal field $\boldsymbol{E}_i$. The internal field is given by

$$E_i = \frac{\tau_P \cos\vartheta}{b^2}\, dS, \tag{10.5}$$

where ϑ is the angle between $\boldsymbol{P}$ and the normal to the spherical surface. Now $\tau_P = P\cos\vartheta$, and the element of surface may be taken as $dS = 2\pi\, b^2 \sin\vartheta\, d\vartheta$, so that

$$E_i = 2\pi P \int \cos^2\vartheta \sin\vartheta\, d\vartheta = \frac{4\pi}{3} P. \tag{10.6}$$

The total effective field on the molecule is therefore

$$\boldsymbol{E}_{\text{eff}} = \boldsymbol{E} + \frac{4\pi}{3}\boldsymbol{P}. \tag{10.7}$$

The dipole moment $\boldsymbol{p}$ produced in the molecule by this field may be written

$$\boldsymbol{p} = \alpha \boldsymbol{E}_{\text{eff}}, \tag{10.8}$$

where α is the molecular polarizability. The form of (10.8) implies that the molecule itself polarizes isotropically, and thus imposes a further restriction on the type of substance to which the theory applies. If (10.8) is valid the dipole moment per unit volume is

$$\boldsymbol{P} = N\boldsymbol{p} = N\alpha\left(\boldsymbol{E} + \frac{4\pi}{3}\boldsymbol{P}\right), \tag{10.9}$$

where N is the number of molecules per unit volume. But $4\pi\boldsymbol{P} = (\varepsilon - 1)\boldsymbol{E}$, and therefore (10.9) leads to the relation

$$\frac{\varepsilon - 1}{\varepsilon + 2} = \frac{4\pi N\alpha}{3}. \tag{10.10}$$

This is known as the Clausius-Mossotti[1] relation. If both numerator and denominator of the fraction on the right are multiplied by the mass per molecule it is seen that the expression varies with the density; this behavior has been checked experimentally for gaseous dielectrics[2]. For dilute gases, where ε is very nearly unity,

$$\varepsilon - 1 \approx \frac{4\pi N m\alpha}{m} = \frac{4\pi g\alpha}{m}, \tag{10.11}$$

in which g is the density and m is the molecular mass. The internal field given by Eq. (10.6) is valid for a lattice point within a cubic crystal as well as for a molecule of a random isotropic distribution, but it is not valid for crystals of lower symmetry.

In some substances the electric susceptibility is temperature dependent, and behaves approximately as $a + b/T$. This temperature dependence is understood microscopically in terms of polar molecules, i.e., molecules which possess intrinsic electric dipole moment. Ordinarily the dipoles are oriented at random, but a field produces preferential orientation, subject to interference from thermal agitation. The subject thus demands statistical treatment, which was supplied by PETER DEBYE (1912)[3]. The contribution to the polarization from the intrinsic dipoles, each of magnitude p_0, is given by

$$P = N p_0\left(\operatorname{Cot} x - \frac{1}{x}\right), \tag{10.12}$$

where $x = p_0 E/kT$. For small x, Eq. (10.12) becomes

$$P \approx \frac{1}{3} N p_0 x = \frac{N p_0^2 E}{3kT}. \tag{10.13}$$

The total susceptibility,

$$\frac{\varepsilon - 1}{4\pi} = N\alpha + \frac{N p_0^2}{3kT}, \tag{10.14}$$

[1] O. F. MOSSOTTI: Mem. mat. fis. Modena **24**, 49 (1850). — R. CLAUSIUS: Die mechanische Wärmetheorie, Vol. 2, pp. 64—97 (1879). Vol. 2 is: Die mechanische Behandlung der Electricität.

[2] See H. FRÖHLICH, Theory of Dielectrics (Dielectric Constant and Dielectric Loss), Oxford, 2nd ed., 1958. When the conditions on the validity of Eq. (10.10) are satisfied its agreement with experiment are often better than might be expected. For N_2, e.g., the quantity

$$\frac{\text{molecular weight}}{\text{density}} \frac{(\varepsilon - 1)}{(\varepsilon + 2)}$$

varies less than 1% in a range of pressures from 1 to 2000 atmospheres. See also Vol. XVII of this Encyclopedia.

[3] P. DEBYE: Phys. Z. **13**, 97 (1912). See also P. DEBYE: Polar Molecules, New York, 1929, more recently reprinted by Dover.

is thus of the form $a+b/T$ if the dielectric constant is not too different from unity. Dependence of the susceptibility on temperature must be taken formally into account in a consistent treatment of field energy and forces.

11. Electrostatic field energy. In the view of FARADAY and MAXWELL the energy of a charge distribution resides in the field, not in the charge itself. The energy should then be expressible as an integral of a function of the field over all space. The expression for energy density may be derived by considering the change in energy associated with the introduction of an infinitesimal increment of charge $\delta\varrho$ in such a way that no work is done on mechanical constraints. If ϕ is the potential from which the field intensity may be derived at any point, this energy increment is

$$\left.\begin{aligned}\delta U &= \int \phi\,\delta\varrho\,dV = \frac{1}{4\pi}\int \phi\,\mathrm{div}\,\delta\boldsymbol{D}\,dV\\ &= \frac{1}{4\pi}\int \mathrm{div}\,[(\delta\boldsymbol{D})\,\phi]\,dV - \frac{1}{4\pi}\int \delta\boldsymbol{D}\cdot\mathrm{grad}\,\phi\,dV.\end{aligned}\right\} \tag{11.1}$$

The first integral in Eq. (11.1) may be converted to a surface integral bounding all space, which vanishes if the charge distribution producing the initial field is finite. Hence

$$\delta U = \int \frac{\boldsymbol{E}\cdot\delta\boldsymbol{D}}{4\pi}\,dV. \tag{11.2}$$

In empty space where $\boldsymbol{D}=\boldsymbol{E}$ Eq. (11.2) may be integrated from 0 to $\boldsymbol{E}$ to give

$$U = \int \frac{E^2}{8\pi}\,dV. \tag{11.3}$$

The vacuum energy density is therefore given by $E^2/8\pi$.

In the presence of dielectrics the integration of Eq. (11.2) may be carried out if there is a linear constitutive relation between $\boldsymbol{D}$ and $\boldsymbol{E}$, so that ε is defined at all points. If it is further assumed that the process is isothermal, to insure constancy of the dielectric constant,

$$U = \int \frac{\boldsymbol{E}\cdot\boldsymbol{D}}{8\pi}\,dV. \tag{11.4}$$

It is clear from the restrictions necessary for performing this integral that $\boldsymbol{E}\cdot\boldsymbol{D}/8\pi$ is the density of thermodynamic free energy, not the total energy density. Because of the dependence of ε on temperature the consideration of thermal effects must take into account this distinction. It can be easily shown from the relations between thermodynamic functions that the total energy is given by

$$U_{\text{total}} = \int \frac{\boldsymbol{E}\cdot\boldsymbol{D}}{8\pi}\,\frac{1}{\varepsilon}\,\frac{d(T\varepsilon)}{dT}\,dV \tag{11.5}$$

where T is the absolute temperature. For most purposes the free energy is the more useful expression, and it is the free energy that is minimized in an equilibrium distribution of charge on a conductor.

The sense in which energy resides in the field is much the same as that in which mechanical energy resides in an elastic medium under stress. It should therefore be possible to define a force density $\boldsymbol{f}$ and a stress tensor of which the force is the divergence, as is customary in the theory of continuous elastic media. This is done by applying the energy principle, assuming a virtual velocity $\boldsymbol{u}$ so small that the situation remains electrostatic. The formal expression for the isothermal time rate of change of field energy in terms of a volume force is

$$\frac{\partial U}{\partial t} = -\int \boldsymbol{f}\cdot\boldsymbol{u}\,dV = \left.\frac{\partial U}{\partial t}\right]_{\varrho} + \left.\frac{\partial U}{\partial t}\right]_{\varepsilon}, \tag{11.6}$$

since the fields are completely determined if ϱ and ε are known everywhere. But $\left.\frac{\partial U}{\partial t}\right]_\varepsilon$ is just $\int \phi \frac{\partial \varrho}{\partial t}\, dV$, and, if the free energy density is given by $D^2/8\pi\,\varepsilon$,

$$\left.\frac{\partial U}{\partial t}\right]_\varrho = \left.\frac{\partial U}{\partial t}\right]_{\operatorname{div}\boldsymbol{D}} = \frac{1}{8\pi}\int D^2 \frac{\partial}{\partial t}\left(\frac{1}{\varepsilon}\right) dV = -\int \frac{E^2}{8\pi}\frac{\partial \varepsilon}{\partial t}\, dV,$$

since the volume integral over all space of the time rate of change of $\boldsymbol{D}$ is zero if div $\boldsymbol{D}$ is held constant. Therefore

$$\frac{dU}{dt} = \int \left(\phi \frac{\partial \varrho}{\partial t} - \frac{1}{8\pi} E^2 \frac{\partial \varepsilon}{\partial t}\right) dV. \tag{11.7}$$

The integrand of Eq. (11.7) may be transformed to the scalar product of the virtual velocity and a vector expression to be identified with the volume force $\boldsymbol{f}$ by means of the continuity equations for the conservation of charge and mass during the virtual process:

$$\frac{\partial \varrho}{\partial t} = -\operatorname{div}(\varrho\, \boldsymbol{u}), \tag{11.8}$$

$$\frac{\partial g}{\partial t} = -\operatorname{div}(g\, \boldsymbol{u}). \tag{11.9}$$

One further assumption is necessary: ε must be a single-valued function of the mass density g. The application of GAUSS's divergence theorem yields vanishing surface integrals, and the final form is

$$\frac{dU}{dt} = \int \left[-\varrho \boldsymbol{E} + \frac{E^2}{8\pi} \operatorname{grad} \varepsilon - \frac{1}{8\pi} \operatorname{grad}\left(E^2 \frac{\partial \varepsilon}{\partial g} g\right)\right] \cdot \boldsymbol{u}\, dV,$$

from which it is seen that

$$\boldsymbol{f} = \varrho \boldsymbol{E} - \frac{1}{8\pi} E^2 \operatorname{grad} \varepsilon + \frac{1}{8\pi} \operatorname{grad}\left(E^2 \frac{\partial \varepsilon}{\partial g} g\right). \tag{11.10}$$

The first term in this expression for $\boldsymbol{f}$ is the electrostatic force on the true charges, the second is a contribution that arises in the presence of dielectric inhomogeneities, and the third is the electrostriction term.

The Maxwell stress tensor T_{ij} is so constructed that

$$f_i = \frac{\partial T_{ij}}{\partial x_j}. \tag{11.11}$$

This condition is satisfied by

$$T_{ij} = \frac{\varepsilon}{4\pi}\left[E_i E_j - \frac{\delta_{ij}}{2}(1-b)\, E_k E_k\right] \tag{11.12}$$

where for convenience we have put $b = \frac{g}{\varepsilon}\frac{\partial \varepsilon}{\partial g}$ and δ_{ij} is the Kronecker symbol defined by $\delta_{ij}=1$ for $i=j$ and $\delta_{ij}=0$ for $i\neq j$; summation over the three components is implied by the repeated subscript k. The physical interpretation of the stress tensor is that of a surface force equivalent to the volume force $\boldsymbol{f}$; i.e., for any component of $\boldsymbol{f}$,

$$\int f_i\, dV = \int T_{ij}\, dS_j, \tag{11.13}$$

where the righthand integral is taken over the bounding surface of the volume under consideration. The tensor is symmetric, and can therefore be reduced to three principal values.

The terms containing b occur only in diagonal elements of the stress tensor, and contribute a hydrostatic pressure at right angles to any surface element considered. If $b=0$ the principal values of the tensor, obtained by solving the secular determinant

$$|T_{ij} - \delta_{ij}\lambda| = 0, \tag{11.14}$$

are

$$\lambda_1 = \frac{\varepsilon}{8\pi} E^2, \quad \lambda_2 = \lambda_3 = -\frac{\varepsilon}{8\pi} E^2. \tag{11.15}$$

The principal axes are so oriented that the coordinate axis corresponding to the single root of the secular determinant, λ_1, is parallel to $\boldsymbol{E}$, while the two axes corresponding to the roots λ_2 and λ_3 are perpendicular to $\boldsymbol{E}$. This fact is often expressed qualitatively by saying that the electric field transmits a tension parallel to the field and a pressure at right angles to the field. (The algebraic sign of $\boldsymbol{E}$ does not affect the stress.) The stress transmitted across a surface element at 45° to the field is a pure shearing stress.

The mechanical ether model implied by the language of elasticity is useful, although the words are now interpreted figuratively, as are "flux" and "displacement current". MAXWELL himself attempted to avoid hypotheses, and KELVIN, who was largely responsible for the mathematical analogies, remarked explicitly that no physical hypothesis follows from the fact of these and similar analogies[1].

III. Magnetostatics.

12. Vacuum fields of steady current distributions. The magnetic induction field is solenoidal and may be derived unambiguously from a vector potential $\boldsymbol{A}$,

$$\boldsymbol{B} = \operatorname{curl} \boldsymbol{A}. \tag{12.1}$$

For steady currents in vacuo $\operatorname{curl} \boldsymbol{B} = 4\pi \boldsymbol{j}$, and therefore $\boldsymbol{A}$ satisfies the equation

$$\operatorname{curl}\operatorname{curl} \boldsymbol{A} = 4\pi \boldsymbol{j}. \tag{12.2}$$

Since it is permissable to assume $\operatorname{div} \boldsymbol{A} = 0$ for steady currents, Eq. (12.2) may be transformed by use of the vector identity

$$\operatorname{curl}\operatorname{curl} \boldsymbol{A} = \operatorname{grad}\operatorname{div} \boldsymbol{A} - \nabla^2 \boldsymbol{A} \tag{12.3}$$

to the form

$$\nabla^2 \boldsymbol{A} = -4\pi \boldsymbol{j}, \tag{12.4}$$

which is the vector Poisson equation. In Cartesian coordinates Eq. (12.4) is a set of three scalar equations to be solved subject to the condition $\operatorname{div} \boldsymbol{A} = 0$. In coordinates other than Cartesian the Laplacian of a vector must be interpreted by means of Eq. (12.3). A particular solution of Eq. (12.4) is

$$\boldsymbol{A} = \int \frac{\boldsymbol{j}}{r}\, dV', \tag{12.5}$$

as an integral over all sources, provided the currents are confined to a finite region of space.

The vector potential arising from a steady current distribution limited to a small volume may be expanded in a manner analogous to the scalar potential expansion of Eq. (7.3). Again let the origin of coordinates be in or near the finite distribution of current whose coordinates are ξ_i. For any field point at a distance greater than the radius of a sphere inclosing all currents, the i-th component of

[1] MARY B. HESSE: Isis **46**, 337 (1955).

the vector potential is then given by

$$\left.\begin{aligned}A_i=\Big\{\frac{1}{R}\int j_i\,dV'+\left[\frac{\partial}{\partial\xi_j}\left(\frac{1}{r}\right)\right]_{\xi=0}\int j_i\xi_j\,dV'+\\+\frac{1}{2!}\left[\frac{\partial^2}{\partial\xi_j\,\partial\xi_k}\left(\frac{1}{r}\right)\right]_{\xi=0}\int j_i\xi_i\xi_k\,dV'+\cdots\Big\}.\end{aligned}\right\}\tag{12.6}$$

The term in this series are called monopole, dipole, etc., by analogy with the electrostatics terminology. There is no magnetic monopole, however, since $\int \boldsymbol{j}\,dV'$ vanishes for steady currents almost by definition.

The leading term in Eq. (12.6), the magnetic dipole, must be transformed to exhibit its physical significance. Since the derivatives with respect to the source coordinates are related to those of the field point by $\partial/\partial\xi_i=-\partial/\partial x_i$,

$$\frac{\partial}{\partial\xi_j}\left(\frac{1}{r}\right)j_i\xi_j=-j_i\xi_j\frac{\partial}{\partial x_j}\left(\frac{1}{r}\right)=\frac{x_j\xi_j j_i}{r^3},\tag{12.7}$$

where the x_i are coordinates of the field point whose position vector is $\boldsymbol{R}$. But

$$\xi_j j_i=\tfrac{1}{2}(\xi_j j_i+\xi_i j_j)+\tfrac{1}{2}(\xi_j j_i-\xi_i j_j),\tag{12.8}$$

and

$$\xi_j j_i+\xi_i j_j=\boldsymbol{j}\cdot\operatorname{grad}\xi_i\xi_j=\operatorname{div}(\boldsymbol{j}\,\xi_i\xi_j)-\xi_i\xi_j\operatorname{div}\boldsymbol{j}.\tag{12.9}$$

The volume integral of the expression (12.9) vanishes, since the complete divergence can be transformed to a surface integral outside the distribution and since $\operatorname{div}\boldsymbol{j}=0$. The dipole term of Eq. (12.6) in vector form thus reduces to

$$\left.\begin{aligned}\boldsymbol{A}&=-\int\frac{1}{2}\left[\boldsymbol{j}\,\boldsymbol{\xi}\cdot\operatorname{grad}\left(\frac{1}{r}\right)-\boldsymbol{\xi}\boldsymbol{j}\cdot\operatorname{grad}\left(\frac{1}{r}\right)\right]dV'\\&=\int\frac{1}{2}\operatorname{curl}\frac{\boldsymbol{\xi}\times\boldsymbol{j}}{r}\,dV'=\operatorname{curl}\int\frac{\frac{1}{2}(\boldsymbol{\xi}\times\boldsymbol{j})}{r}\,dV',\end{aligned}\right\}\tag{12.10}$$

where the vector differential operator acts only on the coordinates of the field point and the resulting expression is to be evaluated at $r=R$. Eq. (12.10) may be written as

$$\boldsymbol{A}=\operatorname{curl}\left(\frac{\boldsymbol{m}}{r}\right)=-\boldsymbol{m}\times\operatorname{grad}\left(\frac{1}{R}\right).\tag{12.11}$$

Here

$$\boldsymbol{m}=\tfrac{1}{2}\int(\boldsymbol{\xi}\times\boldsymbol{j})\,dV'\tag{12.12}$$

is the magnetic moment of the current distribution, which reduces to the elementary expression for a linear current loop J of area $\boldsymbol{S}$ so directed as to agree with the right-hand rule, namely,

$$\boldsymbol{m}=J\,\boldsymbol{S}.\tag{12.13}$$

For any actual current distribution the field is given by the dipole term alone only at very large distances (unless the distribution is spherically symmetric so that the higher terms vanish at all external points). Eq. (12.11) is strictly the potential arising from a point dipole, a current loop whose moment remains constant as the area goes to zero. In the vicinity of a configuration of currents Eq. (12.6) converges slowly, and direct integration of Eq. (12.5) is possible in special cases.

By means of the vector analogue of GREEN's theorem,

$$\left.\begin{aligned}&\int(\boldsymbol{F}\cdot\nabla^2\boldsymbol{G}-\boldsymbol{G}\cdot\nabla^2\boldsymbol{F})\,dV\\&\quad=\int[\boldsymbol{F}\times\operatorname{curl}\boldsymbol{G}+\boldsymbol{F}(\operatorname{div}\boldsymbol{G})-\boldsymbol{G}\times\operatorname{curl}\boldsymbol{F}-\boldsymbol{G}(\operatorname{div}\boldsymbol{F})]\cdot d\boldsymbol{S},\end{aligned}\right\}\tag{12.14}$$

with $\boldsymbol{G}$ appropriately chosen, the vector potential within a finite region may be represented by a surface distribution of current and a magnetic dipole layer on the boundary of the region, in addition to the volume integral over all currents within the boundary. The method is not as a rule particularly useful in the solution of the vector Poisson equation, however, since there is no physical equivalent (at ordinary temperatures) in magnetostatics for the conductor in electrostatics. The fields contributed by magnetic media are best formally described by finding the equivalent distributions of (inaccessible) currents.

13. Fields in magnetic media. The boundary conditions on the magnetic induction field $\boldsymbol{B}$ and the magnetic field intensity $\boldsymbol{H}$ at the interface between two regions of different magnetic properties follow formally from the field equations, just as do those of the electrostatic fields. The total flux of $\boldsymbol{B}$ from any volume is always zero, and the application of the flux theorem to an elementary "pill box" as shown in Fig. 6 (p. 15) leads to

$$\boldsymbol{n}\cdot(\boldsymbol{B}_1-\boldsymbol{B}_2)=0, \tag{13.1}$$

where $\boldsymbol{n}$ is the unit normal to the interface between media 1 and 2. Thus the component of $\boldsymbol{B}$ normal to the interface is continuous. For the tangential component, Eq. (4.1), equivalent to

$$\oint \boldsymbol{H}\cdot d\boldsymbol{s}=4\pi\int \boldsymbol{j}_{\text{true}}\cdot d\boldsymbol{S}, \tag{13.2}$$

may be applied to a loop such as that shown in Fig. 5. It follows that

$$\boldsymbol{n}\times(H_1-H_2)=4\pi\boldsymbol{K}, \tag{13.3}$$

where $\boldsymbol{K}$ is a true surface current density (if any), measured in abamperes per unit length at right angles to the direction of the current. Thus in the absence of the flow of free surface charges on an interface the tangential component of $\boldsymbol{H}$ is continuous across the boundary.

The contribution to the magnetic field due to a material medium possessing no true currents may be formally expressed in terms of the partial field $4\pi\boldsymbol{M}$ of Eq. (4.3):

$$\operatorname{curl}\boldsymbol{B}=\operatorname{curl}\boldsymbol{H}+4\pi\operatorname{curl}\boldsymbol{M}=4\pi(\boldsymbol{j}_{\text{true}}+\boldsymbol{j}_m), \tag{13.4}$$

and thus $4\pi\boldsymbol{M}$ is that part of the field whose circulation (vortex) density is given by the inaccessible currents. But the magnetization $\boldsymbol{M}$ may also be defined as a source of induction field. Consider an elementary volume magnetic moment $\boldsymbol{M}\,dV'$, where $\boldsymbol{M}$ is the magnetic dipole moment per unit volume. The vector potential due to a distribution of magnetic dipole density $\boldsymbol{M}$ is, by Eq. (12.11)

$$\boldsymbol{A}=-\int \boldsymbol{M}\times\operatorname{grad}\left(\frac{1}{r}\right)dV', \tag{13.5}$$

where it is understood that the vector differential operator acts on the coordinates of the field point. But $\partial/\partial x(1/r)=-\partial/x'(1/r)$, and therefore

$$\boldsymbol{A}=-\int\operatorname{curl}\left(\frac{\boldsymbol{M}}{r}\right)dV'+\int\frac{\operatorname{curl}\boldsymbol{M}}{r}\,dV', \tag{13.6}$$

where the differentiation now applies to the source coordinates. The first volume integral may be transformed to a surface integral, and Eq. (13.6) becomes

$$\left.\begin{aligned}\boldsymbol{A}&=\int\frac{\boldsymbol{M}\times d\boldsymbol{S}}{r}+\int\frac{\operatorname{curl}\boldsymbol{M}}{r}\,dV'\\ &=\int\frac{\boldsymbol{M}\times d\boldsymbol{S}}{r}+\int\frac{\boldsymbol{j}_m}{r}\,dV',\end{aligned}\right\} \tag{13.7}$$

where curl $\boldsymbol{M}$ has again been identified with the inaccessible or magnetization current density. The surface integral over the boundary of the magnetized region represents a magnetization surface current density $\boldsymbol{M}\times\boldsymbol{n}$, where $\boldsymbol{n}$ is a unit vector directed outward normal to the boundary. The magnetization, analogous to the electric polarization, plays the dual role of partial field and source of field.

Materials with permanent (or residual) magnetic moment are much more common than electrets. An impressed magnetic field also quite generally produces magnetization in material media, as discovered by FARADAY. Except for ferromagnetic (and related) substances, where the magnetization depends on the history of the sample, the relation between the magnetization and the impressed field is often to a very good approximation linear, and may be written

$$\boldsymbol{M} = \chi_m \boldsymbol{H}, \tag{13.8}$$

where χ_m is the magnetic susceptibility. For historical reasons, based on the concept of magnetic poles as the source of magnetic fields, $\boldsymbol{H}$ appears in Eq. (13.8) instead of $\boldsymbol{B}$, but this makes no difference for ideally permeable media, i.e., so long as the linear relations between fields is valid. The relation between the induction field and the field intensity is

$$\boldsymbol{B} = (1 + 4\pi\chi_m)\,\boldsymbol{H} = \mu\,\boldsymbol{H} \tag{13.9}$$

where $\mu = 1 + 4\pi\chi_m$ is the permeability of the medium in electromagnetic units, the system of units in which B and H are dimensionally the same.

Before OERSTED'S discovery of electromagnetism all magnetic fields were attributed to permanent or residual magnetism. The modern formalism for describing magnetic fields reduces to this view if there are no true currents. Then curl $\boldsymbol{H} = 0$, and the magnetic field intensity is attributed to a source density

$$\operatorname{div}\boldsymbol{H} = 4\pi\varrho_m = -4\pi\operatorname{div}\boldsymbol{M} \tag{13.10}$$

analogous to the charge density of electrostatics. On this view the magnetization may be defined in terms of magnetic charges in complete analogy with electric polarization, and div $\boldsymbol{B} = 0$ is equivalent to the statement that free magnetic charges are not found in nature. The complete equivalence between current loops and magnetic dipole layers (valid except for the nature of the singularity at a point source and hence only for points outside the dipole layer) was first postulated by AMPÈRE. In practice the treatment of permanent magnets in terms of magnetic charges (poles) is often convenient, since the calculation of consequent magnetization is thereby reduced to the formalism of electrostatics. If $\boldsymbol{H}$ is written in terms of a magnetic scalar potential as

$$\boldsymbol{H} = -\operatorname{grad}\phi_m, \tag{13.11}$$

the scalar potential may be expressed in terms of ϱ_m, or in terms of the volume magnetization,

$$\phi_m = \int \frac{\boldsymbol{M}\cdot d\boldsymbol{S}}{r} - \int \frac{\operatorname{div}\boldsymbol{M}}{r}\,dV' \tag{13.12}$$

which is formally analogous to Eq. (9.5) of electrostatics.

14. Magnetic properties of matter (see Vol. XVIII of this Encyclopedia). The identification of macroscopic magnetic fields and currents, including magnetization currents, with the space-time averages of microscopic fields and currents is justified by an averaging process such as was considered in Sect. 10. The currents are of three distinct kinds for stationary matter. "True" currents may be attributed to ionic convection, under the assumption that the atomic distribution in the medium does not change. The magnetization currents are due to spatial inhomogeneity or anisotropy of atomic currents, and are correlated directly with the magnetic properties of matter. We should note that, in order to arrive at the equivalent of MAXWELL'S equations, we must also consider (slow) time variations of external fields, primarily to obtain the polarization current. That is to say, the microscopic fields and currents satisfy the vacuum relation

$$\operatorname{curl} \boldsymbol{b} = 4\pi\, \boldsymbol{j}_a + \frac{1}{c}\frac{\partial \boldsymbol{e}}{\partial t}. \tag{14.1}$$

The average of the microscopic current density is

$$\bar{\boldsymbol{j}}_a = \boldsymbol{j}_{\text{true}} + \frac{1}{c}\frac{\partial \boldsymbol{P}}{\partial t} + \operatorname{curl} \boldsymbol{M} + \text{higher order terms}. \tag{14.2}$$

The space-time average of Eq. (14.1) over a physically infinitesimal interval thus yields

$$\operatorname{curl} \bar{\boldsymbol{b}} = \frac{1}{c}\frac{\partial \boldsymbol{E}}{\partial t} + 4\pi\left(\boldsymbol{j}_{\text{true}} + \frac{1}{c}\frac{\partial \boldsymbol{P}}{\partial t} + \operatorname{curl} \boldsymbol{M}\right) \tag{14.3}$$

to the desired order, and $\boldsymbol{b}$ may be identified with the macroscopic field $\boldsymbol{B}$. The magnetization current density $\boldsymbol{j}_m = \operatorname{curl} \boldsymbol{M}$ need not be computed explicitly on the atomic level; it is often more convenient to find $\boldsymbol{M} = N\,\boldsymbol{m}$, where $\boldsymbol{m}$ is the (effective) magnetic dipole moment per atom or per molecule.

One magnetic property, diamagnetism, is presumably common to all matter, and is a measure of induced magnetic moment produced by an impressed field. Unlike dielectric susceptibility, diamagnetic susceptibility is negative, and the relative permeability of so-called diamagnetic substances is less than unity. Qualitatively this can be understood by considering an adiabatic process of setting up the field and postulating that the inaccessible currents so induced persist as resistanceless circuits. In accord with the law of induction these currents are such as to oppose the field, and thus the diamagnetic susceptibility is negative. This qualitative argument is essentially sound, but it is not consistent with classical statistical mechanics, as was shown independently by NIELS BOHR (1911) and Miss VAN LEEWEN (1919) in their respective doctoral dissertations[1]. Its validity depends on the stability of intrinsic electronic orbits against interactions with each other. Strictly speaking, only a quantum theory of magnetism can be developed consistently (see Refs. [*18*], [*22*]).

Paramagnetic substances show a positive susceptibility which is temperature dependent and may be ascribed to preferential orientation of intrinsic molecular magnetic dipoles in the field, analogous to the orientation of polar molecules in an electric field. The effect is sufficiently small that it can be understood on the assumption that the elementary dipoles do not interact with each other magnetically, although there must be an exchange of energy in accord with statistical laws. The classical theory of paramagnetism was given by PAUL LANGEVIN[2] in 1905; the magnetization density is

$$M = N\, m_0 \left(\operatorname{Cot} x - \frac{1}{x}\right), \tag{14.4}$$

with $x = m_0 H/kT$, where m_0 is the magnitude of the intrinsic atomic dipole moment. For small fields and high temperatures, the usual conditions for observation,

$$M = N\, m_0 \frac{x}{3} = \frac{N\, m_0^2\, H}{3kT}. \tag{14.5}$$

[1] J.H. VAN LEEWAN: Abstract and bibliography, "Problèmes de la théorie électronique du magnétisme". J. Phys. Radium (6) **2**, 361—377 (1921).

[2] P. LANGEVIN: Ann. chim. phys. **5**, 70 (1905).

Saturation, as predicted by (14.4), has been observed for paramagnetic salts only within a fraction of a degree of absolute zero. The inverse first power temperature dependence of paramagnetic susceptibility was discovered by PIERRE CURIE[1]. (Paramagnetic susceptibility, although small, effectively masks diamagnetism.)

Ferromagnetic substances do not in general exhibit a linear relation between magnetization and the impressed field, and they are characterized by both saturation and remanence. For weak fields in ferromagnetic materials which have low remanence it is possible to define a susceptibility, although the magnetization is typically hundreds of times greater than the impressed field. A phenomenological treatment of ferromagnetism was given by P. WEISS[2] on the assumption of a strong internal field proportional to the magnetization, which acts in addition to the external field. Microscopically this internal field represents the interaction between the elementary dipoles, which can be taken properly into account only in quantum mechanical treatment[3]. For every ferromagnetic substance there is a critical temperature, called the Curie point, above which the substance behaves like a simple paramagnetic material, except that

$$\chi_m = \frac{\text{constant}}{T - \Theta}, \tag{14.6}$$

where Θ is the Curie point. The interaction of elementary dipoles is also responsible for anti-ferromagnetism, first observed in manganese oxide, and for ferrite-type magnetism, as in magnetite. Both in anti-ferromagnetic substances and in the ferrites the interaction is such as to produce anti-parallel orientation of neighboring dipoles, but the cancellation is incomplete in the ferrites and the resulting behavior resembles ferromagnetism. For anti-ferromagnetic substances the behavior of the susceptibility above the Curie temperature is described by

$$\chi_m = \frac{\text{constant}}{T + \Theta}. \tag{14.7}$$

For details of the theory of magnetic materials, see Refs. [*2*], [*5—18*], [*22*].

15. Magnetostatic field energy. The expression for the energy density of the magnetostatic field may be computed from a consideration of FARADAY'S law of induction to find the energy that must be supplied in establishing a current in a single circuit. Let the circuit be fixed rigidly in space, so that no work is done against mechanical constraints. An external electromotive force $\mathcal{E}$ is related to the instantaneous current J by the equation

$$\mathcal{E} - \frac{d\Phi}{dt} = R J, \tag{15.1}$$

where $\Phi = \int \boldsymbol{B} \cdot d\boldsymbol{S}$ is the total magnetic flux through the circuit and R is the resistance. The rate at which the external source must supply energy is then given by

$$\mathcal{E} J = R J^2 + J \frac{d\Phi}{dt}. \tag{15.2}$$

The term RJ^2 represents the irreversible Joule heat, and is not of interest here. In accord with the last term of (15.2) the increment of energy supplied to the field

---

[1] P. CURIE: Ann. de Chimie (7) **5**, 289 (1895).

[2] Beginning with: Thèses présentées à la faculté des sci. de Paris (1900) and summarized by P. WEISS and G. FOËX: Le Magnétisme. Paris 1931.

[3] The energy of orientation is actually a quantum mechanical exchange energy of electrostatic origin.

during an infinitesimal change is

$$U_m = J\,\delta\Phi = J\int \delta\boldsymbol{B}\cdot d\boldsymbol{S} = \oint J\,\delta\boldsymbol{A}\cdot d\boldsymbol{s}, \tag{15.3}$$

where the final expression is a closed line integral around the circuit. Eq. (15.3) may be transformed to represent a volume distribution of current; if $\operatorname{div}\boldsymbol{j}=0$, the volume integral over any distribution of density $\boldsymbol{j}$ may be represented by a sum of filamentary currents J, and the sum of all integrands $J\,d\boldsymbol{s}$ is equivalent to $\int \boldsymbol{j}\,dV$. Therefore

$$U_m = \int \delta\boldsymbol{A}\cdot\boldsymbol{j}\,dV. \tag{15.4}$$

The displacement current may be neglected, and $4\pi\boldsymbol{j}=\operatorname{curl}\boldsymbol{H}$, so that

$$\left.\begin{aligned} U_m &= \frac{1}{4\pi}\int \operatorname{curl}\boldsymbol{H}\cdot\delta\boldsymbol{A}\,dV \\ &= \frac{1}{4\pi}\int \operatorname{div}(\boldsymbol{H}\times\delta\boldsymbol{A})\,dV + \frac{1}{4\pi}\int \boldsymbol{H}\cdot\operatorname{curl}\delta\boldsymbol{A}\,dV. \end{aligned}\right\} \tag{15.5}$$

The volume integral of the total divergence is equivalent to a surface integral, and for a finite distribution of steady currents both $\boldsymbol{H}$ and $\boldsymbol{A}$ fall off sufficiently rapidly at large distances that this term vanishes if the integral is taken over all of space. In that case

$$U_m = \frac{1}{4\pi}\int \boldsymbol{H}\cdot\delta\boldsymbol{B}\,dV. \tag{15.6}$$

Eq. (15.6) is quite general, and may be obtained in a variety of ways. The expression can be integrated with respect to $\boldsymbol{B}$ to yield the total energy in the magnetostatic field only if the relation between $\boldsymbol{B}$ and $\boldsymbol{H}$ is linear. If this condition is satisfied,

$$U_m = \int \frac{\boldsymbol{H}\cdot\boldsymbol{B}}{8\pi}\,dV, \tag{15.7}$$

and the energy density is given by $\boldsymbol{H}\cdot\boldsymbol{B}/8\pi$, which is analogous to the electrostatic energy density $\boldsymbol{E}\cdot\boldsymbol{D}/8\pi$. For the consideration of thermal effects it is necessary to remember that $\boldsymbol{H}\cdot\boldsymbol{B}/8\pi$ is the density of the thermodynamic free energy.

If the relation between $\boldsymbol{B}$ and $\boldsymbol{H}$ is not linear, Eq. (15.6) may be integrated only between definitely prescribed states of the material, which in general depend on the previous history of the sample. In ferromagnetic media energy is expended during the application of a cyclic field by an amount

$$\Delta U_m = \frac{1}{4\pi}\int \oint \boldsymbol{H}\cdot\delta\boldsymbol{B}\,dV \tag{15.8}$$

per cycle. Magnetic work is performed on the material, and the energy thus dissipated per unit volume is proportional to the area of a loop traced out during a cycle on a plot of $\boldsymbol{B}$ against $\boldsymbol{H}$, called the hysteresis loop. The hysteresis loop most commonly described carries the material to full magnetic saturation, but hysteresis is also exhibited by any finite cycle even though the initially unmagnetized material is not brought to saturation. It is possible, however, to define a reversible susceptibility at any value of $\boldsymbol{H}$ by means of an infinitesimal loop on the initial magnetization curve. The susceptibility so defined corresponds to changes so small that the equilibrium distribution of magnetization of the substance is not disturbed thereby.

The magnetic volume force may be found in a manner analogous to that used for the derivation of the electrostatic volume force. In the absence of any residual magnetization, and with the assumption of a linear relation between $\boldsymbol{B}$ and $\boldsymbol{H}$, the result is relatively simple. If $\boldsymbol{B}=\mu\boldsymbol{H}$,

$$\frac{dU_m}{dt} = \int \frac{\boldsymbol{H}}{4\pi}\cdot\frac{\partial \boldsymbol{B}}{\partial t}\,dV + \int \frac{H^2}{8\pi}\,\frac{\partial\mu}{\partial t}\,dV. \tag{15.9}$$

With neglect of magnetostriction the effect of a small virtual velocity $\boldsymbol{u}$ on the permeability is given by the relation $\partial\mu/\partial t = -\boldsymbol{u}\cdot\operatorname{grad}\mu$. The first term of Eq. (15.9) may be transformed by means of the field equations, but the time derivative required in the law of induction for moving media must be defined so that

$$\frac{d}{dt}\int \boldsymbol{B}\cdot d\boldsymbol{S} = \int \frac{D\boldsymbol{B}}{Dt}\cdot d\boldsymbol{S}. \tag{15.10}$$

The general relation desired is

$$\frac{D\boldsymbol{B}}{Dt} = \frac{\partial \boldsymbol{B}}{\partial t} - \operatorname{curl}(\boldsymbol{u}\times\boldsymbol{B}) + \boldsymbol{u}\operatorname{div}\boldsymbol{B}, \tag{15.11}$$

of which the last term vanishes. The equations

$$\operatorname{curl}\boldsymbol{H} = 4\pi\boldsymbol{j}, \qquad \operatorname{curl}\boldsymbol{E} = -\frac{D\boldsymbol{B}}{Dt}$$

may be combined to yield

$$\int (\boldsymbol{E}\cdot\operatorname{curl}\boldsymbol{H} - \boldsymbol{H}\cdot\operatorname{curl}\boldsymbol{E})\,dV = \int\left(4\pi\boldsymbol{j}\cdot\boldsymbol{E} + \boldsymbol{H}\cdot\frac{D\boldsymbol{B}}{Dt}\right)dV. \tag{15.12}$$

The integral on the left may be transformed to a surface integral which vanishes for magnetostatic fields. Therefore, with the aid of (15.11) and the substitution of $4\pi\boldsymbol{j}$ for $\operatorname{curl}\boldsymbol{H}$, Eq. (15.9) may be transformed to

$$-\frac{dU_m}{dt} = \int \boldsymbol{j}\cdot\boldsymbol{E}\,dV + \int \boldsymbol{f}_m\cdot\boldsymbol{u}\,dV, \tag{15.13}$$

where

$$\boldsymbol{f}_m = \boldsymbol{j}\times\boldsymbol{B} - \frac{1}{8\pi}H^2\operatorname{grad}\mu. \tag{15.14}$$

The volume integral of $\boldsymbol{j}\cdot\boldsymbol{E}$ which appears here is called the thermochemical activity, and is made up of the irreversible heat loss and the work done against electromotive forces. This energy is generated at the expense of U_m. The last term of the force density (15.14) is entirely analogous to the term $(E^2/8\pi)\operatorname{grad}\varepsilon$ of Eq. (11.10). The first term is the ponderomotive force density; for a charge density ϱ moving with velocity $\boldsymbol{v}$ it becomes $\varrho(\boldsymbol{v}\times\boldsymbol{B})/c$. The total ponderomotive force density on charged matter in the presence of both electric and magnetic fields is

$$\boldsymbol{f} = \varrho\left(\boldsymbol{E} + \frac{1}{c}\boldsymbol{v}\times\boldsymbol{B}\right), \tag{15.15}$$

in Gaussian units. This expression was first derived rigorously by H. A. LORENTZ[1], and (15.15) is called the Lorentz force. Note that the form of this force is not actually inherent in the field equations, but follows from energy considerations, i.e., from application of the principles of mechanics.

For media in which there is a linear relation between $\boldsymbol{B}$ and $\boldsymbol{H}$ a magnetic "stress" tensor may be set up by considerations formally analogous to those

[1] Arch. Néerl. **25**, 432 (1892). The Lorentz force follows from an expression for the mutual energy of two charges first derived by R. CLAUSIUS: Crelle's Journal **82**, 85 (1877). — Phil. Mag. (5), **10**, 255 (1880).

leading to the electrostatic tensor, Eq. (11.12). If the magnetostriction term is omitted,

$$T_{ij} = \frac{1}{4\pi}\left(H_i B_j - \frac{\delta_{ij}}{2} H_k B_k\right), \tag{15.16}$$

with the customary summation convention. FARADAY's interpretation of the field as tension along the lines of flux and pressure at right angles (of magnitude given by $\boldsymbol{H}\cdot\boldsymbol{B}/8\pi$ and $-\boldsymbol{H}\cdot\boldsymbol{B}/8\pi$ respectively) follows immediately, in complete analogy to the electric stresses.

IV. Rapidly variable fields.

16. Energy conservation and POYNTING's theorem. MAXWELL's equations for continuous media (including vacuum) may be written without approximation as

$$\operatorname{div}\boldsymbol{D} = 4\pi\varrho, \tag{16.1}$$

$$\operatorname{div}\boldsymbol{B} = 0, \tag{16.2}$$

$$\operatorname{curl}\boldsymbol{E} = -\frac{1}{c}\frac{\partial\boldsymbol{B}}{\partial t}, \tag{16.3}$$

$$\operatorname{curl}\boldsymbol{H} = \frac{1}{c}\frac{\partial\boldsymbol{D}}{\partial t} + 4\pi\boldsymbol{j} \tag{16.4}$$

in Gaussian units, where ϱ and $\boldsymbol{j}$ are to be interpreted as true charge and current densities (in esu and emu respectively) which satisfy the continuity equation

$$\operatorname{div}\boldsymbol{j} = -\frac{1}{c}\frac{\partial\varrho}{\partial t}.$$

An energy integral of MAXWELL's equations may be obtained by formal combination of the two vector relations: after scalar multiplication of (16.3) by $\boldsymbol{H}$ and of (16.4) by $\boldsymbol{E}$ one of these equations is subtracted from the other. The result is

$$\boldsymbol{H}\cdot\operatorname{curl}\boldsymbol{E} - \boldsymbol{E}\cdot\operatorname{curl}\boldsymbol{H} + \frac{1}{c}\boldsymbol{H}\cdot\dot{\boldsymbol{B}} + \frac{1}{c}\boldsymbol{E}\cdot\dot{\boldsymbol{D}} + 4\pi\boldsymbol{j}\cdot\boldsymbol{E} = 0. \tag{16.5}$$

The first two terms are equal to $\operatorname{div}(\boldsymbol{E}\times\boldsymbol{H})$ by a general vector identity, and Eq. (16.5), when multiplied by $c/4\pi$, becomes

$$\frac{1}{4\pi}(\boldsymbol{H}\cdot\dot{\boldsymbol{B}} + \boldsymbol{E}\cdot\dot{\boldsymbol{D}}) + c\boldsymbol{j}\cdot\boldsymbol{E} + \operatorname{div} c(\boldsymbol{E}\times\boldsymbol{H})/4\pi = 0. \tag{16.6}$$

Eq. (16.6) is POYNTING's theorem[1], and $c(\boldsymbol{E}\times\boldsymbol{H})/4\pi$ (which may be denoted by $\boldsymbol{N}$) is known as the Poynting vector. The term $c\boldsymbol{j}\cdot\boldsymbol{E} = \varrho\boldsymbol{v}\cdot\boldsymbol{E}$ is readily identified as the rate at which the field $\boldsymbol{E}$ does work on the charge ϱ per unit volume. If there are linear constitutive relations between $\boldsymbol{H}$ and $\boldsymbol{B}$, and between $\boldsymbol{E}$ and $\boldsymbol{D}$, the first two terms may be written as the time rate of change of $(\boldsymbol{H}\cdot\boldsymbol{B} + \boldsymbol{E}\cdot\boldsymbol{D})/8\pi$. The meaning of the last term becomes clear if Eq. (16.6) is integrated over any volume and the divergence theorem is applied:

$$\frac{\partial}{\partial t}\int\frac{\boldsymbol{H}\cdot\boldsymbol{B} + \boldsymbol{E}\cdot\boldsymbol{D}}{8\pi}\,dV + \int c\boldsymbol{j}\cdot\boldsymbol{E}\,dV + \int\frac{c(\boldsymbol{E}\times\boldsymbol{H})}{4\pi}\cdot d\boldsymbol{S} = 0. \tag{16.7}$$

Eq. (16.7) expresses the conservation of energy if $(\boldsymbol{H}\cdot\boldsymbol{B} + \boldsymbol{E}\cdot\boldsymbol{D})/8\pi$ is identified with the electromagnetic energy density. The expression

$$c(\boldsymbol{E}\times\boldsymbol{H})/4\pi = \boldsymbol{N}$$

[1] J.H. POYNTING: Phil. Trans. Roy. Soc. Lond. **175**, 343 (1884).

must therefore be interpreted as the flux of energy per unit area through the surface of the volume considered. (It is assumed that no charge crosses the boundary.)

In any region devoid of charge POYNTING's theorem takes the form of a hydrodynamic equation of continuity: the time rate of change of the sum of electric and magnetic field energy densities equals the negative divergence of the energy flux density $\boldsymbol{N}$. The Poynting vector has no definition independent of the above derivation, however, and has no effect on the energy balance if the divergence of $\boldsymbol{N}$ is everywhere equal to zero, as in crossed static fields.

17. Electromagnetic momentum. The identification of $(\boldsymbol{E}\cdot\boldsymbol{D}+\boldsymbol{H}\cdot\boldsymbol{B})/8\pi$ with energy density when no restrictions are imposed on the time variation of the fields suggests that the volume force should be derivable as the divergence of an energy tensor T_{ij} which is the sum of Eqs. (11.12) and (15.16). Actually, with neglect of the electrostriction and magnetostriction terms,

$$\left.\begin{aligned}\frac{\partial T_{ij}}{\partial x_j} &= \frac{1}{4\pi}\Big(E_i\frac{\partial D_j}{\partial x_j}+D_j\frac{\partial E_i}{\partial x_j}-\frac{1}{2}E^2\frac{\partial\varepsilon}{\partial x_i}-D_j\frac{\partial E_j}{\partial x_i}+H_i\frac{\partial B_j}{\partial x_j}+\\ &\quad+B_j\frac{\partial H_i}{\partial x_j}-\frac{1}{2}H^2\frac{\partial\mu}{\partial x_i}-B_j\frac{\partial H_j}{\partial x_i}\Big)\\ &=\left[\varrho\boldsymbol{E}+\boldsymbol{j}\times\boldsymbol{B}-\frac{1}{8\pi}E^2\operatorname{grad}\varepsilon\frac{1}{8\pi}\operatorname{grad}\mu+\frac{\partial}{\partial t}\frac{(\boldsymbol{D}\times\boldsymbol{B})}{4\pi c}\right]_i\end{aligned}\right\}\tag{17.1}$$

where the final formulation has been obtained with the aid of MAXWELL's equations. To trace the significance of this expression it will suffice to consider an unbounded medium of uniform properties, so that the susceptibilities have zero gradient. Under these circumstances it is clear that only the first terms of Eq. (17.1) constitute the volume force as determined by the mechanical motion of the charges, since the fields $\boldsymbol{E}$ and $\boldsymbol{B}$ were defined initially in terms of the mechanical forces $\varrho\,\boldsymbol{E}$ and $\boldsymbol{j}\times\boldsymbol{B}$. In the static case it is consistent to interpret T as a momentum dyadic such that $-\operatorname{div} T$ gives the volume force and $\int T\cdot d\boldsymbol{S}$ is the flow across a surface S of mechanical momentum associated with the charge and current densities ϱ and $\boldsymbol{j}$. Here, however,

$$\int \operatorname{div} T\,dV=\int T\cdot d\boldsymbol{S}=-\frac{d\boldsymbol{p}}{dt}-\frac{\partial}{\partial t}\int\frac{\boldsymbol{D}\times\boldsymbol{B}}{4\pi c}\,dV,\tag{17.2}$$

where $\boldsymbol{p}$ is the mechanical momentum of the charge and current system. In other words,

$$\frac{d\boldsymbol{p}}{dt}=-\int T\cdot d\boldsymbol{S}-\frac{\partial}{\partial t}\int\frac{\boldsymbol{D}\times\boldsymbol{B}}{4\pi c}\,dV\tag{17.3}$$

is the conservation law for momentum which follows from MAXWELL's equations, and it is the time rate of change of mechanical momentum plus that of $\int(\boldsymbol{D}\times\boldsymbol{B})/4\pi c\,dV$ which is equal to $-\int T\cdot d\boldsymbol{S}$, the flow of momentum across the surface bounding the region under consideration. We are thus forced by the theory to attribute momentum to the electromagnetic field itself. For an isolated system, one in which the fields vanish at large distances so that the surface integral in (17.3) is zero, it is the sum of $\boldsymbol{p}$ and $\int(\boldsymbol{D}\times\boldsymbol{B})/4\pi c\,dV$ that vanishes, and not the mechanical momentum alone.

Electromagnetic momentum as represented by the density $(\boldsymbol{D}\times\boldsymbol{B})/4\pi c$ does not depend on the presence of charges, and exists even in vacuum as $(\boldsymbol{E}\times\boldsymbol{H})/4\pi c$. Historically the vacuum term was first interpreted as mechanical momentum of the ether, produced by electromagnetic fields, in the same sense that the vacuum displacement current was considered as motion of the ether, and that vacuum field energy was attributed to ether stresses. In modern theory each of these properties is attributed directly to the fields. We note that in vacuo $(\boldsymbol{D}\times\boldsymbol{B})/4\pi c = \boldsymbol{N}/c^2$, i.e., that the momentum density multiplied by c^2 is equal to the flux of energy density. With the advent of special relativity this relation between momentum and flux of energy was recognized as completely general.

Since it is necessary to attribute linear momentum to an electromagnetic field, it is also necessary to allow for the possibility of angular momentum corresponding to the linear momentum density $(\boldsymbol{D}\times\boldsymbol{B})/4\pi c$; the density of angular momentum is given by $\boldsymbol{r}\times(\boldsymbol{D}\times\boldsymbol{B})/4\pi c$. In an isolated system the sum of mechanical and electromagnetic angular momentum is conserved.

18. The homogeneous wave equation and plane waves. For a homogeneous isotropic medium in which the fields are related by linear constitutive equations MAXWELL's equations may be combined to yield a second order differential equation in a single field variable. By taking the curl of (16.3) and making use of (16.4) one obtains

$$\operatorname{curl}\operatorname{curl}\boldsymbol{E} = -\frac{\mu}{c}\frac{\partial}{\partial t}\operatorname{curl}\boldsymbol{H} = -\frac{\varepsilon\mu}{c^2}\frac{\partial^2 E}{\partial t^2} - \frac{4\pi\mu}{c}\frac{\partial \boldsymbol{j}}{\partial t}. \tag{18.1}$$

If no true charges are present $\operatorname{div}\boldsymbol{E}=0$, and in the absence of external electromotive forces $c\boldsymbol{j}=\sigma\boldsymbol{E}$ by OHM's law, so that

$$\nabla^2\boldsymbol{E} = \frac{\mu\varepsilon}{c^2}\frac{\partial^2\boldsymbol{E}}{\partial t^2} + \frac{4\pi\mu\sigma}{c^2}\frac{\partial\boldsymbol{E}}{\partial t}. \tag{18.2}$$

Eq. (18.2) is known as the general homogeneous wave equation.

The two terms on the right of Eq. (18.2) arise from the displacement current and the conduction current, respectively. In most applications one or the other of these two terms is negligible. Their relative importance depends on the time variation of the field as well as on the properties of the medium. Consider a field that varies harmonically with the time,

$$\boldsymbol{E}(x,y,z,t) = \boldsymbol{E}(x,y,z)\,\mathrm{e}^{-i\omega t}. \tag{18.3}$$

Eq. (18.2) then becomes

$$\nabla^2\boldsymbol{E} + \frac{\mu\varepsilon\omega^2}{c^2}\boldsymbol{E} + \frac{4\pi i\mu\sigma\omega}{c^2}\boldsymbol{E} = \nabla^2\boldsymbol{E} + \left(1+\frac{4\pi\sigma i}{\varepsilon\omega}\right)\frac{\mu\varepsilon\omega^2}{c^2}\boldsymbol{E}. \tag{18.4}$$

The magnitude of the ratio $(4\pi\sigma/\varepsilon):\omega$ determines the nature of the solution. The quantity $\varepsilon/4\pi\sigma$ is known as the relaxation time of the medium, a characteristic time for the establishment of stationary charge distribution. (This may be seen by combining the continuity equation for charge and current, OHM's law, and the source equation for $\boldsymbol{E}$ to obtain a differential equation for the charge density.) For pure metals the relaxation time is of the order of 10^{-14} seconds; thus for all frequencies lower than those of the optical region the displacement current in metals can be neglected. The resulting differential equation is of the same form as that for heat conduction or diffusion. For homogeneous isotropic insulating media, and of course in empty space, Eq. (18.2) becomes the homogeneous equation for wave propagation,

$$\nabla^2\boldsymbol{E} = \frac{\mu\varepsilon}{c^2}\frac{\partial^2\boldsymbol{E}}{\partial t^2}. \tag{18.5}$$

Under the same conditions it can be similarly shown that

$$\nabla^2\boldsymbol{B} = \frac{\mu\varepsilon}{c^2}\frac{\partial^2\boldsymbol{B}}{\partial t^2}, \tag{18.6}$$

but Eqs. (18.5) and (18.6) are not to be solved independently, since the fields must satisfy the first order Maxwell relations.

The meaning of Eq. (18.5) is most readily exhibited on the assumption of a plane wave, i.e., on the assumption that the fields are uniform over a plane, so that the spatial derivatives with respect to two Cartesian coordinates vanish. If the coordinate system is so chosen that x is the independent variable, the plane wave equation becomes

$$\frac{\partial^2 \boldsymbol{E}(x,t)}{\partial x^2} = \frac{\mu\,\varepsilon}{c^2}\,\frac{\partial^2 \boldsymbol{E}(x,t)}{\partial t^2}\,, \tag{18.7}$$

a relation which is satisfied by any vector function of the argument $(x \pm c\,t/\sqrt{\mu\,\varepsilon})$. But $f(x \pm c\,t/\sqrt{\mu\,\varepsilon})$ represents a disturbance traveling along $\mp x$ with a velocity $v = c/\sqrt{\mu\,\varepsilon}$. In free space this velocity is c; it was the agreement between c as the empirical numerical ratio between electromagnetic and electrostatic units of charge on the one hand and the velocity of light on the other which led MAXWELL to conclude from the above equation that "light itself (including radiant heat, and other radiations if any) is an electromagnetic disturbance in the form of waves propagated through the electromagnetic field according to electromagnetic law"[1].

MAXWELL was able to show that electromagnetic waves are transverse: the solutions of Eq. (18.7) are restricted to fields of zero divergence, and thus apart from uniform static fields the components of $\boldsymbol{E}$ in the direction of propagation vanish, as do those of $\boldsymbol{B}$. The relation between $\boldsymbol{B}$ and $\boldsymbol{E}$ also follows from the field equations. Without loss of generality it may be assumed that the solution of Eq. (18.5) is plane polarized, i.e., that $\boldsymbol{E}$ lies in a single plane. Then, if $\boldsymbol{n}$ is a unit normal along positive x, Eq. (16.3) reduces to

$$\boldsymbol{n}\times\frac{\partial \boldsymbol{E}}{\partial x} = -\frac{1}{c}\,\frac{\partial \boldsymbol{B}}{\partial t}\,. \tag{18.8}$$

Moreover, as a formal solution of Eq. (18.7), $\boldsymbol{E}$ satisfies the relation

$$\frac{\partial \boldsymbol{E}}{\partial x} = \pm\frac{\sqrt{\mu\varepsilon}}{c}\,\frac{\partial \boldsymbol{E}}{\partial t}\,, \tag{18.9}$$

and thus

$$\boldsymbol{n}\times\frac{\partial \boldsymbol{E}}{\partial t} = \mp\frac{1}{\sqrt{\mu\varepsilon}}\,\frac{\partial \boldsymbol{B}}{\partial t}\,. \tag{18.10}$$

Except for static fields, Eq. (18.10) reduces to

$$\frac{1}{\sqrt{\mu\varepsilon}}\,\boldsymbol{B} = \mp\,\boldsymbol{n}\times\boldsymbol{E}\,. \tag{18.11}$$

The negative sign in Eq. (1) describes a wave traveling in the direction of positive x; therefore $\boldsymbol{E}$, $\boldsymbol{B}$ and the direction of propagation form a right-handed set of orthogonal vectors. The Poynting vector for a plane wave traveling in the positive x direction is

$$\boldsymbol{N} = \frac{c}{4\pi}\,\boldsymbol{E}\times\boldsymbol{H} = \sqrt{\frac{\varepsilon}{\mu}}\,\frac{c\,E^2}{4\pi}\,\boldsymbol{n}\,. \tag{18.12}$$

From Eq. (18.11) it also follows that the energy density in a progressive wave traversing a dielectric is equally divided between the electric and magnetic fields: $\boldsymbol{E}\cdot\boldsymbol{D} = \boldsymbol{B}\cdot\boldsymbol{H}$. If the total energy density is denoted by W, the Poynting vector may be written as

$$\boldsymbol{N} = \frac{c}{\sqrt{\mu\varepsilon}}\,\frac{\varepsilon\,E^2}{4\pi}\,\boldsymbol{n} = W\,v\,\boldsymbol{n}\,, \tag{18.13}$$

where v is the velocity of the wave, which in free space is simply c.

[1] J.C. MAXWELL: Phil. Trans. Roy. Soc. Lond. **155**, 466 (1865).

The prediction that the velocity of light in a dielectric should be inversely proportional to the square root of the dielectric constant is well substantiated in many insulating materials, although not in those whose molecules are polar. (The permeability of most dielectrics is so nearly unity that its variation may be neglected.) Note, however, that MAXWELL's theory of waves in a continuous medium can give no explanation for dispersion, i.e., for the variation of velocity with the frequency of the electromagnetic disturbance.

19. Radiation pressure. Both the field momentum concept and that of a volume force derived from MAXWELL's stress tensor lead to the conclusion that radiation should exert pressure on being absorbed or reflected. Consider first a perfectly absorbing surface (one in which the fields are absorbed within a short penetration depth) at right angles to the direction of propagation of a plane wave in vacuum, and let the absorber be sufficiently massive that its recoil may be neglected. The radiation in a right cylinder of unit cross section and length $c\,dt$ will be absorbed per unit absorber area in time dt; this corresponds to an amount of field momentum given by $(1/4\pi)\,(\boldsymbol{E}\times\boldsymbol{H})\,dt$. The corresponding force per unit area on the absorber is then

$$\text{Pressure} = \frac{|\boldsymbol{E}\times\boldsymbol{H}|}{4\pi} = \frac{|\boldsymbol{N}|}{c} = W, \tag{19.1}$$

numerically equal to the energy density, if momentum is to be conserved.

The same expression, except for transient terms, may be derived by integrating the ponderomotive volume force implied by (17.3) from the surface of the absorber to a depth in which the fields have been absorbed. For a plane wave traveling in the x direction, whose fields satisfy the relations of Sect. 18,

$$\int_0^\infty \overline{f_x}\,dx = \frac{\overline{\boldsymbol{E}\cdot\boldsymbol{D}+\boldsymbol{B}\cdot\boldsymbol{H}}}{8\pi} + \left[\frac{\partial}{\partial t}\,\frac{(\boldsymbol{E}\times\boldsymbol{H})_x}{4\pi c}\,dx\right]_{\text{av}}. \tag{19.2}$$

The last term has instantaneous values, but its time average is zero for a steady state, or for any finite pulse. This term therefore contributes nothing to the net transfer of momentum, and again the net result of the wave absorption is a pressure equal to the energy density.

Slightly more elaborate computations may be made for totally or partially reflecting surfaces, and for radiation at oblique incidence (see Ref. [*12*]). For normal incidence of radiation on a perfectly reflecting surface there is a reflected wave of amplitude equal to that of the incident wave, and the pressure is given by $|\boldsymbol{E}\times\boldsymbol{H}|/2\pi$. A result of great interest is that for isotropic radiation in a cavity (the ideal blackbody), where the random distribution of radiation fields results in a pressure normal to the walls given by

$$\text{Pressure} = \tfrac{1}{3}\,W. \tag{19.3}$$

This equation forms the basis for deriving the Stefan-Boltzmann law and the Wien displacement law for blackbody radiation.

The prediction of radiation pressure was verified quantitatively by LEBEDEW[1] (1899) and by NICHOLS and HULL[2] (1901). The phenomenon is responsible (at least in part) for the form of comets, as predicted by EULER[3] in 1748 and previously

[1] P. LEBEDEW: Ann. Phys. **6**, 433 (1901).

[2] E.F. NICHOLS and G.F. HULL: Astrophys. J. **17**, 315 (1903). Also Ann. Phys. **12**, 225 (1903).

[3] L. EULER: Histoire Acad. Berlin **2**, 117 (1748), or WHITTAKER **1**, 274.

suggested by KEPLER. Radiation pressure plays a large role in the internal dynamics of stars[1].

20. Plane waves in a dielectric and the effect of dielectric boundaries. The assumption of harmonic dependence of the fields on the time, introduced above to exhibit the relative importance of the displacement current and conduction current in the wave equation (18.2), is justified in considering the general solution of MAXWELL'S equations. Since these equations are linear, particular harmonic solutions may be superposed to constitute any solution required by physical boundary and initial conditions. In other words, no loss of generality is involved in treating separately the Fourier components of the field variables, for each of which the time dependence is indicated by a factor $e^{-i\omega t}$. Fourier analysis of the fields of moving charges is outlined in Sect. 37 for application to particular charge and current configurations; here it is only necessary to admit the validity of this resolution.

The plane wave solution of the homogeneous wave equation (18.7) may therefore be taken as the real part of

$$\boldsymbol{E} = \boldsymbol{E}_0 \, e^{-i(\omega t - \boldsymbol{k}\cdot\boldsymbol{r})}, \tag{20.1}$$

where $\boldsymbol{r}$ is the coordinate vector and $\boldsymbol{k}$ is the propagation vector, a vector in the direction of wave propagation whose magnitude is $\omega/v = \omega\sqrt{\mu\varepsilon}/c = 2\pi/\lambda$. The wave velocity v is also called the phase velocity, since the planes perpendicular to the direction of propagation are surfaces of constant phase of the field (20.1). The ratio c/v is the absolute index of refraction of the medium in optics. In all but ferromagnetic materials μ is very nearly unity, and MAXWELL pointed out that the index of refraction of dielectrics should therefore be equal to the square root of the dielectric constant. The notable departures from this rule (as, for example, water) cannot be understood on the basis of macroscopic theory, but many aspects of wave propagation, including the effects of dielectric boundaries, are accounted for by the electrodynamics of continua. These effects and other aspects of wave transmission are treated fully in Vol. XVII of this Encyclopedia, and are only summarized here as an integral part of classical electrodynamics.

Consider two homogeneous isotropic media having a common plane boundary, and let a plane harmonic wave transmitted by one medium be incident on this boundary. The boundary conditions at the interface (on which there is no true charge or current) demand the continuity of the tangential components of $\boldsymbol{E}$ and $\boldsymbol{H}$ and the continuity of the components of $\boldsymbol{B}$ and $\boldsymbol{D}$ normal to the surface. These conditions cannot be satisfied by a single progressive wave if the two media are characterized by different electromagnetic properties; in general a fraction of the incident energy is reflected, and that transmitted into the second medium is refracted, i.e. is altered in direction and wavelength. The physical situation is represented schematically in Fig. 7, with the direction of the incident plane wave indicated by $\boldsymbol{k}$. The propagation vector $\boldsymbol{k}$ and the unit vector $\boldsymbol{n}$ which specifies the plane boundary together determine what is called the plane of incidence of the wave; $\boldsymbol{k}'$ and $\boldsymbol{k}''$ are the propagation vectors of the transmitted and reflected waves, as yet undetermined. In accord with the relations derived in Sect. 18,

$$\boldsymbol{E} = \boldsymbol{E}_0 \, e^{i(\boldsymbol{k}\cdot\boldsymbol{r} - \omega t)}, \qquad \boldsymbol{H} = \frac{c\,\boldsymbol{k}\times\boldsymbol{E}}{\omega\,\mu_1}. \tag{20.2}$$

The continuity of the tangential fields is possible only if the exponentials are the same at the boundary for all three sets of fields, and thus the angular frequency ω

[1] See, e.g., LAWRENCE H. ALLER: Astrophysics. New York: Ronald, Vol. 1, 1953; Vol. 2, 1954. Also Vol. LI of this Encyclopedia.

of the transmitted and reflected waves must be the same as for the incident wave. With primed symbols for the fields in medium 2 and double primes for the reflected wave, the fields corresponding to refraction and reflection are

$$\boldsymbol{E}' = \boldsymbol{E}_0' \, e^{i(\boldsymbol{k}' \cdot \boldsymbol{r} - \omega t)}, \qquad \boldsymbol{H}' = \frac{c\,\boldsymbol{k}' \times \boldsymbol{E}'}{\omega \mu_2}; \tag{20.3}$$

$$\boldsymbol{E}'' = \boldsymbol{E}_0'' \, e^{i(\boldsymbol{k}'' \cdot \boldsymbol{r} - \omega t)}, \qquad \boldsymbol{H}'' = \frac{c\,\boldsymbol{k}'' \times \boldsymbol{E}''}{\omega \mu_1}. \tag{20.4}$$

Equality of the exponentials further demands that

$$\boldsymbol{k} \cdot \boldsymbol{r} = \boldsymbol{k}' \cdot \boldsymbol{r} = \boldsymbol{k}'' \cdot \boldsymbol{r} \tag{20.5}$$

on the boundary surface, and it is evident that all the propagation vectors are coplanar. Eq. (20.5) also leads to the well known laws of reflection and refraction. If for convenience the origin of the coordinate vector $\boldsymbol{r}$ is located in the boundary plane, $\boldsymbol{n} \cdot \boldsymbol{r} = 0$, and $\boldsymbol{n} \times (\boldsymbol{n} \times \boldsymbol{r}) = -\boldsymbol{r}$. Therefore Eq. (20.5) yields

$$(\boldsymbol{k} - \boldsymbol{k}'') \times \boldsymbol{n} \cdot (\boldsymbol{n} \times \boldsymbol{r}) = 0, \tag{20.6}$$

$$(\boldsymbol{k} - \boldsymbol{k}') \times \boldsymbol{n} \cdot (\boldsymbol{n} \times \boldsymbol{r}) = 0. \tag{20.7}$$

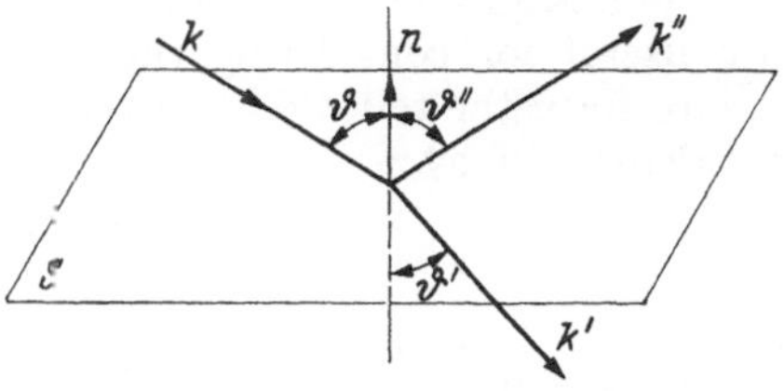

Fig. 7. Plane wave incident on a dielectric boundary.

Since $\boldsymbol{k}$ and $\boldsymbol{k}''$ in the same medium are equal in magnitude, and since $|\boldsymbol{k}|/|\boldsymbol{k}'| = v_2/v_1$, Eqs. (20.6) and (20.7) may be written in terms of the angles of incidence, reflection and refraction indicated in Fig. 7,

$$\vartheta = \vartheta'', \tag{20.8}$$

$$\sin\vartheta / \sin\vartheta' = v_1/v_2, \tag{20.9}$$

in the usual way. Eq. (20.9) is known as SNELL's law[1].

The relations of the reflected and transmitted amplitudes to each other and to those of the incident wave follow from the boundary conditions, which may be written as

$$\boldsymbol{n} \times (\boldsymbol{E} + \boldsymbol{E}'') = \boldsymbol{n} \times \boldsymbol{E}', \tag{20.10}$$

$$\boldsymbol{n} \times (\boldsymbol{k} \times \boldsymbol{E} + \boldsymbol{k}'' \times \boldsymbol{E}'')/\mu_1 = \boldsymbol{n} \times (\boldsymbol{k}' \times \boldsymbol{E}')/\mu_2. \tag{20.11}$$

Both the analysis and the resulting formulas are greatly simplified if the components of the primary vector $\boldsymbol{E}$ at right angles to the plane of incidence and in the plane of incidence are treated separately. The amplitude relations may be written in a variety of forms; perhaps the simplest is

$$\boldsymbol{n} \cdot \boldsymbol{E} = 0, \qquad \left.\begin{matrix} 2E_0 \\ 2E_0'' \end{matrix}\right\} = \left(1 \pm \frac{\mu_1}{\mu_2} \frac{\sin\vartheta \cos\vartheta'}{\sin\vartheta' \cos\vartheta}\right) E_0' \tag{20.12}$$

for the polarization in which the electric field vector is perpendicular to the plane of incidence, and

$$\boldsymbol{n} \cdot \boldsymbol{H} = 0, \qquad \left.\begin{matrix} 2E_0 \\ 2E_0'' \end{matrix}\right\} = \left(\frac{\mu_1}{\mu_2} \frac{\sin\vartheta}{\sin\vartheta'} \pm \frac{\cos\vartheta'}{\cos\vartheta}\right) E_0' \tag{20.13}$$

for the polarization in which the electric field vector lies in the plane of incidence. It is possible to express ϑ' in terms of ϑ and the relative index of refraction by means of SNELL's law, but the ratio of the magnetic permeabilities persists in

[1] First published by RENÉ DESCARTES, Dioptrique, Discours second (1637).

the general formulation. Only if $\mu_1 \approx \mu_2$ do Eqs. (20.12) and (20.13) reduce to the simpler Fresnel formulas:

$$E_0 : E_0' : E_0'' = \sin(\vartheta' + \vartheta) : [\sin(\vartheta' + \vartheta) + \sin(\vartheta' - \vartheta)] : \sin(\vartheta' - \vartheta) \tag{20.14}$$

for $\boldsymbol{E}$ perpendicular to the plane of incidence, and

$$E_0 : E_0' : E_0'' = \tan(\vartheta + \vartheta') : \left[\frac{\tan(\vartheta + \vartheta')}{\cos(\vartheta - \vartheta')} - \frac{\tan(\vartheta - \vartheta')}{\cos(\vartheta + \vartheta')}\right] : \tan(\vartheta - \vartheta') \tag{20.15}$$

for $\boldsymbol{E}$ in the plan of incidence. Relations equivalent to (20.14) and (20.15) were derived by FRESNEL[1] in 1823 on the basis of an elastic ether model for wave propagation.

21. Energy transmission across a boundary; production of polarized light (see also Vol. XXIV of this Encyclopedia). The Fresnel formulas (20.14) and (20.15) are useful for considering the transmission of energy across the boundary. In vector form the reflected and refracted fields at the boundary are related to the incident field by

$$\boldsymbol{E}' = \frac{2\cos\vartheta \sin\vartheta'}{\sin(\vartheta' + \vartheta)} \boldsymbol{E}, \tag{21.1}$$

$$\boldsymbol{E}'' = \frac{\sin(\vartheta' - \vartheta)}{\sin(\vartheta' + \vartheta)} \boldsymbol{E}, \tag{21.2}$$

for the components of $\boldsymbol{E}$ normal to the plane of incidence. For the components of $\boldsymbol{E}$ lying in the plane of incidence,

$$\boldsymbol{k}' \times \boldsymbol{E}' = \frac{2\cos\vartheta \sin\vartheta'}{\sin(\vartheta + \vartheta')\cos(\vartheta - \vartheta')} \sqrt{\frac{\varepsilon_2}{\varepsilon_1}}\, \boldsymbol{k} \times \boldsymbol{E}, \tag{21.3}$$

$$\boldsymbol{k}'' \times \boldsymbol{E}'' = \frac{\tan(\vartheta - \vartheta')}{\tan(\vartheta + \vartheta')} \boldsymbol{k} \times \boldsymbol{E}. \tag{21.4}$$

The time average of the energy flow is given by the real part of the complex Poynting vector, $\overline{\boldsymbol{N}} = \frac{1}{2} \frac{\boldsymbol{E} \times \boldsymbol{H}^*}{4\pi}$, and the normal component of the energy flow across the surface must be continuous. Let us temporarily exclude complex angles of refraction, i.e., exclude angles of incidence for which $\sin\vartheta > v_1/v_2$ in the case that $v_2 > v_1$. The energy principle then simply demands that $\boldsymbol{n} \cdot (\boldsymbol{N} - \boldsymbol{N}'') = \boldsymbol{n} \cdot \boldsymbol{N}'$, which may be written as

$$\sqrt{\varepsilon_1}\, E^2 \cos\vartheta = \sqrt{\varepsilon_2}\, E'^2 \cos\vartheta' + \sqrt{\varepsilon_1}\, E''^2 \cos\vartheta. \tag{21.5}$$

The reflection and transmission coefficients are defined by

$$R = \frac{\boldsymbol{n} \cdot \overline{\boldsymbol{N}}''}{\boldsymbol{n} \cdot \overline{\boldsymbol{N}}} = \frac{E_0''^2}{E_0^2}, \qquad T = \frac{\boldsymbol{n} \cdot \overline{\boldsymbol{N}}'}{\boldsymbol{n} \cdot \overline{\boldsymbol{N}}} = \sqrt{\frac{\varepsilon_2}{\varepsilon_1}} \frac{\cos\vartheta'}{\cos\vartheta} \frac{E_0'^2}{E_0^2}, \tag{21.6}$$

with $R + T = 1$. If $\boldsymbol{E}$ is normal to the plane of incidence,

$$R_\perp = \frac{\sin^2(\vartheta' - \vartheta)}{\sin^2(\vartheta' + \vartheta)}, \qquad T_\perp = \frac{\sin 2\vartheta \sin 2\vartheta'}{\sin^2(\vartheta + \vartheta')}. \tag{21.7}$$

If $\boldsymbol{E}$ lies in the plane of incidence,

$$R_\parallel = \frac{\tan^2(\vartheta - \vartheta')}{\tan^2(\vartheta + \vartheta')}, \qquad T_\parallel = \frac{\sin 2\vartheta \sin 2\vartheta'}{\sin^2(\vartheta + \vartheta')\cos^2(\vartheta - \vartheta')}. \tag{21.8}$$

[1] The manuscript was believed lost, but was found among the papers of FOURIER, and printed in Mém. Acad. Sci. 11, 393 (1832).

The reflection coefficient $R_{\parallel}$ is zero if $\vartheta+\vartheta'=\pi/2$. When this condition is satisfied the reflected and transmitted rays are perpendicular to each other, and $\tan\vartheta_0=\sqrt{\varepsilon_2/\varepsilon_1}$. The angle ϑ_0 is known as BREWSTER'S angle[1] or the polarizing angle; only the component of radiation for which $\boldsymbol{E}$ is normal to the plane of incidence is reflected at that angle.

If the refractive index of the second medium is less than that of the first, a range of incidence angles exists for which $\sqrt{\varepsilon_1/\varepsilon_2}\sin\vartheta>1$; then $\sin\vartheta'>1$, and $\cos\vartheta$ is pure imaginary. The angle ϑ' itself, having reached the real value $\pi/2$ as ϑ is increased to $\arcsin(v_1/v_2)$, may be written for still greater angles of incidence as $\vartheta=\pi/2\pm i\delta$; thus $\sin\vartheta'=\operatorname{Cos}\delta$. The Fresnel formula (20.14) yields

$$-\frac{E_0''}{E_0}=\frac{\sin\left(\frac{\pi}{2}\pm i\,\delta-\vartheta\right)}{\sin\left(\frac{\pi}{2}\pm i\,\delta+\vartheta\right)}=\frac{\cos\left(\vartheta\mp i\,\delta\right)}{\cos\left(\vartheta\pm i\,\delta\right)}=\mathrm{e}^{i\varphi} \tag{21.9}$$

for E perpendicular to the plane of incidence, and from (20.15)

$$-\frac{E_0''}{E_0}=-\frac{\tan\left(\frac{\pi}{2}\pm i\,\delta-\vartheta\right)}{\tan\left(\frac{\pi}{2}\pm i\,\delta+\vartheta\right)}=\frac{\cot\left(\vartheta\mp i\,\delta\right)}{\cot\left(\vartheta\pm i\,\delta\right)}=\mathrm{e}^{i\psi} \tag{21.10}$$

for $\boldsymbol{E}$ in the plane of incidence. The reflection coefficient is unity for both cases, and there is said to be total reflection. The two components are changed in phase by unequal amounts, and as a result plane polarized light is in general totally reflected with elliptical polarization. The angle of incidence for which there is a maximum phase difference between the components of $\boldsymbol{E}$ parallel to and perpendicular to the plane of incidence is given by

$$\sin^2\vartheta_{\max}=\frac{2}{\frac{\varepsilon_1}{\varepsilon_2}+1}=\frac{2}{\left(\frac{v_2}{v_1}\right)^2+1}, \tag{21.11}$$

and the phase difference at this angle is

$$\varphi-\psi=2\operatorname{arc\,tan}\frac{v_1}{2v_2}\left(\frac{v_1^2}{v_2^2}-1\right) \tag{21.12}$$

(see Ref. [20]). Although the transmission coefficient as defined by (21.6) is zero, the instantaneous fields do not vanish in the second medium; since $\cos\vartheta'$ and therefore $\boldsymbol{k}'\cdot\boldsymbol{n}$ are imaginary, it is clear that these fields are confined to the neighborhood of the surface. These fields have been detected, and some transmission obtained, for waves sufficiently long that a second object of large index of refraction may be placed within a few wavelengths of the "totally reflecting" surface.

22. Waves in conducting media and metallic reflection. The propagation of plane waves in a conducting medium can be described in a way that is formally analogous to wave propagation in a non-conductor if the propagation vector (which we shall call $\boldsymbol{k}'$) is permitted to assume complex values. For a wave whose time dependence is given by $\mathrm{e}^{-i\omega t}$ the differential equation (18.2) reduces to

$$\nabla^2\boldsymbol{E}=-\left(\frac{\varepsilon\mu\omega^2+4\pi i\omega\mu\sigma}{c^2}\right)\boldsymbol{E}=k'^2\boldsymbol{E}, \tag{22.1}$$

where

$$k'=\alpha+i\beta=\frac{\omega\sqrt{\mu\varepsilon}}{c}\sqrt{1+\frac{4\pi i\sigma}{\varepsilon\omega}}. \tag{22.2}$$

[1] DAVID BREWSTER: Phil. Trans. Roy. Soc. Lond. **105**, 125 (1815).

The real and imaginary parts of k' are

$$\left.\begin{aligned}\alpha &= \frac{\omega\sqrt{\mu\varepsilon}}{c}\sqrt{\frac{\sqrt{1+\left(\frac{4\pi\sigma}{\varepsilon\omega}\right)^2}+1}{2}}\,;\\ \beta &= \frac{\omega\sqrt{\mu\varepsilon}}{c}\sqrt{\frac{\sqrt{1+\left(\frac{4\pi\sigma}{\varepsilon\omega}\right)^2}-1}{2}}\end{aligned}\right\} \tag{22.3}$$

with the signs of the roots so chosen that the fields do not become infinite. For a plane wave within the conductor propagated in the positive x direction,

$$\left.\begin{aligned}\boldsymbol{E} &= \boldsymbol{E}_0\,\mathrm{e}^{-\beta x}\,\mathrm{e}^{i(\alpha x-\omega t)}\,;\\ \boldsymbol{H} &= c\,\frac{\boldsymbol{k}'\times\boldsymbol{E}}{\mu\omega}\,.\end{aligned}\right\} \tag{22.4}$$

Since $\boldsymbol{k}'$ is complex, the magnetic vector is out of phase with the electric vector, lagging by an angle whose tangent is β/α. The wave is exponentially damped, and β may be called the absorption coefficient for the field amplitudes.

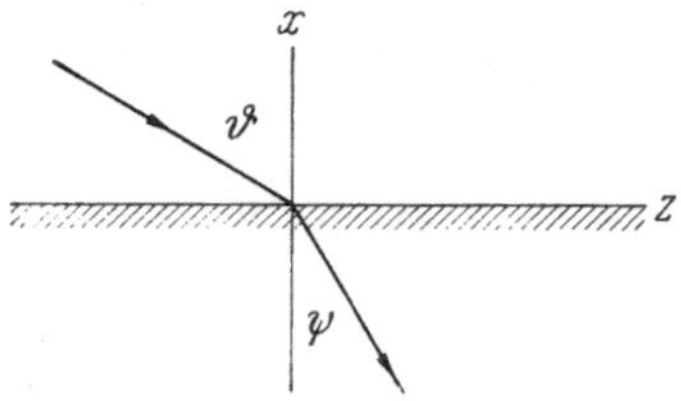

Fig. 8. Plane wave incident on a conductor.

Refraction and reflection at a conducting surface may be determined by a formal application of SNELL's law and the Fresnel formulas, but the physical results are strikingly different from those produced at the boundary between two dielectrics. Suppose that a plane wave traversing a perfect dielectric is incident on the surface of a plane conductor at an angle ϑ, as in Fig. 8. The propagation constant of the incident wave is $k=\omega\sqrt{\omega\varepsilon}/c$ in the dielectric, and in the conductor $k'=\alpha+i\beta$, where α and β are related to the constants of the conductor and to the frequency by (22.3) above. According to SNELL's law,

$$\sin\vartheta' = \frac{k}{\alpha+i\beta}\sin\vartheta\,, \tag{22.5}$$

which defines a complex angle of refraction. In terms of the explicit coordinates of Fig. 8 the phase of the refracted wave is

$$\boldsymbol{k}'\cdot\boldsymbol{r} = (\alpha+i\beta)\,(-x\cos\vartheta'+z\sin\vartheta') = -i\,p\,x - q\,x + k\,z\sin\vartheta\,, \tag{22.6}$$

where q and p are the real and imaginary parts of $(\alpha+i\beta)\cos\vartheta'$ respectively. The amplitude is damped by the factor e^{px}, and is thus a unique function of the distance from the surface. The planes of constant real phase, determined by

$$-q\,x + k\,z\sin\vartheta = \text{const}\,, \tag{22.7}$$

give the direction of propagation in the conductor, and a real angle of refraction is defined if this condition is written as

$$-x\cos\psi + z\sin\psi = \text{const}\,, \tag{22.8}$$

or

$$\sin\psi = \frac{k\sin\vartheta}{\sqrt{q^2+k^2\sin^2\vartheta}}\,.$$

A real index of refraction of the conductor with respect to the dielectric may be defined by a modification of SNELL's law,

$$\frac{v_1}{v_2} = n(\vartheta) = \frac{\sin\vartheta}{\sin\psi} = \frac{\sqrt{q^2+k^2\sin^2\vartheta}}{k}\,. \tag{22.9}$$

Note that the parameters p, q and n depend on the angle of incidence; explicitly,

$$\left.\begin{aligned} p^2 &= \tfrac{1}{2}\left[-\alpha^2+\beta^2+k^2\sin^2\vartheta+\sqrt{4\alpha^2\beta^2+(\alpha^2-\beta^2-k^2\sin^2\vartheta)^2}\right];\\ q^2 &= \tfrac{1}{2}\left[\alpha^2-\beta^2-k^2\sin^2\vartheta+\sqrt{4\alpha^2\beta^2+(\alpha^2-\beta^2-k^2\sin^2\vartheta)^2}\right];\\ k^2 n^2 &= \tfrac{1}{2}\left[\alpha^2-\beta^2+k^2\sin^2\vartheta+\sqrt{4\alpha^2\beta^2+(\alpha^2-\beta^2-k^2\sin^2\vartheta)^2}\right] \end{aligned}\right\} \tag{22.10}$$

For optical frequencies the dependence of the index of refraction on the angle of incidence as given by (22.10) has been confirmed, but the experimental values of the parameters α and β do not correspond to the static or quasi-static values of σ, ε and μ [see Ref. [*21*]).

Within metallic conductors $4\pi\sigma \gg \varepsilon\omega$ for frequencies below the optical range, and

$$\alpha \approx \beta \approx \frac{1}{c}\sqrt{2\pi\mu\sigma\omega}, \qquad \sin\psi \approx \frac{c\,k}{\sqrt{2\pi\mu\sigma\omega}}. \tag{22.11}$$

Thus ψ in good conductors tends to zero for all angles of incidence, and the propagation into the conductor is normal to the surface. For simplicity, let a plane wave in the dielectric be incident normally on the surface of a conductor in the direction of $-x$ (Fig. 8). The fields within the conductor are

$$\left.\begin{aligned} \boldsymbol{E}' &= \boldsymbol{E}_0'\, e^{\frac{\sqrt{2\pi\mu\sigma\omega}}{c}x}\, e^{-i\left(\frac{\sqrt{2\pi\mu\sigma\omega}}{c}x+\omega t\right)};\\ \boldsymbol{H}' &= -\frac{(1+i)}{\sqrt{2}}\sqrt{\frac{4\pi\sigma}{\mu\omega}}\,\boldsymbol{n}\times\boldsymbol{E}'. \end{aligned}\right\} \tag{22.12}$$

Both fields decrease with penetration, and fall to $1/e$ of their values inside the metal surface in a distance

$$\delta = \frac{c}{\sqrt{2\pi\mu\sigma\omega}}, \tag{22.13}$$

which is known as the skin depth. The skin depth goes to zero as the conductivity approaches infinity; it also becomes smaller as the frequency is increased, but it must be noted that Eqs. (22.12) are valid descriptions of the fields only if $4\pi\sigma \gg \omega\varepsilon$. At a distance δ from the surface the phase difference between $\boldsymbol{E}$ and $\boldsymbol{H}$ is 180°.

In general the two polarization components are reflected from a metallic surface with different phases, so that it is possible to produce elliptically polarized light by reflection of a plane polarized wave. Except for normal incidence the general formulas for metallic reflections are extremely complicated and have not been experimentally verified in detail. In the case of normal incidence of a wave characterized by the propagation constant k in medium 1 on a conducting medium 2 in which the wave parameters are α and β, the reflection coefficient is

$$R = \frac{(\mu_1\alpha-\mu_2 k)^2+\mu_1^2\beta^2}{(\mu_1\alpha+\mu_2 k)^2+\mu_1^2\beta^2}. \tag{22.14}$$

If $4\pi\sigma \gg \omega\varepsilon$, this formula reduces to

$$R = 1 - 2\sqrt{\frac{\omega\,\varepsilon\,\mu_2}{2\pi\,\sigma\,\mu_1}} \tag{22.15}$$

if powers of $\omega\varepsilon/4\pi\sigma$ higher than the first are neglected.

In the limit of high conductivity (vanishing skin depth) no oscillatory fields penetrate the conductor. At the surface of a perfect conductor the external electric field is normal and the magnetic field is tangential, accompanied by surface charge and current configurations, in accord with Eqs. (9.1), (9.2), (13.1) and (13.3). For good conductors these conditions lead to excellent representations of external fields, although the skin depth must be considered to account for losses and attenuation.

In the approximation $4\pi\sigma \ll \omega\varepsilon$, valid for radio frequencies incident on relatively poor conductors such as fresh water and dry earth, the reflected wave at all angles behaves very much as if the second medium were also a dielectric, although the refracted wave is absorbed rapidly.

23. The distribution of electromagnetic frequencies within a cavity. Oscillatory solutions of the wave equation exist for a source free finite region bounded by perfectly reflecting walls. This three-dimensional eigenvalue problem has found extensive practical application in the generation of microwave radiation, and the characteristic electromagnetic modes of cavities are treated elsewhere in this Encyclopedia (Vol. XVI). Of more general theoretical interest is the distribution of possible frequencies in the approximation that the dimensions of the cavity are large compared with the wavelength of the radiation. This problem was first attacked by Lord RAYLEIGH[1], and the result was of great historical significance in the consideration of radiation in statistical equilibrium with the cavity walls, the so-called blackbody problem. To this latter point we shall return in Sect. 51; the most elementary treatment yields the frequency distribution which is quite general for wavelengths short in comparison with the cavity dimensions.

It suffices to solve the equation for the electric field vector, subject to the boundary condition that the tangential component of $\boldsymbol{E}$ vanishes at the walls. If the time variation is given by $e^{-i\omega t}$, the wave equation for a vacuum region reduces to

$$\nabla^2 \boldsymbol{E}(x, y, z) + \frac{\omega^2}{c^2} \boldsymbol{E}(x, y, z) = 0. \tag{23.1}$$

The vector wave equation in Cartesian coordinates is a set of three scalar equations, one in each of the rectangular components of the vector. Consider a cubical cavity of length L, having perfectly conducting walls. The solutions of (23.1) which satisfy the boundary conditions are

$$\left.\begin{aligned} E_x &= E_1 \cos\frac{n_1 x}{L} \sin\frac{n_2 y}{L} \sin\frac{n_3 z}{L}, \\ E_y &= E_2 \sin\frac{n_1 x}{L} \cos\frac{n_2 y}{L} \sin\frac{n_3 z}{L}, \\ E_z &= E_3 \sin\frac{n_1 x}{L} \sin\frac{n_2 y}{L} \cos\frac{n_3 x}{L}, \end{aligned}\right\} \tag{23.2}$$

in which n_1, n_2 and n_3 are integers which are related by the condition that

$$\frac{\omega^2}{c^2} = \frac{\pi^2}{L^2}(n_1^2 + n_2^2 + n_3^2) = \left(\frac{2\pi\nu}{c}\right)^2. \tag{23.3}$$

The condition $\operatorname{div} \boldsymbol{E} = 0$ within the boundary imposes one relation between the electric field amplitudes, namely,

$$n_1 E_1 + n_2 E_2 + n_3 E_3 = 0. \tag{23.4}$$

Thus there are two linearly independent waves corresponding to a set of three integers, which correspond to the two possibilities of polarization of a transverse wave. The number of modes of vibration in the cavity within the frequency interval between $\boldsymbol{v}$ and $\boldsymbol{v} + \Delta\nu$ corresponds to the number of sets of positive integers

[1] Lord RAYLEIGH: Phil. Mag. (5) **44**, 539 (1900).

n_1, n_2, n_3 characterizing these frequencies by Eq. (23.3). If n_1, n_2, n_3 are viewed as net points in a three-dimensional coordinate space, all points on a sphere of radius $\sqrt{n_1^2+n_2^2+n_3^2}$ correspond to the same frequency. If this radius is sufficiently large, the required number of modes is then the number of lattice points in the volume of one octant of a spherical shell of radius $2\nu L/c$ and thickness $2L\,\Delta\nu/c$, multiplied by a factor 2 to account for the two polarization possibilities:

$$\Delta N = \frac{8\pi L^3 \nu^2 \Delta\nu}{c^3}.$$

Since L^3 is the volume of the cavity, the number of modes per unit volume is

$$\frac{dN}{d\nu} = \frac{8\pi\nu^2}{c^3}, \tag{23.5}$$

or, in terms of the wavelength of the radiation,

$$\frac{dN}{d\lambda} = \frac{8\pi}{\lambda^4}. \tag{23.6}$$

The generalization of (23.5) or (23.6) to cavities of arbitrary shape may be readily accomplished if separation parameters k_1, k_2, k_3 are introduced explicitly in the solution of the wave equation. (For a cube $k_1 = \pi n_1/L$, etc.) In every case the k's must be so chosen that the boundary conditions are satisfied, and each set of k values is characterized by a set of three integers. The number of modes contained in a frequency interval is then related to the density of allowed points in k space. Since only high frequencies are considered, irregular cavities may be approximated by a sum of rectangular parallelopipeds; it is thus proved quite generally that the number of modes per unit frequency range is proportional to the square of the frequency. This result is also independent of the nature of the boundary conditions; all that is required is that there be such conditions as to restrict the allowed modes to discrete values[1].

24. Spherical waves. Only in Cartesian coordinates, where the coordinate vectors are in the same direction for all points of space, is the vector wave equation (18.5) or (18.6) simply equivalent to three scalar equations for the three components of the field vector. Electromagnetic fields may however be derived from solutions of the homogeneous scalar equation (see Sect. 32), and the method is particularly useful in spherical coordinates. The problem of spherical waves was first treated by Mie and by Debye[2], and the scalar functions from which the fields are derived are often called the Debye potentials. (Equivalent scalar potentials for the electromagnetic fields in spherical coordinates were apparently derived as early as 1899 by Bromwich[3], but were not made available in published form until 1911.) These scalar potentials are special cases of vector superpotentials, as is shown in Sect. 32.

Assume a scalar function Π, which depends on the time through a factor $e^{-i\omega t}$ (not explicitly written), and which satisfies the equation

$$\nabla^2\Pi = \frac{1}{c^2}\frac{\partial^2\Pi}{\partial t^2} = -k^2\Pi. \tag{24.1}$$

---

[1] See, for example, P. Morse and H. Feshbach: Methods of Mathematical Physics. New York 1953.

[2] G. Mie: Ann. Phys. (4) **25**, 377 (1908) and P. Debye: Ann. Phys. (4) **30**, 57 (1909).

[3] T. J. I'A. Bromwich: Phil. Trans. Roy. Soc. Lond. A **220**, 175 (1920), also Phil. Mag. (6) **38**, 143 (1919). According to a footnote, first publication was as a question in part II of the Mathematical Tripos (1910).

Solution of this equation in spherical coordinates by separation of variables leads to

$$\Pi(\boldsymbol{r}; l, m) = z_l(k r)\, Y_l^m(\vartheta, \varphi), \tag{24.2}$$

where l is a positive integer, and $m = l, l-1 \cdots -l$. Here

$$\left.\begin{aligned} Y_l^m(\vartheta,\varphi) &= \frac{(-1)^{l+m}}{2^l l!}\left[(2l+1)\frac{(l-m)!}{(l+m)!}\right]^{\frac{1}{2}}(\sin\vartheta)^m \frac{d^{l+m}}{(d\cos\vartheta)^{l+m}}\left[(\sin\vartheta)^{2l}\right] e^{im\varphi} \\ &= \left[(2l+1)\frac{(l-m)!}{(l+m)!}\right]^{\frac{1}{2}} P_l^m(\cos\vartheta)\, e^{im\varphi}, \end{aligned}\right\} \tag{24.3}$$

so normalized that the mean value of $|Y_l^m|^2$ over all angles is equal to unity. These functions satisfy the relation that

$$Y_l^{-m} = (-1)^m (Y_l^m)^*. \tag{24.4}$$

The spherical Bessel function $z_l(kr)$ is related to the ordinary cylinder function by

$$z_l(k r) = \sqrt{\frac{\pi}{2 k n}}\, Z_{l+\frac{1}{2}}(k r), \tag{24.5}$$

and is to be interpreted as that particular Bessel function which is appropriate to the boundary conditions. If a single solution is to be regular at all points in space $z_l(kr)$ must be chosen as $j_l(kr)$.

If Π is a solution of the scalar wave equation, grad Π is a solution of the vector wave equation, and so is $\boldsymbol{r}\times\operatorname{grad}\Pi$, if the equation is invariant under rotation of the coordinates. For the solution $\boldsymbol{r}\times\operatorname{grad}\Pi = -\operatorname{curl}(\boldsymbol{r}\,\Pi)$ the divergence vanishes, and this solution therefore satisfies a condition on $\boldsymbol{E}$ or $\boldsymbol{H}$ in a uniform source-free medium. A vector field which has no radial components could thus be expressed as a sum over l and m of such solutions, with coefficients appropriate to the nature of the field. Since the operator $\boldsymbol{r}\times\operatorname{grad}$ does not operate on the variable r, it is often convenient to define as vector spherical harmonics the result of operating on $Y_l^m(\vartheta,\varphi)$ with $\boldsymbol{r}\times\operatorname{grad}$. To take full advantage of the simplification inherent in the complex dependence on φ, and to facilitate comparison with the angular momentum operators of quantum mechanics, the operator

$$-i\,\boldsymbol{r}\times\operatorname{grad} \equiv \boldsymbol{L} \tag{24.6}$$

is now sometimes used for defining vector spherical harmonics. (See Refs. [*4*], [*17*].) Note, however, that the normalization here introduced differs by a factor 4π from that in Ref. [*4*], in order that it may conform with that used in the chief references for Sect. 33.) The set of functions

$$\boldsymbol{X}_l^m(\vartheta,\varphi) = \boldsymbol{L}\, Y_l^m(\vartheta,\varphi) = i\operatorname{curl}\left[\boldsymbol{r}\,\Pi(r; l, m)/z_l(k r)\right] \tag{24.7}$$

will suffice for the expansion of the angular dependence of any wholly transverse field[1].

If for any scalar solution Π_1 the magnetic field $\boldsymbol{H}$ is identified with $\boldsymbol{L}\Pi_1$, the corresponding electric field $\boldsymbol{E}$ may be found from the relation $-ik\boldsymbol{E} = \operatorname{curl}\boldsymbol{H}$. (Vacuum fields are here assumed.) A general field is not represented in this way, however, since $\boldsymbol{H}$ so defined has no component along $\boldsymbol{r}$. Another solution is obtained by interchanging the roles of $\boldsymbol{E}$ and $\boldsymbol{H}$, so that $\boldsymbol{E} = \boldsymbol{L}\Pi_2$, where Π_2 is also a solution of Eq. (24.1); $\boldsymbol{H}$ may then be obtained from the relation $ik\boldsymbol{H} = \operatorname{curl}\boldsymbol{E}$. These two kinds of fields may be distinguished by the subscripts E

[1] The properties of these functions, and the expansion of a plane wave in terms of them, are summarized in Sect. 53 of this article.

and M, the significance of which is clarified in Sect. 33. The space dependence of the two varieties of fields may be written as a standard function for each particular l and m:

$$\left.\begin{aligned} \boldsymbol{H}_E(r;l,m) &= -i\,k\,\mathrm{curl}\,(\boldsymbol{r}\,\Pi_1) = k\,z_l(k\,r)\,\boldsymbol{X}_l^m(\vartheta,\varphi),\\ \boldsymbol{E}_E(r;l,m) &= \mathrm{curl}\,\mathrm{curl}\,(\boldsymbol{r}\,\Pi_1) = -i\,\mathrm{curl}\,[z_l(k\,r)\,\boldsymbol{X}_l^m(\vartheta,\varphi)], \end{aligned}\right\} \tag{24.8}$$

and, independently,

$$\left.\begin{aligned} \boldsymbol{H}_M(r;l,m) &= \mathrm{curl}\,\mathrm{curl}\,(\boldsymbol{r}\,\Pi_2) = -i\,\mathrm{curl}\,[z_l(k\,r)\,\boldsymbol{X}_l^m(\vartheta,\varphi)],\\ \boldsymbol{E}_M(r;l,m) &= i\,k\,\mathrm{curl}\,(\boldsymbol{r}\,\Pi_2) = -k\,z_l(k\,r)\,\boldsymbol{X}_l^m(\vartheta,\varphi). \end{aligned}\right\} \tag{24.9}$$

It is clear that these "unit" fields vanish identically for $l=0$. Actual fields may be represented as series of these partial fields, with dimensional amplitude coefficients and with radial functions appropriate to the boundary conditions. For representing the modes of a spherical cavity, for example, $z_l(kr)$ must be specified as $j_l(kr)$, and the possible values of k are restricted by the boundary conditions at the cavity wall. In the expansion of a plane wave $j_l(kr)$ must also be used, since it is the only Bessel function which is finite for all points in space. If on the other hand $z_l(kr)$ is chosen as $h_l^{(1)}(kr)$, which behaves as $i^{-l-1}\,e^{ikr}/r$ for large r, both (24.8) and (24.9) represent radiation from what may be called unit sources at the origin. At large distances from the origin the time average of the Poynting vector is equal to $c|E|^2/8\pi = c|H|^2/8\pi$, and the total rate of radiated energy from each unit source is found by integrating this expression over a sphere. Since it is readily shown that

$$\int \boldsymbol{X}_l^m \cdot \boldsymbol{X}_l^{m*}\,d\Omega = 4\pi\,l(l+1), \tag{24.10}$$

in this notation, each unit field corresponds to an average rate of radiated energy given by

$$\int \overline{\boldsymbol{N}}\cdot d\boldsymbol{S} = \tfrac{1}{2}\,l(l+1)\,c. \tag{24.11}$$

The vector spherical harmonics are orthogonal over a sphere, so that a general radiated field represented by

$$\boldsymbol{E} = \sum_{l,m}\left[a_l^m\,\boldsymbol{E}_E(r;l,m) + b_l^m\,\boldsymbol{E}_M(r;l,m)\right] \tag{24.12}$$

corresponds to an outward flux of energy

$$\int \overline{\boldsymbol{N}}\cdot d\boldsymbol{S} = \tfrac{1}{2}\,c\sum_{l,m} l(l+1)\,(|a_l^m|^2 + |b_l^m|^2). \tag{24.13}$$

It is often convenient to know the coordinate components corresponding to Eqs. (24.8) and (24.9). These are

$$\begin{aligned} (E_E)_r &= l(l+1)\,\frac{1}{r}\,z_l(k\,r)\,Y_l^m(\vartheta,\varphi), & (H_E)_r &= 0,\\ (E_E)_\vartheta &= \frac{1}{r}\,\frac{d}{dr}\,(r\,z_l(k\,r))\,\frac{\partial}{\partial\vartheta}\,Y_l^m(\vartheta,\varphi), & (H_E)_\vartheta &= k\,m\,z_l(kr)\,Y_l^m(\vartheta,\varphi)/\sin\vartheta,\\ (E_E)_\varphi &= i\,m\,\frac{1}{r}\,\frac{d}{dr}\,(r\,z_l(k\,r))\,Y_l^m(\vartheta,\varphi)/\sin\vartheta, & (H_E)_\varphi &= i\,k\,z_l(kr)\,\frac{\partial}{\partial\vartheta}\,Y_l^m(\vartheta,\varphi) \end{aligned}$$

and

$$\begin{aligned} (H_M)_r &= l(l+1)\,\frac{1}{r}\,z_l(k\,r)\,Y_l^m(\vartheta,\varphi), & (E_M)_r &= 0,\\ (H_M)_\vartheta &= \frac{1}{r}\,\frac{d}{dr}\,(r\,z_l(k\,r))\,\frac{\partial}{\partial\vartheta}\,Y_l^m(\vartheta,\varphi), & (E_M)_\vartheta &= -k\,m\,z_l(k\,r)\,Y_l^m(\vartheta,\varphi)/\sin\vartheta,\\ (H_M)_\varphi &= i\,m\,\frac{1}{r}\,\frac{d}{dr}\,(r\,z_l(k\,r))\,Y_l^m(\vartheta,\varphi)/\sin\vartheta, & (E_M)_\varphi &= -i\,k\,z_l(k\,r)\,\frac{\partial}{\partial\vartheta}\,Y_l^m(\vartheta,\varphi). \end{aligned}$$

In Sect. 33 it is shown that the fields designated by subscripts E and M correspond to those produced by electric and magnetic multipoles respectively. In the theory of cavity resonators they are known as transverse magnetic (TM) waves and transverse electric (TE) waves (see Vol. XVI, this Encyclopedia).

25. The "spin" of a vector field; polarization parameters[1]. If for any linear vector field the differential equations and boundary conditions are invariant under rotation of the coordinates, rotation of a solution yields a field which is also a solution. Thus if $\boldsymbol{E}(\boldsymbol{r})$, say, is a solution of the wave equation, a new solution may be obtained by rotating the field through an angle φ about the z axis. The components of the new solution are

$$\left.\begin{aligned} E'_x(x,y,z) &= \cos\varphi\, E_x(x\cos\varphi + y\sin\varphi,\, y\cos\varphi - x\sin\varphi,\, z) - \\ &\quad - \sin\varphi\, E_y(x\cos\varphi + y\sin\varphi,\, y\cos\varphi - x\sin\varphi,\, z); \\ E'_y(x,y,z) &= \sin\varphi\, E_x(x\cos\varphi + y\sin\varphi,\, y\cos\varphi - x\sin\varphi,\, z) + \\ &\quad + \cos\varphi\, E_y(x\cos\varphi + y\sin\varphi,\, y\cos\varphi - x\sin\varphi,\, z); \\ E'_z(x,y,z) &= E_z(x\cos\varphi + y\sin\varphi,\, y\cos\varphi - x\sin\varphi,\, z). \end{aligned}\right\} \tag{25.1}$$

The component E_z, parallel to the axis of rotation, transforms like a scalar function, and for an infinitesimal angle $\delta\varphi$,

$$E'_z = E_z + \left(y\frac{\partial E_z}{\partial x} - x\frac{\partial E_z}{\partial y}\right)\delta\varphi = E_z - i\,\delta\varphi\, L_z E_z, \tag{25.2}$$

where L_z is the z component of the differential operator defined by Eq. (23.5). The behavior of the field components perpendicular to the axis of rotation is more complicated. Formally one may write

$$\boldsymbol{E}' = \boldsymbol{E} - i\,\delta\varphi\,(L_z + S_z)\,\boldsymbol{E}, \tag{25.3}$$

where S_z is an operator defined by

$$S_z\begin{pmatrix} E_x \\ E_y \\ E_z \end{pmatrix} = \begin{pmatrix} -i E_y \\ i E_x \\ 0 \end{pmatrix}, \tag{25.4}$$

or, in matrix form

$$S_z = \begin{pmatrix} 0 & -i & 0 \\ i & 0 & 0 \\ 0 & 0 & 0 \end{pmatrix}. \tag{25.5}$$

Since the difference between two solutions of the field equations is also a solution, $(L_z + S_z)\boldsymbol{E}$ is a solution of the field equations. We have already noted that the operator $\boldsymbol{L}$ may be correlated with the angular momentum operator of quantum mechanics; this suggests that S_z may be similarly interpreted as a "spin" operator. Since S_z satisfies the identity $S_z(S_z^2 - 1) = 0$, it follows that the "eigen values" of S_z are $1, 0, -1$. In this sense it may be said that a classical vector field has unit spin. This formal analogy must not be taken to mean that angular momentum is necessarily associated with all classical vector fields, or that it

[1] See, e.g., A. I. AKHIEZER and V. B. BERESTETSKY, Quantum Electrodynamics, Moscow 1953, or J. M. JAUCH and F. ROHRLICH, The Theory of Photons and Electrons, Addison-Wesley Cambridge (Mass.), 1955, which includes a treatment of the Stokes parameters. The rotation operators are also given by J. BLATT and V. F. WEISSKOPF, Theoretical Nuclear Physics, New York 1952. See also R. H. GOOD jr., Phys. Rev. **105**, 1914 (1957) and Ann. Phys. **1**, 213 (1957) on particle aspect of electromagnetic fields.

will be possible to distinguish physically between "spin" and "orbital" angular momentum in a vector field.

The quantum mechanical description of spin is, however, closely related to the classical description of polarized light introduced by STOKES[1]. An electromagnetic wave in a definite state of polarization is defined by two transverse amplitude components and a phase difference, i.e.,

$$\left.\begin{aligned} E_1 &= \varepsilon_1 \cos \omega t, \\ E_2 &= \varepsilon_2 \cos (\omega t + \alpha) \end{aligned}\right\} \tag{25.6}$$

where directions 1 and 2 are orthogonal to each other and to the direction of propagation. An equivalent description is given by

$$E = c + c^*, \tag{25.7}$$

where

$$E = \begin{pmatrix} E_1 \\ E_2 \end{pmatrix}, \quad c = \begin{pmatrix} c_1 \\ c_2 \end{pmatrix} \tag{25.8}$$

$$c_1 = \tfrac{1}{2}\varepsilon_1 e^{i\omega t}, \quad c_2 = \tfrac{1}{2}\varepsilon_2 e^{i(\omega t+\alpha)}. \tag{25.9}$$

Now the Pauli spin matrices are

$$\sigma_1 = \begin{pmatrix} 0 & 1 \\ 1 & 0 \end{pmatrix}, \quad \sigma_2 = \begin{pmatrix} 0 & -i \\ i & 0 \end{pmatrix}, \quad \sigma_3 = \begin{pmatrix} 1 & 0 \\ 0 & -1 \end{pmatrix}, \tag{25.10}$$

and the quantities

$$\left.\begin{aligned} s_1 &= (c^*, \sigma_1 c) = \tfrac{1}{2}\varepsilon_1 \varepsilon_2 \cos\alpha, \\ s_2 &= (c^*, \sigma_2 c) = \tfrac{1}{2}\varepsilon_1 \varepsilon_2 \sin\alpha, \\ s_3 &= (c^*, \sigma_3 c) = \tfrac{1}{4}(\varepsilon_1^2 - \varepsilon_2^2) \end{aligned}\right\} \tag{25.11}$$

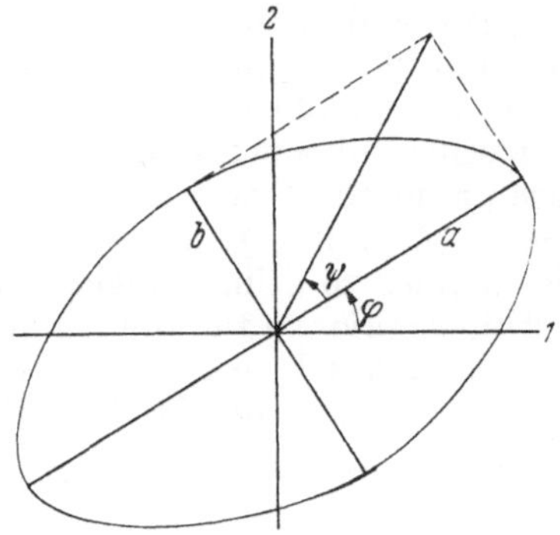

Fig. 9. Polarization ellipse.

are the three independent Stokes parameters for the description of polarized light. STOKES also introduced the dependent parameter

$$s_0 = \sqrt{s_1^2 + s_2^2 + s_3^2} = \tfrac{1}{4}(\varepsilon_1^2 + \varepsilon_2^2), \tag{25.12}$$

which is clearly proportional to the average intensity of the wave.

The angles characterizing the polarization ellipse (Fig. 9) are very simply related to the Stokes parameters. If φ is the angle of inclination of the major axis with the original direction 1,

$$\tan 2\varphi = s_1/s_3. \tag{25.13}$$

The angle ψ, defined by

$$\tan \psi = b/a, \tag{25.14}$$

where a and b are the semi-axes of the polarization ellipse as shown in Fig. 9, satisfies the relation

$$\sin 2\psi = s_2/s_0. \tag{25.15}$$

It is of interest to note that for a plane polarized plane wave $s_2 = 0$, and for circularly polarized radiation $s_1 = 0 = s_3$. In this sense s_2 is the "spin" in the direction of propagation. In the quantum theory of radiation these considerations take on added significance.

A principal advantage of the Stokes parameters in classical physics arises in the description of partially polarized light. Two waves may be said to be *oppositely* polarized if their three independent Stokes parameters are proportional

[1] G. G. STOKES: Trans. Cambridge Phil. Soc. **9**, 399 (1852); reprinted in Mathematical and Physical Papers, Cambridge University Press, London **3**, 233 (1901).

in magnitude but have opposite signs. "Natural" light may be regarded as the incoherent superposition of any two oppositely polarized beams whose Stokes parameters are equal and opposite. Any partially polarized beam may then be considered as a superposition of natural light and a polarized beam characterized by a particular set of Stokes parameters. This method has been used extensively by CHANDRASEKHAR in treating the problem of radiative transfer through atmospheres[1].

V. The relation of fields to macroscopic sources.

26. The inhomogeneous wave equation and retarded potentials; gauge. The wave equation, (18.2) or (18.5), was derived for a charge free region, and gives no indication of the relation between fields and their sources. The existence of charges, and of currents produced by some external agency, must be taken into account in order to describe the origin of fields. For this purpose it is convenient, as in the static case, to introduce potentials. In anticipation of electron theory (in which all sources are "accessible") we may assume initially that the fields are produced in vacuum. (The effect of material media is included explicitly in Sects. 30 to 33.) It is also assumed that the distribution of charge and current is confined to a finite region, and that the charges and currents are known for all times prior to the time of observation.

The fields $\boldsymbol{B}$ and $\boldsymbol{E}$ can in general be derived from the vector potential $\boldsymbol{A}$ and the scalar potential ϕ by the relations

$$\boldsymbol{B} = \operatorname{curl} \boldsymbol{A}, \tag{26.1}$$

$$\boldsymbol{E} = -\operatorname{grad} \phi - \frac{1}{c}\frac{\partial \boldsymbol{A}}{\partial t}. \tag{26.2}$$

MAXWELL'S equations (16.2) and (16.3) are automatically satisfied by these expressions; in terms of the potentials, (16.1) and (16.4) become

$$\nabla^2 \phi + \operatorname{div} \frac{1}{c}\frac{\partial \boldsymbol{A}}{\partial t} = -4\pi \varrho, \tag{26.3}$$

$$\operatorname{curl}\operatorname{curl} \boldsymbol{A} + \frac{1}{c^2}\frac{\partial^2 \boldsymbol{A}}{\partial t^2} + \frac{1}{c}\frac{\partial}{\partial t}\operatorname{grad}\phi = 4\pi \boldsymbol{j}. \tag{26.4}$$

It simplifies the notation to introduce the d'Alembertian operator,

$$\Box \equiv \nabla^2 - \frac{1}{c^2}\frac{\partial^2}{\partial t^2}. \tag{26.5}$$

Eqs. (26.3) and (26.4) may be written

$$\Box \phi + \frac{1}{c}\frac{\partial}{\partial t}\left(\operatorname{div} \boldsymbol{A} + \frac{1}{c}\frac{\partial \phi}{\partial t}\right) = -4\pi \varrho; \tag{26.6}$$

$$\Box \boldsymbol{A} - \operatorname{grad}\left(\operatorname{div} \boldsymbol{A} + \frac{1}{c}\frac{\partial \phi}{\partial t}\right) = -4\pi \boldsymbol{j}. \tag{26.7}$$

The divergence of $\boldsymbol{A}$ is thus far unspecified; if it is so chosen that

$$\operatorname{div} \boldsymbol{A} + \frac{1}{c}\frac{\partial \phi}{\partial t} = 0, \tag{26.8}$$

then ϕ and $\boldsymbol{A}$ satisfy separate but symmetrical second order differential equations of the form called the inhomogeneous wave equation. Eq. (26.8) is the *Lorentz condition* on what is known as the gauge of the potentials. In general the fields

[1] S. CHANDRASEKHAR: Radiative Transfer. Oxford 1950.

defined by (26.1) and (26.2) are not affected by the so-called *gauge transformation*,

$$\left.\begin{aligned} \boldsymbol{A}' &= \boldsymbol{A} - \operatorname{grad}\psi; \\ \phi' &= \phi + \frac{1}{c}\frac{\partial\psi}{\partial t}, \end{aligned}\right\} \tag{26.9}$$

where ψ is an arbitrary function of the coordinates and the time. Then

$$\operatorname{div}\boldsymbol{A}' + \frac{1}{c}\frac{\partial\phi'}{\partial t} = \operatorname{div}\boldsymbol{A} + \frac{1}{c}\frac{\partial\phi}{\partial t} - \Box\psi. \tag{26.10}$$

The Lorentz condition restricts the gauge: only functions ψ for which $\Box\psi=0$ are permitted in transformations (26.9) under the Lorentz condition.

If the Lorentz condition is satisfied the potentials are solutions of the equations

$$\Box\phi = -4\pi\varrho; \tag{26.11}$$

$$\Box\boldsymbol{A} = -4\pi\boldsymbol{j} \tag{26.12}$$

The free wave solutions (i.e., solutions of the homogeneous equation) must be excluded, since the fields sought are those excited by the charge and current distributions ϱ and $\boldsymbol{j}$. It was first shown by L. LORENZ[1] that (26.11) and (26.12) are satisfied by the particular solutions

$$\phi(x, y, z, t) = \int \frac{\varrho(x', y', z', t - r/c)}{r}\, dV', \tag{26.13}$$

$$\boldsymbol{A}(x, y, z, t) = \int \frac{\boldsymbol{j}(x', y', z', t - r/c)}{r}\, dV', \tag{26.14}$$

where $r=\sqrt{(x-x')^2+(y-y')^2+(z-z')^2}$. The volume integral is taken over the source coordinates x', y', z', with the integrand evaluated at the "retarded time" $t-r/c$. There are many methods for obtaining (26.13) and (26.14), but it will suffice here to show by simple substitution that (26.13) is a solution of (26.11). The coordinates of the field point are involved in the integrand twice, once in the denominator r, again in the time $t'=t-r/c$ at which the numerator is evaluated. The Laplacian is symmetrical in source and field points, so that the differentiation variables may be interchanged. For a function independent of the angular variables,

$$\nabla^2 f(r) = \frac{1}{r}\frac{d^2}{dr^2}\big(r f(r)\big); \qquad \text{moreover,} \qquad \frac{\partial^2 f(t-r/c)}{\partial r^2} = \frac{1}{c^2}\frac{\partial^2 f(t-r/c)}{\partial t^2}.$$

Therefore

$$\left.\begin{aligned} \nabla^2\phi &= \int \varrho\nabla^2\left(\frac{1}{r}\right) dV' + \frac{1}{c^2}\frac{\partial^2}{\partial t^2}\int \frac{\varrho(x', y', z', t-r/c)}{r}\, dV' \\ &= -4\pi\int \delta(\boldsymbol{r})\,\varrho\, dV' - \frac{1}{c^2}\frac{\partial^2 \phi}{\partial t^2} = -4\pi\varrho + \frac{1}{c^2}\frac{\partial^2\phi}{\partial t^2}, \end{aligned}\right\} \tag{26.15}$$

and the validity of the solution is established.

The quantities expressed by (26.13) and (26.14) are called the *retarded potentials*; they correspond to contributions from the source that originate sufficiently earlier than the observation time to allow for transmission of the fields to the observation point with velocity c. Mathematically, functions of $(t+r/c)$, instead of $(t-r/c)$, are equally valid solutions of (26.11) and (26.12), and give rise to the advanced potentials. No physical meaning was attached to the advanced potentials in classical electrodynamics. DIRAC and others have investigated the possible

[1] L. LORENZ: Pogg. Ann. Phys. u. Chem. **131**, 243 (1867). — Phil. Mag. (4) **84**, 287 (1867).

contribution of these potentials to the radiation field; see Ref. [*15*] and this volume, Special Relativity, Sect. 36.

The solutions ϕ and $\boldsymbol{A}$ as given by (26.13) and (26.14) automatically satisfy the Lorentz condition (26.8), since relation (26.8) was used in obtaining (26.11) and (26.12). It is of interest to note that the Lorentz condition is consistent with the continuity of charge and current; for charge and current densities that satisfy (26.11) and (26.12),

$$-4\pi\left(\operatorname{div}\boldsymbol{j}+\frac{1}{c}\frac{\partial\varrho}{\partial t}\right)=\Box\left(\operatorname{div}\boldsymbol{A}+\frac{1}{c}\frac{\partial\phi}{\partial t}\right). \tag{26.16}$$

The Lorentz condition also assures a relativistically covariant form for the potentials.

It is sometimes convenient to use a gauge for which the Lorentz condition is not satisfied. From the second of Eqs. (26.9) it is clear that ψ may be so chosen that the scalar potential is zero. (It is not in general possible to make the vector potential zero, since to do so would require three auxiliary conditions instead of one.) Another important gauge is that in which $\operatorname{div}\boldsymbol{A}=0$, so that

$$\left.\begin{aligned}\Box\boldsymbol{A}-\frac{1}{c}\operatorname{grad}\frac{\partial\phi}{\partial t}&=-4\pi\boldsymbol{j},\\ \nabla^2\phi&=-4\pi\varrho.\end{aligned}\right\} \tag{26.17}$$

Eqs. (26.17) define what is called the *Coulomb gauge*, since the equation for ϕ is the same as in electrostatics, although the electric field as given by Eq. (26.2) is of course the same as for any other gauge.

The field variables describe directly measurable quantities, whereas the potentials do not; the mathematical ambiguity in the potentials whose derivatives lead to the same fields does not correspond to a physically significant interaction of charges and currents in classical electromagnetic theory[1]. Quantities which remain unchanged by the transformations of Eqs. (26.9) are called *gauge invariant*, and are physically observable in the sense of the field formulation of electrodynamics.

27. The Hertz potential. The fact that $\boldsymbol{A}$ and ϕ are related by the Lorentz condition (or that some alternative arbitrary condition may be imposed) indicates that they may be derived from a single vector superpotential. Historically the vector superpotential was introduced by HERTZ, and is called the *Hertz vector*. If the scalar potential is given in terms of a vector $\boldsymbol{\Pi}$ by the relation

$$\phi=-\operatorname{div}\boldsymbol{\Pi} \tag{27.1}$$

then the Lorentz condition is satisfied if

$$\boldsymbol{A}=\frac{1}{c}\frac{\partial\boldsymbol{\Pi}}{\partial t}. \tag{27.2}$$

Formally the charge and current densities may be represented by a single vector $\mathfrak{P}$ such that the continuity condition is identically satisfied:

$$\varrho=-\operatorname{div}\mathfrak{P};\quad \boldsymbol{j}=\frac{1}{c}\frac{\partial\mathfrak{P}}{\partial t}. \tag{27.3}$$

With this definition of $\mathfrak{P}$, the inhomogeneous wave equation for $\boldsymbol{\Pi}$,

$$\Box\boldsymbol{\Pi}=-4\pi\mathfrak{P} \tag{27.4}$$

[1] In the quantum domain there exist effects of the potentials on charged particles: see Y. AHARONOV and D. BOHM: Phys. Rev. **115**, 485 (1959).

implies Eqs. (26.11) and (26.12) for ϕ and $\boldsymbol{A}$. The name polarization potential, sometimes used for the Hertz vector, reflects the formal similarity of $\mathfrak{P}$ and the dielectric polarization density $\boldsymbol{P}$, but $\mathfrak{P}$ is a representation of the true charge and current densities and is not to be interpreted as a partial field. The vector $\mathfrak{P}$ is in fact the stream potential for the charge and current.

It is necessary to impose a further condition if a retarded solution of (27.4), analogous to the solutions of similar equations of Sect. 26, is to be written down. It has been noted that the source density must be confined to a finite region of space; this condition is satisfied by $\mathfrak{P}$ as defined only if the total charge of the system is zero. That this is true may be seen from the first of Eqs. (27.3), since

$$\int \varrho\, dV' = -\int \operatorname{div} \mathfrak{P}\, dV' = -\int \mathfrak{P} \cdot d\boldsymbol{S}, \tag{27.5}$$

where the last integration may be taken over any surface enclosing the charge. If it is assumed, however, that the total charge involved is zero, it is valid to write

$$\boldsymbol{\Pi}(x, y, z, t) = \int \frac{\mathfrak{P}(x', y', z', t - r/c)}{r}\, dV' \tag{27.6}$$

by analogy with (26.14). Again r is the distance from the field point to the source point.

In terms of the Hertz vector the fields are

$$\boldsymbol{B} = \operatorname{curl} \boldsymbol{A} = \frac{1}{c}\frac{\partial}{\partial t} \operatorname{curl} \boldsymbol{\Pi}; \tag{27.7}$$

$$\boldsymbol{E} = -\operatorname{grad} \phi - \frac{1}{c}\frac{\partial \boldsymbol{A}}{\partial t} = \operatorname{grad} \operatorname{div} \boldsymbol{\Pi} - \frac{1}{c^2}\frac{\partial^2 \boldsymbol{\Pi}}{\partial t^2} = \operatorname{curl} \operatorname{curl} \boldsymbol{\Pi} - 4\pi \mathfrak{P}. \tag{27.8}$$

Outside the source (where $\operatorname{div} \boldsymbol{E} = 0$) the last term of (27.8) vanishes, and the electric field intensity is simply curl curl $\boldsymbol{\Pi}$.

In the historic paper of Hertz (Ref. [7]) $\int \mathfrak{P}\, dV' = \boldsymbol{p}$ was identified with an infinitesimal linear oscillating dipole at the origin, whose fixed direction is taken as the axis of polar coordinates. The vector $\boldsymbol{\Pi}$ is then in the same direction as $\boldsymbol{p}$, and is given by

$$\boldsymbol{\Pi} = \frac{\boldsymbol{p}(t - r/c)}{r}. \tag{27.9}$$

The fields may be computed by (27.7) and (27.8); when differentiation with respect to r is transformed into that with respect to t, by virtue of the argument $(t - r/c)$ of $\boldsymbol{p}$, the results are

$$H_r = H_\vartheta = E_\varphi = 0, \tag{27.10}$$

$$H_\varphi = \frac{\sin\vartheta}{r}\left(\frac{\ddot{p}}{c^2} + \frac{\dot{p}}{c\,r}\right), \tag{27.11}$$

$$E_r = \frac{\cos\vartheta}{r}\left(\frac{2\dot{p}}{r\,c} + \frac{2p}{r^2}\right), \tag{27.12}$$

$$E_\vartheta = \frac{\sin\vartheta}{r}\left(\frac{\ddot{p}}{c^2} + \frac{\dot{p}}{r\,c} + \frac{p}{r^2}\right). \tag{27.13}$$

Here it is assumed that $\boldsymbol{p}$ is along the axis of polar coordinates. These expressions for the fields, with the time dependence unspecified, do not exhibit the relative importance of the various terms, but in the limit as $r \to \infty$ it is clear that the two leading terms are:

$$H_\varphi = \frac{\sin\vartheta}{c^2 r}\ddot{p}, \tag{27.14}$$

$$E_\vartheta = \frac{\sin\vartheta}{c^2 r}\ddot{p}. \tag{27.15}$$

In this "distant zone" the vectors $\boldsymbol{E}$ and $\boldsymbol{H}$ are thus to a very good approximation perpendicular to each other and to the radius vector $\boldsymbol{r}$ from the origin. The Poynting vector is

$$|\boldsymbol{N}| = N_r = \frac{c\,(\boldsymbol{E}\times\boldsymbol{H})_r}{4\pi} = \frac{c\,E_\vartheta H_\varphi}{4\pi} = \frac{1}{4\pi c^3}\,\frac{\sin^2\vartheta}{r^2}\,\ddot{p}^2, \tag{27.16}$$

which gives the angular distribution of the radiated energy. The total energy radiated per unit time is given by the integral of the Poynting vector over the surface of a sphere of radius r,

$$-\frac{dW}{dt} = 2\pi r^2 \int N_r \sin\vartheta\, d\vartheta = \frac{2}{3}\,\frac{\ddot{p}^2}{c^3}. \tag{27.17}$$

Note, however, that the time involved in the second time derivative of $\boldsymbol{p}$ is the retarded time; the energy received at distance r at time t was lost by the dipole at time $t - r/c$.

If $\boldsymbol{p}$ is interpreted physically as the dipole moment of a moving charge $+e$ in the immediate neighborhood of a stationary charge $-e$, $\ddot{\boldsymbol{p}} = e\dot{\boldsymbol{v}}$, where $\dot{\boldsymbol{v}}$ is the acceleration of $+e$ at time $t - r/c$. The rate of radiation is then given by

$$-\frac{dW}{dt} = \frac{2}{3}\,\frac{e^2\,\dot{v}^2}{c^3}. \tag{27.18}$$

The radiation law in terms of the acceleration of the moving charge was first given by J. J. LARMOR[1].

A vector expression for the fields of an arbitrarily oriented infinitesimal dipole is often useful. If $\hat{\boldsymbol{r}}$ is a unit vector in the direction of $\boldsymbol{r}$, these fields may be written

$$\boldsymbol{H} = \frac{\ddot{\boldsymbol{p}}\times\hat{\boldsymbol{r}}}{c^2 r} + \frac{\dot{\boldsymbol{p}}\times\hat{\boldsymbol{r}}}{c\,r^2}; \tag{27.19}$$

$$\boldsymbol{E} = \frac{\hat{\boldsymbol{r}}\times(\hat{\boldsymbol{r}}\times\ddot{\boldsymbol{p}})}{c^2 r} + \frac{2\hat{\boldsymbol{r}}(\hat{\boldsymbol{r}}\cdot\dot{\boldsymbol{p}}) + \hat{\boldsymbol{r}}\times(\hat{\boldsymbol{r}}\times\dot{\boldsymbol{p}})}{c\,r^2} + \frac{2\hat{\boldsymbol{r}}(\hat{\boldsymbol{r}}\cdot\boldsymbol{p}) + \hat{\boldsymbol{r}}\times(\hat{\boldsymbol{r}}\times\boldsymbol{p})}{r^3}. \tag{27.20}$$

Note again that $\boldsymbol{p}$ is $\boldsymbol{p}(t - r/c)$, i.e., the dipole and its derivatives are to be evaluated at the retarded time.

28. Multipole expansion of the Hertz potential. Only for very special cases is it possible to perform exactly the integration indicated in (27.6) and proceed to obtain therefrom the electromagnetic fields. For a source distribution of finite extent this potential may be expanded in a manner similar to the method of expanding the static scalar and vector potentials given in Sects. 7 and 12. Again the origin of coordinates is located inside the region occupied by the sources, and the fields are ordinarily computed at an observation point whose distance from the origin is large compared with the maximum dimension of the source distribution. The resulting approximations give rise to relatively simple equations for the fields, although no restrictions are necessary in the general formulation. The only additional complication in the present problem is due to the appearance of r, the distance from the source point to the field point, in the retarded time; thus the expansion depends on the nature of the time variation of the source density $\mathfrak{P}$.

Since any solutions of the inhomogeneous wave equation may be constructed from harmonic components, it is necessary to consider only $\mathfrak{P}(x', y', z', t') = \mathfrak{P}(\xi)\, e^{-i\omega t'}$, where t' is the retarded time. In terms of the time of observation at

[1] See Sir JOSEPH LARMOR: Aether and Matter, Cambridge 1900, or Phil. Mag. (5) **44**, 503 (1897).

the field point, the solution of (27.6) becomes

$$\boldsymbol{\Pi}(x, y, z, t) = \int \frac{\mathfrak{P}(\xi)\, e^{i(kr-\omega t)}}{r}\, dV' = e^{-i\omega t} \int \mathfrak{P}(\xi)\, \frac{e^{ikr}}{r}\, dV', \tag{28.1}$$

where $k = \omega/c$. The expansion of e^{ikr}/r about the origin of R and ξ, for $R > \xi$, is

$$\frac{e^{ikr}}{r} = i\,k \sum_{n=0}^{\infty} (2n+1)\, P_n(\cos\Theta)\, j_n(k\,\xi)\, h_n^{(1)}(k\,R). \tag{28.2}$$

Here Θ is the angle between ξ and $\boldsymbol{R}$, and $P_n(\cos\Theta)$ is the Legendre polynomial of order n; $j_n(k\xi)$ and $h_n^{(1)}(kR)$ are the spherical Bessel and Hankel functions, defined in terms of ordinary cylinder functions by Eq. (24.5). By means of the addition theorem of spherical harmonics $P_n(\cos\Theta)$ can be written in terms of the angular coordinates ϑ', φ' of the source and ϑ, φ of the field:

$$P_n(\cos\Theta) = \sum_{m=-n}^{n} (-1)^m P_n^m(\cos\vartheta)\, P_n^{-m}(\cos\vartheta')\, e^{im(\varphi-\varphi')}. \tag{28.3}$$

Therefore the Hertz potential can be written as

$$\left.\begin{aligned} \boldsymbol{\Pi} = i\,k\, e^{-i\omega t} \sum_n \sum_m (2n+1)\,(-1)^m h_n^{(1)}(k\,R)\, P_n^m(\cos\vartheta)\, e^{im\varphi} \times \\ \times \int \mathfrak{P}(\xi)\, j_n(k\,\xi)\, P_n^{-m}(\cos\vartheta')\, e^{-im\varphi'}\, dV'. \end{aligned}\right\} \tag{28.4}$$

This series is exact. (Here n is used as an index for the Bessel and Legendre functions; the index l is reserved for the somewhat different multipole expansion of Sect. 33.)

The ordinary multipole potentials are obtained by replacing $j_n(k\xi)$ by the first term of its series expansion valid for $k\xi \ll 1$,

$$j_n(k\,\xi) \approx \frac{2^n\, n!}{(2n+1)!}\, (k\xi)^n. \tag{28.5}$$

Since $k = \omega/c = 2\pi/\lambda$, the condition for validity of this expansion is that the wavelength of the emitted radiation be large compared with the maximum dimension of the source. The spherical Hankel function may be written in terms of elementary functions, although this procedure is practical only for the first few values of n,

$$h_0^{(1)}(k\,R) = j_0(k\,R) + i\, n_0(k\,R) = \frac{\sin kR}{kR} - i\,\frac{\cos kR}{kR} = -\,i\,\frac{e^{ikR}}{kR}, \tag{28.6}$$

and, similarly,

$$h_1^{(1)}(k\,R) = -\frac{1}{kR}\left(1 + \frac{i}{kR}\right) e^{ikR}. \tag{28.7}$$

The radiation fields (although not those in the near zone for $n > 0$) are correctly given by the asymptotic behavior of $h_n^{(1)}(kR)$ for $kR \gg 1$,

$$h_n^{(1)}(k\,R) \approx (-i)^{n+1}\, \frac{e^{ikR}}{kR}. \tag{28.8}$$

Usually the successive values of n become important only if lower order moments of the distribution vanish; if two or more moments contribute sensibly to the radiation, the approximation (28.5) is invalid. In that case (28.4) may be evaluated with $j_n(k\xi)$ in its exact form, or further terms in the expansion of $j_n(k\xi)$ may be taken into account. The integrals over the source coordinates are then not moments in the elementary sense, although the nomenclature is retained.

If approximation (28.5) is justified for $n=0$, $\int \mathfrak{P}\, dV' = \boldsymbol{p}_1$, and $\boldsymbol{p}=\boldsymbol{p}_1\, e^{-i\omega t}$, the electric dipole moment of the distribution. (The vector $\boldsymbol{p}_1$ is constant.) Eq. (28.4) reduces to the Hertz potential for an oscillating dipole, and, if the axis of polar coordinates is taken along $\boldsymbol{p}_1$, the corresponding fields are given by

$$H_r = H_\vartheta = E_\varphi = 0; \tag{28.9}$$

$$H_\varphi = -i\,k\,p_1 \sin\vartheta \left(\frac{1}{R^2} - \frac{i\,k}{R}\right) e^{i(kR-\omega t)}; \tag{28.10}$$

$$E_r = \frac{2p_1 \cos\vartheta}{R^2}\left(\frac{1}{R} - i\,k\right) e^{i(kR-\omega t)}; \tag{28.11}$$

$$E_\vartheta = \frac{p_1 \sin\vartheta}{R}\left(\frac{1}{R^2} - \frac{i\,k}{R} - k^2\right) e^{i(kR-\omega t)}. \tag{28.12}$$

These expressions may be compared with (27.10) to (27.13); the harmonic dependence on the time clarifies the distinction between the induction or near zone, in which the fields diminish rapidly with distance from the source, and the radiation or distant zone (also called the wave zone) in which the fields are proportional to $1/R$.

The total rate of radiated energy may be found by integrating the time average of the Poynting vector over the surface of a large sphere. For an oscillating dipole of constant amplitude the radiation fields alone are

$$\boldsymbol{E}_{\text{rad}} = \frac{e^{i(kR-\omega t)}}{R}\,[(\boldsymbol{p}_1 \times \boldsymbol{k}) \times \boldsymbol{k}], \tag{28.13}$$

$$\boldsymbol{H}_{\text{rad}} = \frac{k\, e^{i(kR-\omega t)}}{R}\,(\boldsymbol{p}_1 \times \boldsymbol{k}), \tag{28.14}$$

where $\boldsymbol{k}$, the propagation vector, is along $\boldsymbol{R}$. The time average of the Poynting vector is therefore $\overline{|\boldsymbol{N}|} = \frac{1}{2} c\, |\boldsymbol{E}\times\boldsymbol{H}^*|/4\pi = c(\boldsymbol{H}\cdot\boldsymbol{H}^*)/8\pi$. The average rate at which energy is radiated is

$$-\frac{dW}{dt} = \frac{R^2 c}{8\pi}\int \boldsymbol{H}\cdot\boldsymbol{H}^*\, d\Omega = \frac{\omega^4\, \boldsymbol{p}_1\cdot\boldsymbol{p}_1^*}{3c^3} \tag{28.15}$$

which is the time average of (27.18) for sinusoidal time variation.

At very large distance from the source distribution any finite portion of the radiated wave is essentially plane. Because the radiation fields are transverse it might appear that no angular momentum is radiated, and indeed for an infinite plane wave the angular momentum is zero. Actually a plane wave of finite extent (unless it is linearly polarized) involves small longitudinal components which give rise to non-vanishing angular momentum. The concept of angular momentum is most useful as applied to spherical waves, however. The angular momentum density $\boldsymbol{L}$ depends on the total field, but that part of the angular momentum which is radiated will be given by the terms independent of R in the expression

$$\boldsymbol{L} = R^2 \int \frac{\boldsymbol{R}\times(\boldsymbol{D}\times\boldsymbol{B})\, d\Omega\, c\, dt}{4\pi c}. \tag{28.16}$$

The average rate of radiated angular momentum is

$$\frac{dL}{dt} = \frac{R^2}{8\pi}\int \boldsymbol{R}\times(\boldsymbol{D}\times\boldsymbol{B}^*)\, d\Omega = R^2 \int \frac{\boldsymbol{E}(\boldsymbol{R}\cdot\boldsymbol{H}^*) - \boldsymbol{H}^*(\boldsymbol{R}\cdot\boldsymbol{E})}{8\pi}\, d\Omega \tag{28.17}$$

in empty space. For an electric dipole field $(\boldsymbol{E}\cdot\boldsymbol{R})$ has a term that varies as $1/R$, but the integral (28.17) vanishes for a linearly polarized dipole. A rotating dipole, which may be represented by

$$\frac{\hat{\boldsymbol{x}} + i\,\hat{\boldsymbol{y}}}{\sqrt{2}}\, p_1\, e^{-i\omega t}, \tag{28.18}$$

does radiate angular momentum; on substitution of the fields the magnitude of (28.17) becomes

$$\frac{dL_z}{dt} = \frac{\omega^3 p_1^2}{3c^3} \tag{28.19}$$

which is $1/\omega$ times the rate of radiated energy. For a dipole of arbitrary orientation and phase, represented by

$$e^{i\omega t}\boldsymbol{p} = a\,e^{i\alpha}\,\hat{\boldsymbol{x}} + b\,e^{i\beta}\,\hat{\boldsymbol{y}} + c\,e^{i\gamma}\,\hat{\boldsymbol{z}}, \tag{28.20}$$

the relations between the mean rate of radiated angular momentum and that of energy is given by

$$\frac{dL_x}{dt} = \frac{2bc\sin(\gamma-\beta)}{a^2+b^2+c^2}\,\frac{1}{\omega}\,\frac{dW}{dt},$$

$$\frac{dL_y}{dt} = \frac{2ca\sin(\alpha-\gamma)}{a^2+b^2+c^2}\,\frac{1}{\omega}\,\frac{dW}{dt},$$

$$\frac{dL_z}{dt} = \frac{2ab\sin(\beta-\alpha)}{a^2+b^2+c^2}\,\frac{1}{\omega}\,\frac{dW}{dt},$$

where dW/dt is the total radiation rate. A more general relation between the radiation of angular momentum and the radiation of energy follows from the expansion of Sect. 33.

29. Computation of magnetic dipole and electric quadrupole fields from the polarization potential. With approximation (27.5) the Hertz vector becomes, for $n=1$,

$$\left.\begin{aligned} \boldsymbol{\Pi}_{(1)} &= \frac{e^{i(kR-\omega t)}}{R}\left(1+\frac{i}{kR}\right)\int k\,\xi\,\boldsymbol{\mathfrak{P}}(\boldsymbol{\xi})\cos\Theta\,dV' \\ &= \frac{e^{i(kR-\omega t)}}{R^2}\left(1+\frac{i}{kR}\right)\int \boldsymbol{\mathfrak{P}}(\boldsymbol{\xi})\,(\boldsymbol{\xi}\cdot\boldsymbol{R})\,dV' \end{aligned}\right\} \tag{29.1}$$

where Θ is the angle between $\boldsymbol{\xi}$ and $\boldsymbol{R}$. In terms of the components ξ_i of $\boldsymbol{\xi}$ and x_i of $\boldsymbol{R}$, the j-th component of the integrand of (29.1) is $x_i\,\mathfrak{P}_j\,\xi_i$, and $\mathfrak{P}_j\,\xi_i$ is a tensor function of the source distribution. Ths tensor may be separated into its symmetric and antisymmetric parts, as was done in the expansion of the static vector potential in Sect. 12. The antisymmetric part is

$$\boldsymbol{\Pi}_a = \frac{k\,e^{i(kR-\omega t)}}{R^2}\left(1+\frac{i}{kR}\right)\boldsymbol{R}\times\frac{1}{2}\int\boldsymbol{\mathfrak{P}}(\boldsymbol{\xi})\times\boldsymbol{\xi}\,dV'. \tag{29.2}$$

Since $\boldsymbol{\mathfrak{P}}$ is related to the current density by $\boldsymbol{\mathfrak{P}} = -\boldsymbol{j}/ik$, and the magnetic moment of the distribution is $\boldsymbol{m} = \frac{1}{2}\int\boldsymbol{\xi}\times\boldsymbol{j}\,dV'$, the polarization potential corresponding to the antisymmetric part of the integrand of (23.1) may be written

$$\boldsymbol{\Pi}_a = \frac{e^{i(kR-\omega t)}}{R^2}\left(1+\frac{i}{kR}\right)(\boldsymbol{m}\times\boldsymbol{R}). \tag{29.3}$$

The field components, found by use of (27.7) and (27.8), are

$$E_R = E_\vartheta = H_\varphi = 0; \tag{29.4}$$

$$E_\varphi = \frac{k^2}{R}\left(1+\frac{i}{kR}\right)\sin\Theta\,|\boldsymbol{m}|\,e^{i(kR-\omega t)}; \tag{29.5}$$

$$H_R = \left(\frac{1}{R^3}-\frac{ik}{R^2}\right)\cos\Theta\,|\boldsymbol{m}|\,e^{i(kR-\omega t)}; \tag{29.6}$$

$$H_\vartheta = \left(\frac{1}{R^3}-\frac{ik}{R^2}-\frac{k^2}{R}\right)\sin\Theta\,|\boldsymbol{m}|\,e^{i(kR-\omega t)}, \tag{29.7}$$

where ϑ is the angle between $\boldsymbol{m}$ and $\boldsymbol{R}$. The geometrical structure of these fields is identical with that of the electric dipole except that the roles of $\boldsymbol{E}$ and $\boldsymbol{H}$ are reversed. An oscillating magnetic dipole moment produces a magnetic field that lies in a meridian plane through the dipole, as does a static magnetic moment. The lines of $\boldsymbol{E}$, whose existence depends on the oscillation, are concentric circles about the dipole axis.

The symmetric part of the quantity $\mathfrak{P}(\boldsymbol{\xi}\cdot\boldsymbol{R})$ is

$$\frac{\mathfrak{P}(\boldsymbol{\xi}\cdot\boldsymbol{R})+\boldsymbol{\xi}(\boldsymbol{R}\cdot\mathfrak{P})}{2}. \tag{29.8}$$

In (7.6) the electric quadrupole moment of a charge distribution was defined as

$$p_{ij}=\int \xi_i\,\xi_j\,\varrho\,dV' = -\int \xi_i\,\xi_j \operatorname{div}\mathfrak{P}\,dV'. \tag{29.9}$$

By the divergence theorem,

$$\int \psi \operatorname{div}\mathfrak{P}\,dV' = -\int \mathfrak{P}\cdot\operatorname{grad}\psi\,dV' + \int \psi\,\mathfrak{P}\cdot d\boldsymbol{S} \tag{29.10}$$

where ψ is any scalar function of the coordinates. Since $\mathfrak{P}$ is zero outside the charge distribution the surface integral may be made to vanish, and hence the i-th component of the symmetric integral is

$$\tfrac{1}{2}\int \mathfrak{P}_i[(\boldsymbol{\xi}\cdot\boldsymbol{R})+\xi_i(\mathfrak{P}\cdot\boldsymbol{R})]\,dV' = \tfrac{1}{2}R_j\int(\xi_j\,\mathfrak{P}_i+\xi_i\,\mathfrak{P}_j)\,dV' = \tfrac{1}{2}x_j\,p_{ij}. \tag{29.11}$$

The i-th component of the symmetrical part of the Hertz potential is thus

$$\Pi_i = -\frac{i\,k\,\mathrm{e}^{i(kR-\omega t)}}{2R}\left(\frac{1}{R}+\frac{i}{k\,R^2}\right)x_j\,p_{ji}. \tag{29.12}$$

This polarization potential represents the contribution of an oscillating electric quadrupole. The curl of (29.12) has no radial component, and therefore $\boldsymbol{H}$ has no component along $\boldsymbol{R}$. If the coordinate axes are chosen to coincide with the principal axes of the quadrupole tensor, and the z axis is identified with the polar axis for the field point coordinates R, ϑ, φ, the radiation fields alone are found to be

$$E_\vartheta = \frac{k^3\,\mathrm{e}^{i(kR-\omega t)}}{8R}\sin 2\vartheta\,[p_{xx}+p_{yy}-2p_{zz}-(p_{yy}-p_{xx})\cos 2\varphi]; \tag{29.13}$$

$$E_\varphi = \frac{k^3\,\mathrm{e}^{i(kR-\omega t)}}{4R}\sin\vartheta\,\sin 2\varphi\,(p_{yy}-p_{xx}); \tag{29.14}$$

$$H_\vartheta = -\frac{k^3\,\mathrm{e}^{i(kR-\omega t)}}{4R}\sin\vartheta\,\sin 2\varphi\,(p_{yy}-p_{xx}); \tag{29.15}$$

$$H_\varphi = \frac{k^3\,\mathrm{e}^{i(kR-\omega t)}}{8R}\sin 2\vartheta\,[p_{xx}+p_{yy}-2p_{zz}-(p_{yy}-p_{xx})\cos 2\varphi]. \tag{29.16}$$

The distribution of quadrupole radiation has two nodal cones on which there is zero field, compared with the single nodal line of a dipole distribution.

If the quadrupole moment is referred to principal axes, and expressed in terms of Q_{ij} as defined in Eq. (7.14), it may be described by two quantities in addition to those giving the orientation of the axes. If the charge distribution has an axis of symmetry, which we may take as the z axis, it is describable by a single parameter,

$$Q = p_{xx} - p_{zz}, \tag{29.17}$$

often called *the* quadrupole moment. The radiation fields are then given simply by

$$H_\varphi = \frac{k^3\,\mathrm{e}^{i(kR-\omega t)}}{4R}\,Q\sin 2\vartheta, \tag{29.18}$$

together with the corresponding equation for E_ϑ.

The term in the expansion of the polarization potential for which $n=2$ corresponds to magnetic quadrupole and electric octupole radiation. The general characteristics of these fields are more immediately evident in the expansion of radial superpotentials, Sect. 33.

30. Generalized superpotentials and stream potentials. In 1901 RIGHI[1] introduced as an alternate to the Hertz vector what might be called the magnetization potential $\boldsymbol{\Pi}_m$, which satisfies the equation

$$\Box \boldsymbol{\Pi}_m = -4\pi \boldsymbol{M}. \tag{30.1}$$

The infinitesimal magnetic dipole, for which $\boldsymbol{M}=\boldsymbol{m}\,\delta(\boldsymbol{r})\,\mathrm{e}^{-i\omega t}$, corresponds to

$$\boldsymbol{\Pi}_m = \frac{\boldsymbol{m}}{r}\,\mathrm{e}^{i(kr-\omega t)}, \tag{30.2}$$

in complete analogy to the relation of the polarization potential (which we now designate as $\boldsymbol{\Pi}_e$) and the infinitesimal electric dipole oscillating with angular frequency ω. The fields in empty space are derived from $\boldsymbol{\Pi}_m$ by use of (27.7) and (27.8) with the roles of $\boldsymbol{E}$ and $\boldsymbol{H}$ interchanged. The full generalization of the vector superpotentials (for media of arbitrary properties), together with their relations to such scalar potentials as those of DEBYE, has been given by NISBET[2] [*14*], whose account will be followed in this and succeeding sections.

It is convenient to introduce $\boldsymbol{\Pi}_m$ and $\boldsymbol{\Pi}_e$ together, as related to the ordinary scalar and vector potentials by

$$\phi = -\operatorname{div}\boldsymbol{\Pi}_e, \qquad \boldsymbol{A} = \frac{1}{c}\dot{\boldsymbol{\Pi}}_e + \operatorname{curl}\boldsymbol{\Pi}_m \tag{30.3}$$

which automatically satisfy the Lorentz condition. The fields $\boldsymbol{E}$ and $\boldsymbol{B}$ are thus

$$\boldsymbol{E} = \operatorname{grad}\operatorname{div}\boldsymbol{\Pi}_e - \frac{1}{c^2}\ddot{\boldsymbol{\Pi}}_e - \frac{1}{c}\operatorname{curl}\dot{\boldsymbol{\Pi}}_m, \tag{30.4}$$

$$\boldsymbol{B} = \frac{1}{c}\operatorname{curl}\dot{\boldsymbol{\Pi}}_e + \operatorname{curl}\operatorname{curl}\boldsymbol{\Pi}_m. \tag{30.5}$$

The wave equations for $\boldsymbol{\Pi}_e$ and $\boldsymbol{\Pi}_m$ are simplified if the charge and current densities are expressed in terms of two stream potentials, one of which is analogous to $\mathfrak{P}$ introduced in Sect. 27; here they will be called $\boldsymbol{Q}_e$ and $\boldsymbol{Q}_m$, such that

$$\varrho = -\operatorname{div}\boldsymbol{Q}_e, \qquad \boldsymbol{j} = \frac{1}{c}\dot{\boldsymbol{Q}}_e + \operatorname{curl}\boldsymbol{Q}_m. \tag{30.6}$$

Then $\boldsymbol{\Pi}_e$ and $\boldsymbol{\Pi}_m$ satisfy the equations

$$\Box \boldsymbol{\Pi}_e = -4\pi(\boldsymbol{P}+\boldsymbol{Q}_e), \tag{30.7}$$

$$\Box \boldsymbol{\Pi}_m = -4\pi(\boldsymbol{M}+\boldsymbol{Q}_m). \tag{30.8}$$

The stream potentials are thus just the electric and magnetic polarization equivalents to the true charges and currents.

The expressions for $\boldsymbol{D}$ and $\boldsymbol{H}$ follow from their definitions as partial fields and from Eqs. (30.4), (30.5), (30.7) and (30.8):

$$\boldsymbol{D} = -4\pi\boldsymbol{Q}_e - \frac{1}{c}\operatorname{curl}\dot{\boldsymbol{\Pi}}_m + \operatorname{curl}\operatorname{curl}\boldsymbol{\Pi}_e, \tag{30.9}$$

$$\boldsymbol{H} = 4\pi\boldsymbol{Q}_m - \frac{1}{c^2}\ddot{\boldsymbol{\Pi}}_m + \operatorname{grad}\operatorname{div}\boldsymbol{\Pi}_m + \frac{1}{c}\operatorname{curl}\dot{\boldsymbol{\Pi}}_m. \tag{30.10}$$

---

[1] A. RIGHI: Nuovo Cim. (5) **2**, 104 (1901).

[2] The general properties of these superpotentials and their gauge transformations have been given in tensor form by W. H. MCCREA: Proc. Roy. Soc. Lond. A **240**, 447 (1957). This treatment is more concise than that of NISBET, but is of course entirely equivalent when translated into ordinary space-time coordinates.

Full symmetry between $\boldsymbol{E}$ and $\boldsymbol{B}$ on the one hand and $\boldsymbol{D}$ and $\boldsymbol{H}$ on the other can be achieved by the introduction of magnetic stream potentials, $\boldsymbol{R}_e$ and $\boldsymbol{R}_m$:

$$\boldsymbol{B} = -4\pi \boldsymbol{R}_m + \frac{1}{c} \operatorname{curl} \dot{\boldsymbol{\Pi}}_e + \operatorname{curl} \operatorname{curl} \boldsymbol{\Pi}_m, \tag{30.11}$$

$$\boldsymbol{E} = 4\pi \boldsymbol{R}_e + \operatorname{grad} \operatorname{div} \boldsymbol{\Pi}_e - \frac{1}{c^2} \ddot{\boldsymbol{\Pi}}_e - \frac{1}{c} \operatorname{curl} \dot{\boldsymbol{\Pi}}_m. \tag{30.12}$$

The wave equations (30.7) and (30.8) then require the addition on the right of $4\pi \boldsymbol{R}_e$ and $4\pi \boldsymbol{R}_m$ respectively. Note that the functions $\boldsymbol{R}_e$ and $\boldsymbol{R}_m$ need not be zero, although the corresponding magnetic current and charge densities,

$$\varrho_m = -\operatorname{div} \boldsymbol{R}_m, \quad \boldsymbol{j}_m = -\frac{1}{c} \dot{\boldsymbol{R}}_m + \operatorname{curl} \boldsymbol{R}_e \tag{30.13}$$

have zero values. The inclusion of $\boldsymbol{R}_e$ and $\boldsymbol{R}_m$ is necessary for full generality of superpotential gauge transformations, as shown in Sect. 31.

The theory as here given holds for all media. If the medium is such that ϱ, $\boldsymbol{j}$, $\boldsymbol{P}$ and $\boldsymbol{M}$ may be prescribed (i.e., the medium is non-conducting and non-polarizable) the inhomogeneous terms in the wave equations for $\boldsymbol{\Pi}_e$ and $\boldsymbol{\Pi}_m$ are given functions of the coordinates and the time, and the time, and the solutions may be written down at once. Otherwise the inhomogeneous terms are functions of the electromagnetic field variables, and integro-differential equations for $\boldsymbol{\Pi}_e$ and $\boldsymbol{\Pi}_m$ result.

31. Gauge transformations of the superpotentials and stream potentials. The transformations of the stream potentials and of the $\boldsymbol{\Pi}$'s that are possible without affecting the fields are of course not the same as the gauge transformations of $\boldsymbol{A}$ and ϕ. First, it is clear that ϱ and $\boldsymbol{j}$ are invariant when $\boldsymbol{Q}_e$ and $\boldsymbol{Q}_m$ are replaced by $\boldsymbol{Q}_e^0$ and $\boldsymbol{Q}_m^0$ in accordance with

$$\left.\begin{aligned} \boldsymbol{Q}_e &= \boldsymbol{Q}_e^0 + \operatorname{curl} \boldsymbol{G}, \\ \boldsymbol{Q}_m &= \boldsymbol{Q}_m^0 - \frac{1}{c} \dot{\boldsymbol{G}} - \operatorname{grad} g, \end{aligned}\right\} \tag{31.1}$$

where g and $\boldsymbol{G}$ are arbitrary functions of the coordinates and the time. Similarly the magnetic stream potentials still give zero magnetic charge and current densities under the transformation

$$\left.\begin{aligned} \boldsymbol{R}_e &= \boldsymbol{R}_e^0 - \frac{1}{c} \dot{\boldsymbol{L}} - \operatorname{grad} l, \\ \boldsymbol{R}_m &= \boldsymbol{R}_m^0 - \operatorname{curl} \boldsymbol{L}, \end{aligned}\right\} \tag{31.2}$$

where the functions l and $\boldsymbol{L}$ are arbitrary. The correspondingly general form of writing the fields is

$$\boldsymbol{E} = \operatorname{grad} \operatorname{div} \boldsymbol{\Pi}_e - \frac{1}{c^2} \ddot{\boldsymbol{\Pi}}_e - \frac{1}{c} \operatorname{curl} \dot{\boldsymbol{\Pi}}_m + 4\pi \boldsymbol{R}_e^0 - \frac{4\pi}{c} \dot{\boldsymbol{L}} - 4\pi \operatorname{grad} l, \tag{31.3}$$

$$\boldsymbol{B} = \frac{1}{c} \operatorname{curl} \dot{\boldsymbol{\Pi}}_e + \operatorname{curl} \operatorname{curl} \boldsymbol{\Pi}_m - 4\pi \boldsymbol{R}_m^0 + 4\pi \operatorname{curl} \boldsymbol{L}, \tag{31.4}$$

$$\boldsymbol{D} = -\frac{1}{c} \operatorname{curl} \dot{\boldsymbol{\Pi}}_m + \operatorname{curl} \operatorname{curl} \boldsymbol{\Pi}_e - 4\pi \boldsymbol{Q}_e^0 - 4\pi \operatorname{curl} \boldsymbol{G}, \tag{31.5}$$

$$\boldsymbol{H} = \operatorname{grad} \operatorname{div} \boldsymbol{\Pi}_m - \frac{1}{c^2} \ddot{\boldsymbol{\Pi}}_m + \frac{1}{c} \operatorname{curl} \dot{\boldsymbol{\Pi}}_e + 4\pi \boldsymbol{Q}_m^0 - \frac{4\pi}{c} \dot{\boldsymbol{G}} - 4\pi \operatorname{grad} g, \tag{31.6}$$

with

$$\square \boldsymbol{\Pi}_e = -4\pi \left\{ \boldsymbol{P} + \boldsymbol{Q}_e^0 + \boldsymbol{R}_e^0 + \operatorname{curl} \boldsymbol{G} - \frac{1}{c} \dot{\boldsymbol{L}} - \operatorname{grad} l \right\}, \tag{31.7}$$

$$\square \boldsymbol{\Pi}_m = -4\pi \left\{ \boldsymbol{M} + \boldsymbol{Q}_m^0 + \boldsymbol{R}_m^0 - \frac{1}{c} \dot{\boldsymbol{G}} - \operatorname{grad} g - \operatorname{curl} \boldsymbol{L} \right\}. \tag{31.8}$$

Thus the auxiliary functions of the gauge transformations appear as sources for the corresponding vector superpotentials.

The vectors $\boldsymbol{\Pi}_e$ and $\boldsymbol{\Pi}_m$ may be transformed by

$$\left.\begin{aligned}\boldsymbol{\Pi}_e &= \boldsymbol{\Pi}_e^0 + \operatorname{curl}\boldsymbol{\Gamma} - \frac{1}{c}\dot{\boldsymbol{\Lambda}} - \operatorname{grad}\lambda,\\ \boldsymbol{\Pi}_m &= \boldsymbol{\Pi}_m^0 - \frac{1}{c}\dot{\boldsymbol{\Gamma}} - \operatorname{grad}\gamma - \operatorname{curl}\boldsymbol{\Lambda},\end{aligned}\right\} \tag{31.9}$$

where γ, $\boldsymbol{\Gamma}$, λ, and $\boldsymbol{\Lambda}$ are arbitrary functions. Substitution of these equations into Eqs. (31.3) for ϕ and $\boldsymbol{A}$ shows that if γ and $\boldsymbol{\Gamma}$ are arbitrary but λ and $\boldsymbol{\Lambda}$ satisfy

$$\Box\lambda = \frac{1}{c}\dot{\zeta}, \qquad \Box\boldsymbol{\Lambda} = -\operatorname{grad}\zeta, \tag{31.10}$$

where ζ is an arbitrary function, the gauge of the corresponding ϕ and $\boldsymbol{A}$ is also changed, but remains a Lorentz gauge. The gauge function for ϕ and $\boldsymbol{A}$ [function ψ of Eq. (26.9)] is

$$\psi = \zeta + \frac{1}{c}\dot{\lambda} + \operatorname{div}\boldsymbol{\Lambda}. \tag{31.11}$$

Substitution of Eq. (31.9) in the expressions for the fields and comparison with the gauge transformations of the stream potentials shows that the equations satisfied by the fields and the superpotentials, Eqs. (26.9) to (26.12) and the inhomogeneous wave equations for the $\boldsymbol{\Pi}$'s, remain invariant under the transformations (31.9) with arbitrary gauge functions, provided that the stream potentials are at the same time transformed with functions that satisfy

$$4\pi g = \frac{1}{c}\dot{\xi} - \Box\gamma, \qquad 4\pi\boldsymbol{G} = -\Box\boldsymbol{\Gamma} - \operatorname{grad}\xi, \tag{31.12}$$

$$4\pi l = \frac{1}{c}\dot{\zeta} - \Box\lambda, \qquad 4\pi\boldsymbol{L} = -\Box\boldsymbol{\Lambda} - \operatorname{grad}\zeta \tag{31.13}$$

with arbitrary ξ and ζ. Thus if a given electromagnetic field can be represented by the potentials $\boldsymbol{\Pi}_e^0$ and $\boldsymbol{\Pi}_m^0$ for a given choice of stream potentials, then other stream potentials exist for which the representation of the same fields is given by the potentials $\boldsymbol{\Pi}_e$ and $\boldsymbol{\Pi}_m$ of Eqs. (30.9).

The gauge functions, in addition to providing generality to include well known special superpotentials in a single formulation, also serve to trace these potentials to their sources. The generalization to account for both material media and true sources also leads to expressions for the fields within the source distribution.

32. Scalar superpotentials. In general the charge and current distributions and the polarizations are functions of the fields, and the equations for the superpotentials are integro-differential equations. If, however, ϱ, $\boldsymbol{j}$, $\boldsymbol{P}$ and $\boldsymbol{M}$ may be regarded as given, great simplification is possible. In particular, it is possible to derive a general electromagnetic field from two scalar functions, which are really components of the vector superpotentials, with proper choices of the gauge functions. Apparently WHITTAKER[1] was the first to prove this theorem, for the special case of the potentials $\mathcal{F}$ and $\mathcal{G}$,

$$\Pi_e = \mathcal{F}\hat{\boldsymbol{z}}, \qquad \Pi_m = \mathcal{G}\hat{\boldsymbol{z}}, \tag{32.1}$$

in vacuo, at points away from the sources. (Here $\hat{\boldsymbol{z}}$ is a unit vector in a fixed direction.) The method is now well known in the treatment of transverse electric

[1] E.T. WHITTAKER: Proc. Lond. Math. Soc. 1, 367 (1903).

(TE) and transverse magnetic (TM) modes of a cylindrical cavity or a wave guide. NISBET has shown that this result may be extended within the source distribution (assumed given) by a suitable choice of stream functions. Substitution of (32.1) in the general wave equations for the superpotentials, (31.7) and (31.8), with

$$\boldsymbol{Q}_e^0 = c \int \boldsymbol{j}\, dt, \quad \boldsymbol{Q}_m^0 = \boldsymbol{R}_e^0 = \boldsymbol{R}_m^0 = 0, \tag{32.2}$$

yields six scalar equations in which G_x, G_y, L_x and L_y may be set equal to zero:

$$\left.\begin{aligned} -\frac{\partial G_z}{\partial y} + \frac{\partial l}{\partial x} &= P_x + \int j_x c\, dt, & \frac{\partial L_z}{\partial y} + \frac{\partial g}{\partial x} &= M_x, \\ \frac{\partial G_z}{\partial x} + \frac{\partial l}{\partial y} &= P_y + \int j_y c\, dt, & -\frac{\partial L_z}{\partial x} + \frac{\partial g}{\partial y} &= M_y, \end{aligned}\right\} \tag{32.3}$$

which relate the gauge functions to the sources, and

$$\left.\begin{aligned} \Box \mathscr{F} &= -4\pi \left\{P_z + \int j_z c\, dt - \frac{1}{c} \dot{L}_z - \frac{\partial l}{\partial z}\right\}, \\ \Box \mathscr{G} &= -4\pi \left\{M_z - \frac{1}{c} \dot{G}_z - \frac{\partial g}{\partial z}\right\}, \end{aligned}\right\} \tag{32.4}$$

as the wave equations for the scalar superpotentials. Note that, if $\boldsymbol{M}=0$, L_z and g may be taken as zero, and still further simplifications are possible in special cases. The fields $\boldsymbol{E}$ and $\boldsymbol{B}$ are given explicitly by

$$\left.\begin{aligned} \boldsymbol{E} &= \operatorname{grad} \operatorname{div} (\mathscr{F} \hat{\boldsymbol{z}}) - \frac{1}{c^2} \ddot{\mathscr{F}} \hat{\boldsymbol{z}} - \frac{1}{c} \operatorname{curl} (\dot{\mathscr{G}} \hat{\boldsymbol{z}}) - \frac{4\pi}{c} \dot{L}_z \hat{\boldsymbol{z}} - 4\pi \operatorname{grad} l; \\ \boldsymbol{B} &= \frac{1}{c} \operatorname{curl} (\dot{\mathscr{F}} \hat{\boldsymbol{z}}) + \operatorname{curl} \operatorname{curl} (\mathscr{G} \hat{\boldsymbol{z}}) + 4\pi \operatorname{curl} (L_z \hat{\boldsymbol{z}}). \end{aligned}\right\} \tag{32.5}$$

The *Debye potentials*, Π_1 and Π_2, and the equivalent *Bromwich potentials* U and V (equal to $r\Pi_1$ and $r\Pi_2$ respectively), are essentially radial components of $\boldsymbol{\Pi}_e$ and $\boldsymbol{\Pi}_m$. The fields and wave equations are given by Eqs. (31.3) to (31.8) with the quantities $\boldsymbol{P}$, $\boldsymbol{M}$, $\boldsymbol{Q}_e^0$, $\boldsymbol{Q}_m^0$, $\boldsymbol{R}_e^0$, $\boldsymbol{R}_m^0$, $\boldsymbol{G}$ and $\boldsymbol{L}$ all equal to zero, but with

$$2\pi l = \Pi_1, \quad 2\pi g = \Pi_2, \tag{32.6}$$

$$\boldsymbol{\Pi}_e = \boldsymbol{r}\, \Pi_1, \quad \boldsymbol{\Pi}_m = \boldsymbol{r}\, \Pi_2, \tag{32.7}$$

where $\boldsymbol{r}$ is the radius vector from the origin.

The radial superpotentials are solutions of MAXWELL's equations within the source distribution with the choice of functions

$$G_\vartheta = G_\varphi = L_\vartheta = L_\varphi = 0, \quad 2\pi l = \Pi_1 + 2\pi l^1, \; 2\pi g = \Pi_2 + 2\pi g', \tag{32.8}$$

and stream functions given by Eq. (32.2). Then

$$\left.\begin{aligned} \boldsymbol{E} = {} & \operatorname{grad} \operatorname{div} (\boldsymbol{r} \Pi_1) - \frac{1}{c^2} \boldsymbol{r} \ddot{\Pi}_1 - \frac{1}{c} \operatorname{curl} (\boldsymbol{r} \Pi_2) - 2 \operatorname{grad} \Pi_1 - \\ & - \frac{4\pi}{c} \hat{\boldsymbol{r}} \dot{L}_r - 4\pi \operatorname{grad} l', \end{aligned}\right\} \tag{32.9}$$

$$\boldsymbol{B} = \frac{1}{c} \operatorname{curl} (\boldsymbol{r} \dot{\Pi}_1) + \operatorname{curl} \operatorname{curl} (\boldsymbol{r} \Pi_2) + 4\pi \operatorname{curl} (\boldsymbol{r} L_r), \tag{32.10}$$

where

$$\left.\begin{aligned} \Box \Pi_1 &= -\frac{4\pi}{r} \left\{P_r + \int j_r c\, dt - \frac{1}{c} \dot{L}_r - \frac{\partial l'}{\partial r}\right\}, \\ \Box \Pi_2 &= -\frac{4\pi}{r} \left\{M_r - \frac{1}{c} \dot{G}_r - \frac{\partial g'}{\partial r}\right\}, \end{aligned}\right\} \tag{32.11}$$

from Eqs. (31.7) and (31.8). The gauge functions are particular integrals of

$$\left.\begin{aligned}
-\frac{1}{r\sin\vartheta}\frac{\partial G_r}{\partial\varphi}+\frac{1}{r}\frac{\partial l'}{\partial\vartheta} &= P_\vartheta+\int j_\vartheta\, c\, dt;\\
\frac{1}{r}\frac{\partial G_r}{\partial\vartheta}+\frac{1}{r\sin\vartheta}\frac{\partial l'}{\partial\varphi} &= P_\varphi+\int j_\varphi\, c\, dt;\\
\frac{1}{r\sin\vartheta}\frac{\partial L_r}{\partial\varphi}+\frac{1}{r}\frac{\partial \mathfrak{g}'}{\partial\vartheta} &= M_\vartheta;\\
-\frac{1}{r}\frac{\partial L_r}{\partial\vartheta}+\frac{1}{r\sin\vartheta}\frac{\partial \mathfrak{g}'}{\partial\varphi} &= M_\varphi,
\end{aligned}\right\}\tag{32.12}$$

from which G_r and L_r also satisfy

$$\left.\begin{aligned}
\frac{1}{r^2\sin\vartheta}\frac{\partial}{\partial\vartheta}\left(\sin\vartheta\frac{\partial G_r}{\partial\vartheta}\right)+\frac{1}{r^2\sin^2\vartheta}\frac{\partial^2 G_r}{\partial\varphi^2} &= (\operatorname{curl}\boldsymbol{P})_r+\int(\operatorname{curl}\boldsymbol{j})_r\, c\, dt;\\
\frac{1}{r^2\sin\vartheta}\frac{\partial}{\partial\vartheta}\left(\sin\vartheta\frac{\partial L_r}{\partial\vartheta}\right)+\frac{1}{r^2\sin^2\vartheta}\frac{\partial^2 L_r}{\partial\varphi^2} &= -(\operatorname{curl}\boldsymbol{M})_r.
\end{aligned}\right\}\tag{32.13}$$

In general, the particular integral (i.e., the stream potentials) of the inhomogeneous equations may be so chosen that the complementary function can be expressed in terms of only two scalars, which are components of the vector superpotentials. The partial differential equations from which the stream potentials can be determined in terms of the source distribution arise from the Eqs. (31.7) and (31.8) for the vector superpotentials, but are not themselves wave equations. This follows from the fact that there are four non-redundant scalar components of the arbitrary gauge functions in Eq. (31.9), and these can be chosen to reduce to zero four components of the two vector superpotentials. The possibilities include two-component $\boldsymbol{\Pi}_e$, two-component $\boldsymbol{\Pi}_m$, or one component each of $\boldsymbol{\Pi}_e$ and $\boldsymbol{\Pi}_m$. The Whittaker and the Debye-Bromwich potentials are special cases of this third possibility.

33. General multipole expansion in terms of the Debye potentials. Multipole expansions for the Debye potentials give a most convenient representation for the electromagnetic fields produced by radiating charges and currents (see Ref. [*14*][1]). Consider the electromagnetic field produced by a given localized volume distribution of current density $\boldsymbol{j}$, varying harmonically in the time with frequency $\omega=ck$, in otherwise free space. The field equations then become

$$\left.\begin{aligned}
\operatorname{curl}\boldsymbol{E}-ik\boldsymbol{H}&=0, &\quad \operatorname{div}\boldsymbol{H}&=0,\\
\operatorname{curl}\boldsymbol{H}+ik\boldsymbol{E}&=4\pi\boldsymbol{j}, &\quad \operatorname{div}\boldsymbol{E}&=4\pi\varrho
\end{aligned}\right\}\tag{33.1}$$

with $ik\varrho=\operatorname{div}\boldsymbol{j}$. With the choice of $\boldsymbol{\Pi}_e=\boldsymbol{r}\Pi_1$, $\boldsymbol{\Pi}_m=\boldsymbol{r}\Pi_2$, the fields may be expressed as

$$\boldsymbol{E}=\frac{4\pi}{ik}\boldsymbol{j}-4\pi\operatorname{curl}(\hat{\boldsymbol{r}}G)+\operatorname{curl}\operatorname{curl}(\boldsymbol{r}\Pi_1)+ik\operatorname{curl}(\boldsymbol{r}\Pi_2);\tag{33.2}$$

$$\boldsymbol{H}=-ik\operatorname{curl}(\boldsymbol{r}\Pi_1)+\operatorname{curl}\operatorname{curl}(\boldsymbol{r}\Pi_2)\tag{33.3}$$

with

$$\left.\begin{aligned}
(\nabla^2+k^2)(\boldsymbol{r}\Pi_1)&=\frac{4\pi}{ik}\boldsymbol{j}-4\pi\operatorname{curl}(\hat{\boldsymbol{r}}G)+2\operatorname{grad}\Pi_1+4\pi\operatorname{grad}l;\\
(\nabla^2+k^2)(\boldsymbol{r}\Pi_2)&=-4\pi ik\hat{\boldsymbol{r}}G+2\operatorname{grad}\Pi_2,
\end{aligned}\right\}\tag{33.4}$$

[1] See also C. J. Bouwkamp and H. B. G. Casimir: Physica **20**, 539 (1954).

where $\boldsymbol{r}$ and $\hat{\boldsymbol{r}}$ are the radius vector and radial unit vector respectively, as before. The vector equations (33.4) correspond to the scalar wave equations for Π_1 and Π_2 and those which relate G and l to the source current:

$$\left.\begin{aligned} (\nabla^2 + k^2)\,\Pi_1 &= \frac{4\pi}{r}\left\{\frac{j_r}{i\,k} + \frac{\partial l}{\partial r}\right\},\\ (\nabla^2 + k^2)\,\Pi_2 &= -\,4\pi\,i\,k\,G/r, \end{aligned}\right\} \tag{33.5}$$

and

$$\left.\begin{aligned} -\frac{1}{r\sin\vartheta}\,\frac{\partial G}{\partial\varphi} + \frac{1}{r}\,\frac{\partial l}{\partial\vartheta} &= -\,\frac{j_\vartheta}{i\,k},\\ \frac{1}{r}\,\frac{\partial G}{\partial\vartheta} + \frac{1}{r\sin\vartheta}\,\frac{\partial l}{\partial\varphi} &= -\,\frac{j_\varphi}{i\,k}. \end{aligned}\right\} \tag{33.6}$$

Formally,

$$\left.\begin{aligned} \Pi_1(\boldsymbol{r}) &= -\int \frac{1}{r'}\left\{\frac{j_r'}{i\,k} + \frac{\partial l'}{\partial r'}\right\}\frac{e^{ik|\boldsymbol{r}-\boldsymbol{r}'|}}{|\boldsymbol{r}-\boldsymbol{r}'|}\,dV',\\ \Pi_2(\boldsymbol{r}) &= \int \frac{i\,k\,G'}{r'}\,\frac{e^{ik|\boldsymbol{r}-\boldsymbol{r}'|}}{|\boldsymbol{r}-\boldsymbol{r}'|}\,dV', \end{aligned}\right\} \tag{33.7}$$

which are valid at all points of space. Integration is to be carried out over the source coordinates, here designated by primes.

Eqs. (33.6) may be combined to obtain second order differential equations for l and G. These two functions may be expanded in spherical harmonics, from which may be found the expansions for the integrands of Eqs. (33.7). The standard functions involved are those defined in Sect. 24. For the expansion of Π_1,

$$\frac{i\,k}{r}\left(\frac{j_r}{i\,k} + \frac{\partial l}{\partial r}\right) = \sum_{l,m} L_l^m(r)\,Y_l^m(\vartheta,\varphi), \tag{33.8}$$

where

$$\left.\begin{aligned} 4\pi\,L_l^m &= \frac{(-1)^m}{r\,l(l+1)}\iint\left\{\frac{1}{\sin\vartheta}\,\frac{\partial}{\partial\vartheta}\left(\sin\vartheta\,\frac{\partial(r\,j_\vartheta)}{\partial r}\right) + \frac{1}{\sin\vartheta}\,\frac{\partial^2(r\,j_\varphi)}{\partial r\,\partial\varphi} +\right.\\ &\qquad\left. +\,l(l+1)\,j_r\right\}Y_l^{-m}(\vartheta,\varphi)\sin\vartheta\,d\vartheta\,d\varphi\\ &= \frac{(-1)^m}{r\,l(l+1)}\iint\left\{\frac{1}{\sin\vartheta}\,\frac{\partial}{\partial\vartheta}\left(\sin\vartheta\,\frac{\partial(r\,j_\vartheta)}{\partial r}\right) + \frac{1}{\sin\vartheta}\,\frac{\partial^2(r\,j_\varphi)}{\partial r\,\partial\varphi} -\right.\\ &\qquad\left. -\,\frac{1}{\sin\vartheta}\,\frac{\partial}{\partial\vartheta}\left(\sin\vartheta\,\frac{\partial j_r}{\partial\vartheta}\right) - \frac{1}{\sin^2\vartheta}\,\frac{\partial^2 j_r}{\partial\varphi^2}\right\}Y_l^{-m}(\vartheta,\varphi)\sin\vartheta\,d\vartheta\,d\varphi\\ &= \frac{(-1)^m}{l(l+1)}\iint(\boldsymbol{r}\cdot\operatorname{curl}\operatorname{curl}\boldsymbol{j})\,Y_l^{-m}(\vartheta,\varphi)\sin\vartheta\,d\vartheta\,d\varphi. \end{aligned}\right\} \tag{33.9}$$

[There need be no confusion between the index l and the function $l(r,\vartheta,\varphi)$ which does not appear in the final formulation.] Since for $r > r'$,

$$\frac{e^{ik|\boldsymbol{r}-\boldsymbol{r}'|}}{|\boldsymbol{r}-\boldsymbol{r}'|} = i\,k\sum_{l,m}(-1)^m\,h_l^{(1)}(k\,r)\,Y_l^m(\vartheta,\varphi)\,j_l(k\,r')\,Y_l^{-m}(\vartheta',\varphi'), \tag{33.10}$$

the first of Eqs. (33.7) becomes

$$\left.\begin{aligned} \Pi_1(\boldsymbol{r}) &= -\iiint\sum_{l,m} L_l^m(r')\,Y_l^m(\vartheta',\varphi')\sum_{l',m'}(-1)^{m'}\,h_{l'}^{(1)}(k\,r)\,Y_{l'}^{m'}(\vartheta,\varphi)\,\times\\ &\qquad\times j_{l'}(k\,r')\,Y_{l'}^{-m'}(\vartheta',\varphi')\,r'^2\sin\vartheta'\,dr'\,d\vartheta'\,d\varphi'\\ &= -\,4\pi\sum_{l,m} h_l^{(1)}(k\,r)\,Y_l^m(\vartheta,\varphi)\int L_l^m(r')\,j_l(k\,r')\,r'^2\,dr'. \end{aligned}\right\} \tag{33.11}$$

In the nomenclature of Sect. 24,

$$\Pi_1(\boldsymbol{r}) = \sum_{l,m} a_l^m h_l^{(1)}(k r)\, Y_l^m(\vartheta, \varphi) = \sum_{l,m} a_l^m \Pi_1(\boldsymbol{r}; l, m), \tag{33.12}$$

where

$$\left.\begin{aligned} a_l^m &= \frac{(-1)^{m+1}}{l(l+1)} \int j_l(k r')\, Y_l^{-m}(\vartheta' \varphi') \left[\boldsymbol{r}' \cdot \operatorname{curl}\operatorname{curl}\boldsymbol{j}(\boldsymbol{r}')\right] dV', \\ &= \frac{(-1)^{m+1}}{l(l+1)} \int \boldsymbol{j}(\boldsymbol{r}') \cdot \operatorname{curl}\operatorname{curl}\left[\boldsymbol{r}' j_l(k r')\, Y_l^{-m}(\vartheta', \varphi')\right] dV', \end{aligned}\right\} \tag{33.13}$$

with the omission of a vanishing surface integral.

Similarly it may be shown that

$$G(r, \vartheta, \varphi) = \sum_{l,m} G_l^m(r)\, Y_l^m(\vartheta, \varphi), \tag{33.14}$$

where

$$4\pi G_l^m = \frac{r}{i k l(l+1)} \iint (\boldsymbol{r} \cdot \operatorname{curl}\boldsymbol{j})\, (-1)^m\, Y_l^{-m}(\vartheta, \varphi) \sin\vartheta\, d\vartheta\, d\varphi, \tag{33.15}$$

leading to

$$\Pi_2(\boldsymbol{r}) = \sum_{l,m} b_l^m h_l^{(1)}(k r)\, Y_l^m(\vartheta, \varphi) = \sum_{l,m} b_l^m \Pi_2(\boldsymbol{r}; l, m), \tag{33.16}$$

with

$$b_l^m = \frac{(-1)^m i k}{l(l+1)} \int \boldsymbol{j}(\boldsymbol{r}') \cdot \operatorname{curl}\left[\boldsymbol{r}' j_l(k r')\, Y_l^{-m}(\vartheta', \varphi')\right] dV'. \tag{33.17}$$

The partial fields are just those of Sect. 24, and the rate of radiated energy is given by Eq. (24.13).

The partial fields $\boldsymbol{E}_E(\boldsymbol{r}; l, m)$, $\boldsymbol{H}_E(\boldsymbol{r}; l, m)$ derived from $\Pi_1(\boldsymbol{r}; l, m)$ and $\boldsymbol{E}_M(\boldsymbol{r}; l, m)$, $\boldsymbol{H}_M(\boldsymbol{r}; l, m)$ derived from $\Pi_2(\boldsymbol{r}; l, m)$ are called *electric and magnetic multipole fields* respectively, although the sources correspond to elementary multipoles only if it is justified to substitute for $j_l(kr')$ the first term in its power series expansion. In this formulation the expansion is symmetrical in both kinds of multipoles, i.e., the order of the electric multipole is 2^l as derived from $\Pi_1(\boldsymbol{r}; l, m)$ and that of the magnetic multiple is 2^l as derived from $\Pi_2(\boldsymbol{r}; l, m)$.

The z component of the rate of radiation of angular momentum corresponding to any term in the expansion of the Debye potentials is readily computed from Eq. (28.17) and the fields of (24.14) or (24.15). For a unit electric multipole the z component of (28.17) reduces to

$$\frac{r^2}{8\pi} \int H_z^*(\boldsymbol{r} \cdot \boldsymbol{E})\, d\Omega = \frac{k m l (l+1) r^2}{8\pi} \int (h_l^{(1)} Y_l^m)^* h_l^{(1)} Y_l^m\, d\Omega. \tag{33.18}$$

Since (apart from a phase factor) $h_e^{(1)}(kr)$ behaves as e^{ikr}/kr as kr goes to infinity, the integration over a sphere at great distance from the source gives

$$\frac{r^2}{8\pi} \int H_z^*(r E_r)\, d\Omega = \frac{1}{2} l(l+1) \frac{m}{k}, \tag{33.19}$$

which is just (m/ω) times the average rate of radiated energy. This is consistent with the result obtained in Sect. 28 for a rotating dipole, for which the angular dependence was equivalent to $e^{i\varphi}$. Exactly the same result is found for the z component of angular momentum radiated by a unit magnetic multipole.

VI. Classical electron fields.

34. Potentials corresponding to a point charge. If, as postulated by Lorentz, all electromagnetic fields are to be traced to elementary charges, and for computing the fields due to charged particles in vacuo, it is necessary to know the retarded potentials that correspond to a moving electron. It is not possible simply to substitute the retarded distance into the formal solutions of the inhomogeneous wave equation as given by Eqs. (26.13) and (26.14), on the assumption that the charge is concentrated at a point. The inconsistency in such a procedure lies in the fact that the integral of the retarded charge density over space, $\int \varrho(t-r/c)\,dV$, is not in general the total charge. Account must be taken of the fact that the retarded times for different parts of the charge distribution are different even in going to the limit of a point electron. This may be understood physically by considering the effect of retardation as equivalent to information converging on the observation point with a finite velocity c; the information on charge density obtained by mere counting will depend on whether there is a net flux of charge toward or away from the observer.

There are many ways of arriving at the correct potentials for a highly concentrated charge, including the quantitative formulation of the physical process of collecting information at the finite speed c. A consistent formal method of integrating the retarded potential expressions is to represent the position of the point electron $\boldsymbol{r}_e(t')$ as a function of the retarded time, and to transform the time variable so as to account for this dependence. If

$$\varrho(\boldsymbol{r}', t') = e\,\delta\big(\boldsymbol{r}' - \boldsymbol{r}_e(t')\big) \tag{34.1}$$

and

$$\boldsymbol{j}(\boldsymbol{r}', t') = \varrho\,\boldsymbol{v}(t') = e\,\boldsymbol{v}(t')\,\delta\big(\boldsymbol{r}' - \boldsymbol{r}_e(t')\big), \tag{34.2}$$

where e is the charge on an electron moving with velocity $\boldsymbol{v}$ and δ is the Dirac delta function, the potentials of point $\boldsymbol{r}$ may be written

$$\phi(\boldsymbol{r}, t) = \iint \frac{e\,\delta(\boldsymbol{r}' - \boldsymbol{r}(t'))\,\delta\left(t - t' - \frac{|\boldsymbol{r} - \boldsymbol{r}'|}{c}\right)}{|\boldsymbol{r} - \boldsymbol{r}'|}\,dV'\,dt', \tag{34.3}$$

$$\boldsymbol{A}(\boldsymbol{r}, t) = \iint \frac{e\,\boldsymbol{v}(t')\,\delta(\boldsymbol{r} - \boldsymbol{r}_e(t'))\,\delta\left(t - t' - \frac{|\boldsymbol{r} - \boldsymbol{r}'|}{c}\right)}{|\boldsymbol{r} - \boldsymbol{r}'|}\,dV'\,dt'. \tag{34.4}$$

It will suffice to carry out the integration in detail for the scalar potential. The space integration over $\boldsymbol{r}'$ may be performed at once, to yield

$$\phi(\boldsymbol{r}, t) = \iint \frac{e\,\delta\left(t - t' - \frac{|\boldsymbol{r} - \boldsymbol{r}_e(t')|}{c}\right)}{|\boldsymbol{r} - \boldsymbol{r}_e(t')|}\,dt'. \tag{34.5}$$

Here the δ-function over the time is not in correct form for immediate integration, since $\boldsymbol{r}_e$ involves t'. A change of variables to

$$t''(t') = t' + \frac{|\boldsymbol{r} - \boldsymbol{r}_e(t')|}{c} \tag{34.6}$$

transforms the integral so that it may be integrated at once.

$$\left.\begin{aligned} dt'' &= dt' + \frac{1}{c}\,\frac{\partial}{\partial x_{ei}}\,|\boldsymbol{r} - \boldsymbol{r}_e|\,\frac{\partial x_{ei}}{\partial t'}\,dt' \\ &= dt'\left(1 - \frac{(\boldsymbol{r} - \boldsymbol{r}_e)\cdot\boldsymbol{v}(t')}{c\,|\boldsymbol{r} - \boldsymbol{r}_e|}\right). \end{aligned}\right\} \tag{34.7}$$

Therefore, if we let $\boldsymbol{r} - \boldsymbol{r}_e = \boldsymbol{R}$, the radius vector from the electron to the point of observation,

$$\phi(\boldsymbol{r}, t) = \int \frac{e\,\delta(t - t'')}{R - \boldsymbol{R}\cdot\boldsymbol{v}/c}\,dt''. \tag{34.8}$$

The time $t=t''$ defines a particular value of $t'=t-|\boldsymbol{r}-\boldsymbol{r}_e|/c$, which is just the retarded time of the electron. Then

$$\phi(\boldsymbol{r},t)=\left[\frac{e}{R-\boldsymbol{R}\cdot\boldsymbol{v}/c}\right], \tag{34.9}$$

where the square bracket is used as a reminder that the expression is to be evaluated at the retarded time. Similarly,

$$\boldsymbol{A}(\boldsymbol{r},t)=\left[\frac{e\,\boldsymbol{v}/c}{R-\boldsymbol{R}\cdot\boldsymbol{v}/c}\right]. \tag{34.10}$$

The expressions (34.9) and (34.10) are called the *Liénard-Wiechert potentials*[1]; the retarded denominator $R-\boldsymbol{R}\cdot\boldsymbol{v}/c$ is often denoted by s, so that the potentials take the simple form $\phi=[e/s]$, $\boldsymbol{A}=[e\boldsymbol{v}/cs]$. Note that the Liénard-Wiechert potentials are expressed in terms of the retarded positions and velocities of the electron; in computing the fields it is necessary to take into account the difference between the retarded time t' and the time t at which the fields are observed.

It must be pointed out that the Liénard-Wiechert solutions could not be rigorously justified for high velocity charges before the advent of relativity theory. The fields $\boldsymbol{E}$ and $\boldsymbol{B}$ were initially defined in terms of ponderomotive forces on stationary charges and steady currents; in the derivation above it is implied that no change in the basic equations would result if the charge system moves at high velocity. Unlike the laws of classical mechanics, MAXWELL's equations are not invariant under a transformation to a uniformly moving coordinate system with the same flow of time (a Galilean transformation), and thus the equations could be assumed valid only in a particular reference frame, one which was called the ether frame. These difficulties, together with their solutions, are discussed in detail in the second article of this volume (Special Relativity). Pre-relativistically it was necessary to assume that the frame of the observer was the uniquely valid frame, and to interpret the velocity that appears in the Liénard-Wiechert potentials as that of the electron relative to this frame. These restrictions are removed by relativity theory, according to which MAXWELL's equations are equally valid in all unaccelerated frames; the velocity $\boldsymbol{v}$ is then simply that of the electron with respect to the observer, without regard to any hypothetical ether.

35. Fields of a moving charge. The fields are computed from the Liénard-Wiechert potentials by the formulas of Sect. 26,

$$\boldsymbol{B}=\operatorname{curl}\boldsymbol{A}, \tag{35.1}$$

$$\boldsymbol{E}=-\operatorname{grad}\phi-\frac{1}{c}\frac{\partial A}{\partial t}, \tag{35.2}$$

where the differentiations are to be performed with respect to the coordinates and time of the observation. Since the potentials are expressed as functions of the retarded time t', knowledge of $\partial t'/\partial t$ and $\operatorname{grad} t'$ is required. Since $R(t')=c(t-t')$,

$$\frac{\partial R}{\partial t}=\frac{\partial R}{\partial t'}\frac{\partial t'}{\partial t}=-\frac{\boldsymbol{R}\cdot\boldsymbol{v}}{R}\frac{\partial t'}{\partial t}=c\left(1-\frac{\partial t'}{\partial t}\right),$$

or

$$\frac{\partial t'}{\partial t}=\frac{1}{1-\dfrac{\boldsymbol{R}\cdot\boldsymbol{v}}{Rc}}=\frac{R}{s}. \tag{35.3}$$

[1] A. LIÉNARD: Eclairage élect. **16**, 5, 53, 106 (1898). — E. WIECHERT: Arch. Néerl. (2) **5**, 549 (1900).

Similarly,

$$\operatorname{grad} t' = -\frac{1}{c}\operatorname{grad} R(t') = -\frac{1}{c}\left(\frac{\partial R}{\partial t'}\operatorname{grad} t' + \frac{\boldsymbol{R}}{R}\right),$$

and hence

$$\operatorname{grad} t' = -\frac{\boldsymbol{R}}{c(R-\boldsymbol{R}\cdot\boldsymbol{v}/c)} = -\frac{\boldsymbol{R}}{c\,s}. \tag{35.4}$$

With these formulas the calculation of the fields is straightforward. The results are

$$\left.\begin{aligned}\boldsymbol{E} &= \frac{e}{s^2}\operatorname{grad} s - e\frac{\partial}{\partial t}\left(\frac{\boldsymbol{v}}{s\,c^2}\right),\\ &= \frac{e}{s^3}\left(1-\frac{v^2}{c^2}\right)\left(\boldsymbol{R}-\frac{R\,\boldsymbol{v}}{c}\right)+\frac{e}{c^2 s^3}\left\{\boldsymbol{R}\times\left[\left(\boldsymbol{R}-\frac{R\,\boldsymbol{v}}{c}\right)\times\dot{\boldsymbol{v}}\right]\right\}\end{aligned}\right\} \tag{35.5}$$

and

$$\left.\begin{aligned}\boldsymbol{B} &= \frac{e}{c}\operatorname{curl}\frac{\boldsymbol{v}}{s},\\ &= \frac{e\left(1-\frac{v^2}{c^2}\right)}{c\,s^3}\boldsymbol{v}\times\boldsymbol{R}+\frac{e}{c^2 s^3}\frac{\boldsymbol{R}}{R}\times\left\{\boldsymbol{R}\times\left[\left(\boldsymbol{R}-\frac{R\,\boldsymbol{v}}{c}\right)\times\dot{\boldsymbol{v}}\right]\right\}.\end{aligned}\right\} \tag{35.6}$$

It is seen that

$$\boldsymbol{B} = \frac{\boldsymbol{R}\times\boldsymbol{E}}{R}. \tag{35.7}$$

Thus the magnetic field is perpendicular to the electric field everywhere. Both fields are composed of two distinct parts, one of which is independent of the acceleration while the other depends on the acceleration for its existence.

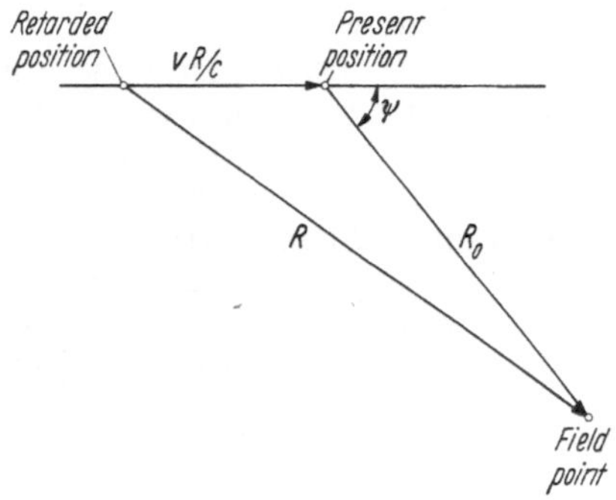

Fig. 10. Coordinate vectors for moving electron.

The part of the electric field (35.5) that is independent of the acceleration consists of a static (Coulomb) term and one which is velocity dependent. These terms may be combined by the recognition that $\boldsymbol{R}-R\boldsymbol{v}/c=\boldsymbol{R}_0$ is just the radius vector from the "present" position of the electron to the observation point in the absence of acceleration. The field

$$\boldsymbol{E}_1 = \frac{e\,\boldsymbol{R}_0}{s^3}\left(1-\frac{v^2}{c^2}\right) \tag{35.8}$$

falls off as $1/R^2$ at large distances, and is called the induction field or the convection field, since it is carried along by the electron. The quantity s may also be expressed in terms of R_0 by noting that since $\boldsymbol{R}\times\boldsymbol{v}=\boldsymbol{R}_0\times\boldsymbol{v}$,

$$R_0^2 - \frac{(\boldsymbol{R}_0\times\boldsymbol{v})^2}{c^2} = \left(R-\frac{\boldsymbol{R}\cdot\boldsymbol{v}}{c}\right)^2 = s^2. \tag{35.9}$$

In terms of the angle ψ of Fig. 10,

$$\boldsymbol{E}_1 = \frac{e\,\boldsymbol{R}_0(1-v^2/c^2)}{R_0^3\left(1-\frac{v^2}{c^2}\sin^2\psi\right)^{\frac{3}{2}}}. \tag{35.10}$$

For very high velocities this field tends to be concentrated in the plane at right angles to $\boldsymbol{v}$. The magnetic field,

$$\boldsymbol{B}_1 = \frac{\boldsymbol{R}\times\boldsymbol{E}_1}{R} = \frac{\boldsymbol{v}\times\boldsymbol{E}_1}{c} \tag{35.11}$$

is at right angles to both $\boldsymbol{E}$ and $\boldsymbol{R}_0$ (as well as to $\boldsymbol{R}$). For low velocities the magnetic field $\boldsymbol{B}_{\text{induction}}$ goes over into the Biot-Savart expression, but for high velocities both $\boldsymbol{E}$ and $\boldsymbol{B}$ are transverse vectors very nearly confined to a plane. It should be noted, however, that these convection fields produce no radiation: the integral of the corresponding Poynting vector over an infinitely distant surface vanishes.

An interesting consequence of the form of the convection field was pointed out by HEAVISIDE[1]. Consider the force exerted by one electron on another moving parallel to it with the same velocity $\boldsymbol{v}$. The force on this second electron is presumably

$$\boldsymbol{F} = e\left(\boldsymbol{E} + \frac{1}{c}\,\boldsymbol{v} \times \boldsymbol{B}\right), \tag{35.12}$$

which, with the substitution of (35.8) and (35.11), may be written as

$$\boldsymbol{F} = -\,e^2 \operatorname{grad}\left(\frac{1 - v^2/c^2}{s}\right), \tag{35.13}$$

or $\boldsymbol{F} = -\operatorname{grad}\phi_1$, where

$$\phi_1 = \frac{e^2\,(1 - v^2/c^2)}{s}, \tag{35.14}$$

was called by HEAVISIDE the *convection potential*. The surfaces of constant s are ellipsoids of revolution (not spheres), with the minor axis in the direction of motion. Since $\boldsymbol{F}$ is perpendicular to the surfaces of constant s, this means that the inverse force of the second electron on the first would not be collinear with that of the first on the second, although the two forces would be equal and opposite. In this interpretation of the equations it is implicitly assumed that there is a single frame of reference in which the field equations are valid. The first electron actually has no velocity with respect to the second, and their interaction is described by the static Coulomb law if the velocity is interpreted as that of the source relative to the field point. The latter interpretation was justified by special relativity theory; the whole problem is obviously a variant of that investigated experimentally by TROUTON and NOBLE (see Sect. 2 of Special Relativity).

The induction fields have here been derived from the Liénard-Wiechert expressions for a point charge, but the form of the potentials for a uniformly moving charge distribution is the same as (34.9) and (34.10). This statement is readily proved by solving the inhomogeneous wave equation under the conditions that the fields are carried convectively, i.e., that $\partial/\partial t = -\boldsymbol{v}\cdot\operatorname{grad}$. Then, if the velocity is taken in the x direction, a change of variables in which x is replaced by $x/\sqrt{1 - v^2/c^2}$ reduces the wave equation to POISSON's equation. In the original variables the solutions are

$$\phi = \int \frac{\varrho\, dV}{s}, \tag{35.15}$$

$$\boldsymbol{A} = \int \frac{\varrho\,\boldsymbol{v}\, dV}{c\, s}, \tag{35.16}$$

where the integral is to be performed over the coordinates of the charge distribution.

With the generalization given by (35.15) and (35.16), the most interesting consequence of Eq. (35.13) arose in connection with attempts by LORENTZ[2] and by ABRAHAM[3] to formulate a detailed theory of the electron. The question of electromagnetic mass will be treated in Sect. 43, but it may be noted here that if

[1] O. HEAVISIDE: Phil. Mag. (5) **27**, 324 (1889). See also O. HEAVISIDE: Electrical Papers.
[2] H. A. LORENTZ: Theory of Electrons.
[3] M. ABRAHAM: Ann. der Phys. **10**, 105–179 (1903).

the field momentum is traced to the electron itself, electromagnetic mass must be attributed to a charge. The form of Eq. (35.13) implied not only a contraction factor $\sqrt{1-v^2/c^2}$ in the direction of motion as a condition for tangential equilibrium of charge, but also led to the conclusion that electromagnetic mass increases with velocity. These arguments arose, of course, prior to the formulation of special relativity.

36. Radiation from an accelerated charge. The terms in Eqs. (35.5) and (35.6) that depend on the acceleration $\dot{\boldsymbol{v}}$ constitute the radiation field. For large values of R and s they vary as $1/R$, and hence predominate sufficiently that the induction field may be neglected in computing the energy radiated. To calculate the rate of radiation it is necessary to evaluate dW/dt', whereas the Poynting vector measures the energy flux as a function of the "present" time t. The rate of energy loss into an elementary solid angle $d\Omega$ is given, in terms of the Poynting vector $\boldsymbol{N}=c\boldsymbol{E}\times\boldsymbol{H}/4\pi$, by

$$-\frac{dW(\vartheta,\varphi)}{dt'}\,d\Omega=|\boldsymbol{N}|\,\frac{dt}{dt'}\,R^2\,d\Omega=|\boldsymbol{N}|\,\frac{s}{R}\,R^2\,d\Omega, \tag{36.1}$$

or

$$-\frac{dW(\vartheta,\varphi)}{dt'}\,d\Omega=\frac{c\,E^2}{4\pi}\,s\,R\,d\Omega=\frac{c\,H^2}{4\pi}\,s\,R\,d\Omega, \tag{36.2}$$

in view of the relation between $\boldsymbol{E}$ and $\boldsymbol{H}$. The general formula for the angular distribution of radiant energy from an accelerated electron is

$$-\frac{dW(\vartheta,\varphi)}{dt'}\,d\Omega=\frac{e^2 R}{4\pi c^3 s^5}\,\{\boldsymbol{R}\times[(\boldsymbol{R}-R\,\boldsymbol{v}/c)\times\dot{\boldsymbol{v}}]\}^2\,d\Omega. \tag{36.3}$$

This expression reduces to simple form only in special cases, but it may be noted that the radiation vanishes in the direction for which $(\boldsymbol{R}-R\boldsymbol{v}/c)$ is parallel or antiparallel to $\dot{\boldsymbol{v}}$. The general radiation pattern in space thus has two nodal lines, which lie in the plane of $\boldsymbol{v}$ and $\dot{\boldsymbol{v}}$. The total rate of radiation is found by integrating Eq. (36.3) over the total solid angle. The result of this integration is

$$-\frac{dW}{dt'}=\frac{2e^2}{3c^3}\,\frac{\dot{\boldsymbol{v}}^2-(\boldsymbol{v}\times\dot{\boldsymbol{v}})^2/c^2}{(1-v^2/c^2)^3} \tag{36.4}$$

without approximation.

Special or approximate forms of Eqs. (36.3) and (36.4) are often useful. If the velocity is sufficiently small that the term in v/c may be neglected,

$$-\frac{dW}{dt'}\,d\Omega=\frac{e^2\,\dot{v}^2\sin^2\vartheta}{4\pi c^3}\,d\Omega \tag{36.5}$$

and

$$-\frac{dW}{dt'}=\frac{2e^2\,\dot{v}^2}{3c^3}. \tag{36.6}$$

These are just the dipole radiation expressions, and (36.6) is identical with Eq. (27.18). Here ϑ is the angle between $\dot{\boldsymbol{v}}$ and the line of observation, but $\dot{\boldsymbol{v}}$ is not constrained to any fixed direction. In the case that $\boldsymbol{v}$ and $\dot{\boldsymbol{v}}$ are directed along the same line, Eq. (36.3) reduces to a relatively simple formula with no approximations. If $\boldsymbol{v}$ is parallel to $\dot{\boldsymbol{v}}$,

$$-\frac{dW}{dt'}\,d\Omega=\frac{e^2\,\dot{v}^2}{4\pi c^3}\,\frac{\sin^2\vartheta}{\left(1-\frac{v}{c}\cos\vartheta\right)^5}, \tag{36.7}$$

where ϑ is now the angle between $\boldsymbol{v}$ and the line of observation. The effect of the velocity dependent denominator is to increase the radiated energy in the forward direction, as indicated in Fig. 11. The total rate of radiation corresponding to Eq. (36.7) is

$$-\frac{dW}{dt'} = \frac{2e^2\dot{v}^2}{3c^3}\,\frac{1}{(1-v^2/c^2)^3}\,. \tag{36.8}$$

Eqs. (36.7) and (36.8) are often applied to compute the rate of Bremsstrahlung, it being assumed that the direction of motion remains unchanged during the slowing down of the particle (see Ref. [*15*]).

The formula for the radiation from a charge in uniform circular motion is of historical interest in connection with the Rutherford atom, and finds important modern applications in the theory of the betatron and the synchrotron. If ϑ is the polar angle between the instantaneous velocity $\boldsymbol{v}$ and the line of observation, and φ is the azimuthal angle of $\boldsymbol{R}$ measured from the plane of $\boldsymbol{v}$ and $\dot{\boldsymbol{v}}$, $\boldsymbol{v}\cdot\boldsymbol{R} = vR\cos\vartheta$ and $\dot{\boldsymbol{v}}\cdot\boldsymbol{R} = \dot{v}R\sin\vartheta\cos\varphi$. Then

$$\left.\begin{aligned} -\frac{dW}{dt'}\,d\Omega &= \frac{e^2\dot{v}^2}{4\pi c^3}\times \\ &\times\frac{\left(1-\frac{v}{c}\cos\vartheta\right)^2-\left(1-\frac{v^2}{c^2}\right)\sin^2\vartheta\cos^2\varphi}{\left(1-\frac{v}{c}\cos\vartheta\right)^5}\,d\Omega\,. \end{aligned}\right\} \tag{36.9}$$

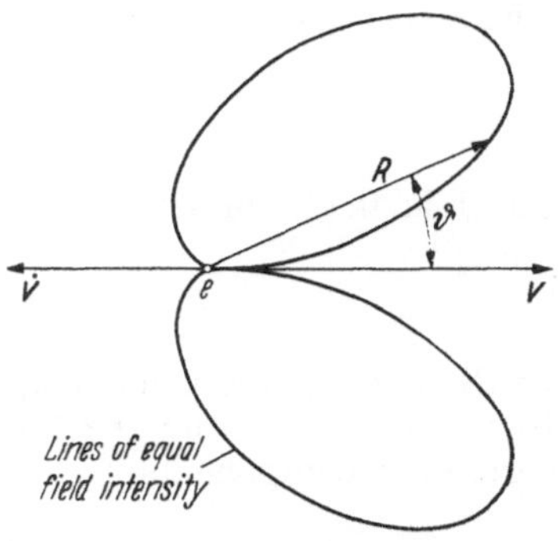

Fig. 11. Radiation pattern of a decelerating charge.

The radiation pattern has nodal lines in the plane of the circle at $\vartheta = \arccos(v/c)$. For high velocities the radiation is chiefly concentrated in the forward direction, and becomes a sharp forward ray as v approaches c. The general expression for the total rate of radiation by a charge e moving in a circular orbit with constant angular velocity is

$$-\frac{dW}{dt'} = \frac{2e^2\dot{v}^2}{3c^3}\,\frac{1}{\left(1-\frac{v^2}{c^2}\right)^2}\,. \tag{36.10}$$

It may be desirable to set up the problem of radiation from a circularly moving charge in fixed coordinates and then find the average intensity during a period of the motion. Let the origin of coordinates be at the center of the circle, with the circle in the xy plane. The polar axis lies along z, and ϑ is now the polar angle of the vector $\boldsymbol{R}$ of the observation point. The time average over a period is equivalent to an average over the azimuthal angle of the rotating charge, φ'. (The average radiation will be symmetrical about the polar axis, so that the azimuthal angle of the vector $\boldsymbol{R}$ is not explicitly involved.) For the orientation of the variable of integration, the yz plane may be taken as that of $\boldsymbol{R}$ and the polar axis, so that $\boldsymbol{v}\cdot\boldsymbol{R} = vR\sin\vartheta\cos\varphi'$, $\dot{\boldsymbol{v}}\cdot\boldsymbol{R} = \dot{v}R\sin\vartheta\sin\varphi'$. Then the average rate of radiation may be found by integrating

$$-\frac{\overline{dW}}{dt'}\,d\Omega = \frac{e^2\dot{v}^2\,d\Omega}{8\pi^2c^3}\int\limits_0^{2\pi}\frac{\left(1-\frac{v^2}{c^2}\right)\cos^2\vartheta+\left(\frac{v}{c}-\sin\vartheta\cos\varphi'\right)^2}{\left(1-\frac{v}{c}\sin\vartheta\cos\varphi'\right)^5}\,d\varphi'\,. \tag{36.11}$$

The result is

$$-\frac{\overline{dW}}{dt'}\,d\Omega = \frac{e^2\dot{v}^2}{8\pi c^3}\left[\frac{2+\frac{v^2}{c^2}\sin^2\vartheta}{\left(1-\frac{v^2}{c^2}\sin^2\vartheta\right)^{\frac{5}{2}}}-\frac{\left(1-\frac{v^2}{c^2}\right)\left(4+\frac{v^2}{c^2}\sin^2\vartheta\right)\sin^2\vartheta}{4\left(1-\frac{v^2}{c^2}\sin^2\vartheta\right)^{\frac{7}{2}}}\right]d\Omega\,. \tag{36.12}$$

It is clear that the radiation tends to be concentrated in the plane of the orbit, especially for high velocities. As the velocity goes to zero, the ratio of the intensity per unit solid angle perpendicular to the plane of the orbit to that at $\vartheta=\pi/2$ approaches $\frac{1}{2}$.

37. Spectral resolution. It is often desirable to express the fields of moving charges in terms of monochromatic waves. As is well known, a real scalar or vector field which is periodic with period T can be expanded as a Fourier series,

$$f(t) = \sum_{n=-\infty}^{\infty} f_n \, \mathrm{e}^{-i n \omega_0 t} = \sum_{\omega} f_\omega \, \mathrm{e}^{-i\omega t}, \tag{37.1}$$

where $\omega_0=2\pi/T$ and $\omega=n\omega_0$. The Fourier components f_n are given by

$$f_n = f_\omega = \frac{1}{T} \int_{-T/2}^{T/2} f(t) \, \mathrm{e}^{i n \omega_0 t} \, dt. \tag{37.2}$$

The real field component corresponding to frequency $\omega=n\omega_0$ is

$$f(\omega, t) = \frac{1}{2} \left(f_\omega \, \mathrm{e}^{-i\omega t} + f_\omega^* \, \mathrm{e}^{i\omega t}\right). \tag{37.3}$$

In the average intensity, i.e., the time mean square of the field, oscillatory factors average out, and the result is just the sum of the intensities of the monochromatic components:

$$\overline{f^2} = f_0^2 + 2 \sum_{n=1}^{\infty} |f_n|^2. \tag{37.4}$$

If the field is multiply periodic with a series of fundamental frequencies $\omega_0^{(1)}, \omega_0^{(2)}, \ldots, \omega_0^{(s)}$ the second form of f in Eq. (37.1) is valid, with ω interpreted as

$$\omega = \sum_{l=1}^{s} n^{(l)} \omega_0^{(l)} . \tag{37.5}$$

where the $n^{(l)}$ are positive and negative integers.

An aperiodic field for which $\int_{-\infty}^{\infty} f \, \mathrm{e}^{i\omega t} \, dt$ is finite may be expanded as a Fourier integral:

$$f(t) = \int_{-\infty}^{\infty} f_\omega \, \mathrm{e}^{-i\omega t} \, d\omega, \tag{37.6}$$

where

$$f_\omega = \frac{1}{2\pi} \int_{-\infty}^{\infty} f(t) \, \mathrm{e}^{i\omega t} \, dt. \tag{37.7}$$

The total radiation associated with an aperiodic field may be written as $\int_{-\infty}^{\infty} f^2 \, dt$ and expressed in terms of the intensities of the monochromatic components.

$$\left.\begin{aligned} \int_{-\infty}^{\infty} f^2 \, dt &= \int_{-\infty}^{\infty} f \, dt \int_{-\infty}^{\infty} f_\omega \, \mathrm{e}^{-i\omega t} \, d\omega \\ &= \int_{-\infty}^{\infty} f_\omega \, d\omega \int_{-\infty}^{\infty} f \, \mathrm{e}^{-i\omega t} \, dt \\ &= 2\pi \int_{-\infty}^{\infty} f_\omega f_{-\omega} \, d\omega = 4\pi \int_{0}^{\infty} |f_\omega|^2 \, d\omega, \end{aligned}\right\} \tag{37.8}$$

since $f_{-\omega}=f_\omega^*$.

Fields which cannot strictly be expanded into a Fourier integral, for example those for which the average intensity approaches a finite limit (not zero) as $t\to\infty$, may be handled in terms of the limit of a Fourier series over a very long period T.

Since T is very large, the interval between successive frequencies is very small, and the sum over n in Eq. (37.4) may be written as an integral over $d\omega/\omega_0$. Thus,

$$\overline{f^2} = \frac{1}{T}\int_{-T/2}^{T/2} f^2\, dt = 2\int_0^\infty |f_n|^2 \frac{d\omega}{\omega_0} = \frac{1}{\pi T}\int_0^\infty \left|\int_{-T/2}^{T/2} f\, e^{i\omega t}\, dt\right|^2 d\omega. \tag{37.9}$$

By means of this formula the intensity corresponding to the frequency interval $d\omega$ may be computed. In view of the fact that $\overline{f^2}$ approaches a finite limit as T goes to infinity it is clear that the integral $\int_{-T/2}^{T/2} f\, e^{i\omega t}\, dt$ behaves as $\sqrt{T}$ for large T.

The charge and current densities producing the field can be expanded in terms of Fourier components, each of which is responsible for the corresponding monochromatic component of the field. Thus for the retarded scalar potential,

$$\phi_\omega\, e^{-i\omega t} = \int \frac{\varrho_\omega\, e^{-i\omega(t-r/c)}}{r}\, dV', \tag{37.10}$$

or, with the substitution $k=\omega/c$,

$$\phi_\omega = \int \frac{\varrho_\omega\, e^{ikr}}{r}\, dV'. \tag{37.11}$$

Here r is of course the distance from source point to field point. Similarly,

$$\boldsymbol{A}_\omega = \int \frac{\boldsymbol{j}_\omega\, e^{ikr}}{r}\, dV'. \tag{37.12}$$

The components of a Fourier integral representation of the source are found by applying Eq. (37.7), and on substitution in (37.11) give

$$\phi_\omega = \frac{1}{2\pi}\int_{-\infty}^{\infty} dt \int \frac{\varrho}{r}\, e^{i(\omega t+kr)}\, dV'. \tag{37.13}$$

For particle velocities sufficiently smaller than that of light, these formulas may be applied to find the fields produced by a single point charge by substituting the δ-function forms of ϱ and $\boldsymbol{j}$ given in Eqs. (24.1) and (34.2). In this approximation the space integration is simply equivalent to the reinterpretation of r as the distance of the point charge from the field point, a distance which is assumed known as a function of the time. Therefore, for a point charge e moving with velocity $\boldsymbol{v}(t)$,

$$\phi_\omega = \frac{e}{2\pi}\int_{-\infty}^{\infty} \frac{1}{r(t)}\, e^{i[\omega t+kr(t)]}\, dt, \tag{37.14}$$

and

$$\boldsymbol{A}_\omega = \frac{e}{2\pi c}\int_{-\infty}^{\infty} \frac{\boldsymbol{v}(t)}{r(t)}\, e^{i[\omega t+kr(t)]}\, dt. \tag{37.15}$$

For very large distances compared with the range of accelerated charge motion it is convenient to introduce a fixed origin, chosen anywhere within the region where this motion occurs. If $\boldsymbol{r}'$ is the coordinate vector of the electron and $\boldsymbol{R}$ that of the field point, then for large R we may write $kr=kR-\boldsymbol{k}\cdot\boldsymbol{r}'$. In the denominator r may be replaced by R, so that

$$\boldsymbol{A}_\omega = \frac{e}{2\pi c}\,\frac{e^{ikR}}{R}\int_{-\infty}^{\infty} \boldsymbol{v}(t)\, e^{i(\omega t-\boldsymbol{k}\cdot\boldsymbol{r}')}\, dt \tag{37.16}$$

will give the correct radiation field. Since $\boldsymbol{v}(t)=d\boldsymbol{r}'/dt$, this expression may be transformed to a line integral along the trajectory of the charge:

$$\boldsymbol{A}_\omega = \frac{e}{2\pi c}\,\frac{e^{ikR}}{R}\int e^{i(\omega t-\boldsymbol{k}\cdot\boldsymbol{r}')}\,d\boldsymbol{r}'. \tag{37.17}$$

The computation of the radiation fields alone from the vector potential is very simple for a monochromatic component, since the term that varies as $1/R$ is given by

$$\boldsymbol{B}_\omega = i\boldsymbol{k}\times\boldsymbol{A}_\omega = \dot{\boldsymbol{A}}_\omega\times\boldsymbol{n}/c, \tag{37.18}$$

where $\boldsymbol{n}$ is a unit vector along the propagation vector $\boldsymbol{k}$. The electric field in the radiation zone is

$$\boldsymbol{E}_\omega = \boldsymbol{B}_\omega\times\boldsymbol{n} \tag{37.19}$$

to a sufficiently good approximation for the computation of the radiated energy.

The formulas (37.14) through (37.17) above are appropriate only to aperiodic motions such as collisions or the motion of an electron through an external field. The total energy per frequency range $d\omega$ may be found by integrating Eq. (36.2) over the time, using the Fourier resolution of $\boldsymbol{B}$. (Note that $s\approx R$.) This is equivalent to the integration carried out in (37.8) above, and the answer is that given by replacing B^2 by $|\boldsymbol{B}_\omega|^2$ on the right side of (36.2) and multiplying by 4π. Thus

$$d W_\omega\, d\Omega = c\,|B_\omega|^2 R^2\, d\Omega \tag{37.20}$$

gives the radiation into the element of solid angle $d\Omega$. In accord with Eqs. (37.17) and (37.18) $\boldsymbol{B}_\omega$ may be written as an integral over the trajectory of the particle:

$$\boldsymbol{B}_\omega = \frac{e}{2\pi c}\,\frac{i\omega\, e^{ikR}}{R}\int e^{i(\omega t-\boldsymbol{k}\cdot\boldsymbol{r}')}\,\boldsymbol{n}\times d\boldsymbol{r}'. \tag{37.21}$$

If the motion of the charge is periodic the field must be expanded in a Fourier series. Eqs. (37.15) and (37.16) must be replaced by

$$\boldsymbol{A}_n = \frac{e}{cT}\int_0^T \frac{\boldsymbol{v}(t)}{r(t)}\, e^{i[n\omega_0 t + kr(t)]}\,dt \tag{37.22}$$

and

$$\boldsymbol{A}_n = \frac{e}{cT}\,\frac{e^{ikR}}{R}\int_0^T \boldsymbol{v}(t)\, e^{i(n\omega_0 t-\boldsymbol{k}\cdot\boldsymbol{r}')}\,dt, \tag{37.23}$$

where (37.23) represents the approximation valid for large distances from the range of motion of the charge. The frequency is now given by $\omega = n\omega_0 = 2\pi n/T$, and in the radiation zone the field $\boldsymbol{B}_n$ may be written either as

$$\boldsymbol{B}_n = \frac{2\pi n e}{T^2}\,\frac{e^{ikR}}{R}\int_0^T e^{i(n\omega_0 t-\boldsymbol{k}\cdot\boldsymbol{r}')}\,\boldsymbol{n}\times\boldsymbol{v}\,dt \tag{37.24}$$

or as

$$\boldsymbol{B}_n = \frac{2\pi n e}{T^2}\,\frac{e^{ikR}}{R}\oint e^{i(r\omega_0 t-k\cdot r')}\,\boldsymbol{n}\times d\boldsymbol{r}'. \tag{37.25}$$

In (37.25) the closed line integral is taken over the closed orbit of the charged particle. In accord with Eq. (37.4) the intensity of radiation with frequency $n\omega_0$ is obtained from the usual formula by replacing the square of the field by twice the square of the n-th Fourier component. Thus

$$\left(\frac{dW}{dt}\right)_n = \frac{c}{2\pi} |B_\omega|^2 R^2 \tag{37.26}$$

gives the monochromatic radiation intensity per unit solid angle.

38. Čerenkov radiation. Whereas a charge moving uniformly in vacuo does not radiate the situation may be different in a dielectric medium for which $\varepsilon > 1$. Within such a medium the inhomogeneous equation for the vector potential becomes

$$\nabla^2 \boldsymbol{A} - \frac{\mu\,\varepsilon}{c^2}\frac{\partial^2 \boldsymbol{A}}{\partial t^2} = -4\pi \boldsymbol{j}. \tag{38.1}$$

For simplicity let us assume $\mu = 1$, and for convenience adopt the notation $n = \sqrt{\varepsilon}$ to signify the index of refraction of the medium. Within the medium the vacuum solutions for the vector potentials are valid if c/n is substituted for c as the velocity with which the fields are propagated. Under these conditions the retardation denominator of the Liénard-Wiechert potentials, $r - \boldsymbol{r}\cdot\boldsymbol{v}\,n/c = r\,[1 - (n v/c)\cos\vartheta]$, may vanish, and in the direction for which $1 - (n v/c)\cos\vartheta = 0$ the field intensities become infinite. In a cone defined by the angle $\vartheta = \operatorname{arc\,cos}\,(c/n v)$ with the direction of motion of a fast particle we might therefore expect to find radiation. Such radiation was first observed by P. A. ČERENKOV in 1934, and explained theoretically by FRANK and TAMM[1]. The effect has since been widely utilized in the design of particle counters.

The current density corresponding to a charge e moving with velocity $\boldsymbol{v}$ in the x direction can be represented by

$$\boldsymbol{j} = \frac{e}{c}\,\boldsymbol{v}\,\delta(x' - v\,t)\;\delta(y')\;\delta(z'). \tag{38.2}$$

A Fourier component of $\boldsymbol{j}$ corresponding to the frequency ω is

$$\boldsymbol{j}_\omega = \frac{1}{2\pi}\int \boldsymbol{j}\,\mathrm{e}^{i\omega t}\,dt = \frac{e}{2\pi\,c}\,\delta(y')\,\delta(z')\,\mathrm{e}^{i\omega x'/v}\,\hat{\boldsymbol{x}}, \tag{38.3}$$

where $\hat{\boldsymbol{x}}$ is a unit vector in the direction of motion. Since for distances large in comparison with the range of source coordinates,

$$\boldsymbol{H}_\omega = -\frac{i\,\mathrm{e}^{ikR}}{R}\int (\boldsymbol{j}_\omega \times \boldsymbol{k})\,\mathrm{e}^{-i\boldsymbol{k}\cdot\boldsymbol{r}'}\,dV', \tag{38.4}$$

the Fourier component of the magnetic field corresponding to Eq. (38.3) is

$$H_\omega = -\frac{i\,e\,k\sin\vartheta}{2\pi\,c\,R}\,\mathrm{e}^{ikR}\int \mathrm{e}^{i\left(\frac{\omega x'}{v} - k x'\cos\vartheta\right)}\,dx', \tag{38.5}$$

when the integrals over y' and z' have been performed. Here $k = n/c$, and ϑ is the angle between the propagation vector $\boldsymbol{k}$ and $\boldsymbol{v}$; the direction of H_ω is clearly azimuthal. In the wave zone the relation between the magnitudes of $\boldsymbol{E}$ and $\boldsymbol{H}$ is

$$|\boldsymbol{E}| = \frac{|\boldsymbol{H}|}{n}, \tag{38.6}$$

and hence the energy loss in a frequency band $d\omega$ is

$$\delta W d\omega = \frac{c}{n}\int |H_\omega|^2 R^2\,d\Omega\,d\omega = \frac{e^2\,n\,\omega^2}{4\pi^2 c^3}\int \left|\int_{-\infty}^{\infty} \mathrm{e}^{i\left(\frac{\omega x'}{v} - k x'\cos\vartheta\right)}\,dx'\right|^2 \sin^2\vartheta\,d\Omega\,d\omega. \tag{38.7}$$

[1] I. FRANK and I. TAMM: C. R. Acad. Sci. USSR. **14**, 109 (1937).

An infinite path is not available to the electron, and the integral over x' can be taken from $-X$ to X:

$$\int_{-X}^{X} e^{i\omega\left(1-\frac{nv}{c}\cos\vartheta\right)\frac{x'}{v}}\,dx' = \frac{2\sin\left[\left(1-\frac{nv}{c}\cos\vartheta\right)\frac{X\omega}{v}\right]}{\omega\left(1-\frac{nv}{c}\cos\vartheta\right)\Big/v}. \tag{38.8}$$

Therefore (38.7) may be written as

$$\delta W = \frac{2e^2 n\,\omega^2}{\pi c^3}\int_{-1}^{1}\sin^2\vartheta\,\frac{\sin^2\left[\left(1-\frac{vn}{c}\cos\vartheta\right)\frac{X\omega}{v}\right]}{\omega^2\left(1-\frac{nv}{c}\cos\vartheta\right)^2\Big/v^2}\,d(\cos\vartheta). \tag{38.9}$$

The angular integrand has a sharp maximum at $\cos\vartheta = c/nv$ and is negligibly small elsewhere. Therefore the integral in (38.9) may be almost exactly approximated by

$$\left.\begin{aligned}&\left(1-\frac{c^2}{n^2v^2}\right)\int_{-1}^{1}\frac{\sin^2\left[\left(1-\frac{nv}{c}\cos\vartheta\right)\frac{X\omega}{v}\right]}{\omega^2\left(1-\frac{nv}{c}\cos\vartheta\right)^2\Big/v^2}\,d(\cos\vartheta)\\ &\quad\approx\left(1-\frac{c^2}{n^2v^2}\right)\frac{cX}{\omega n}\int_{-\infty}^{\infty}\frac{\sin^2 x}{x^2}\,dx = \frac{cX\pi}{\omega n}\left(1-\frac{c^2}{n^2v^2}\right).\end{aligned}\right\} \tag{38.10}$$

The radiation loss in the frequency interval $d\omega$ is then

$$\delta W d\omega = \frac{2e^2X}{c^2}\left(1-\frac{c^2}{n^2v^2}\right)\omega\,d\omega. \tag{38.11}$$

The corresponding energy loss per unit length of path becomes

$$\frac{\delta W}{\delta L}\,d\omega = \frac{e^2}{c^2}\left(1-\frac{c^2}{n^2v^2}\right)\omega\,d\omega, \tag{38.12}$$

since the total path considered is $2X$.

It is necessary in the interpretation of Eq. (38.12) to bear in mind that the index of refraction is not a constant but is in fact a function of the frequency. The velocity of propagation for very high frequency electromagnetic fields in a dielectric approaches the vacuum value, so that $n < c/v$ for sufficiently large values of ω in all media. The total energy loss predicted by Eq. (38.12) is therefore finite and in fact small. It is instructive to write the number of quanta of energy $\hbar\omega$ that are lost per unit length of path:

$$\frac{\delta N}{\delta L}\,d\omega = \frac{e^2}{\hbar c}\left(1-\frac{c^2}{n^2v^2}\right)\frac{d\omega}{c} = \alpha\left(1-\frac{c^2}{n^2v^2}\right)\frac{d\omega}{c}, \tag{38.13}$$

where $\alpha = e^2/\hbar c$ is the fine structure constant, approximately $1/137$. Eq. (38.13) is independent of the frequency except insofar as the index of refraction depends on the frequency, but the total number of quanta is finite since there is in practice an upper limit on the radiated frequency.

39. Multipole radiation from point charge systems. If the distances between members of a system of radiating charges are sufficiently small compared with a wavelength of the emitted radiation, it is possible, as in the case of continuous currents, to make a multipole expansion of the vector potential, or to introduce the superpotentials. For point charges it is instructive to consider expansion of the vector potential, and the general multipole expansion is not often necessary. In the expression

$$\boldsymbol{A} = \int \frac{\boldsymbol{j}(t - r/c)}{r} \, dV' \tag{39.1}$$

the distance r from any source point to the field point must be expressed in terms of the source and field coordinates. Let the coordinate origin be taken within the region where the charges are accelerated; if $\boldsymbol{\xi}$ is the coordinate vector of a source point, and $\boldsymbol{R} = R\boldsymbol{n}$ is that of the field point in terms of a unit radial vector $\boldsymbol{n}$, $r \approx R - \boldsymbol{\xi} \cdot \boldsymbol{n}$. In computing the radiation fields alone the retardation effect may be neglected in the denominator, and the current density is expanded in powers of $\boldsymbol{\xi} \cdot \boldsymbol{n}/c$, the time required for fields to be communicated from one part of the source to another. (The validity of this expansion depends on the assumption that the charge system changes very little during this time, i.e., that the periods associated with the radiation be long in comparison with $\boldsymbol{\xi} \cdot \boldsymbol{n}/c$ for all the charges. This is just the condition that the dimensions of the system be small compared with a wavelength of the emitted radiation, and it also implies that $v \ll c$.) Thus

$$\left.\begin{aligned} \boldsymbol{A} &= \frac{1}{R}\left[\int \boldsymbol{j}(t')\, dV' + \frac{1}{c}\frac{\partial}{\partial t'}\int (\boldsymbol{\xi}\cdot\boldsymbol{n})\,\boldsymbol{j}(t')\, dV'\right] \\ &= \frac{1}{R}\left[\Sigma e\boldsymbol{v} + \frac{1}{c}\frac{\partial}{\partial t'}\Sigma e\boldsymbol{v}(\boldsymbol{\xi}\cdot\boldsymbol{n})\right] \end{aligned}\right\} \tag{39.2}$$

for a system of point charges. Here $t' = t - R/c$ is the same for all charges, in the approximation assumed. Since

$$\left.\begin{aligned} \boldsymbol{v}(\boldsymbol{\xi}\cdot\boldsymbol{n}) &= \frac{1}{2}\frac{\partial}{\partial t'}\boldsymbol{\xi}(\boldsymbol{\xi}\cdot\boldsymbol{n}) + \frac{1}{2}\boldsymbol{v}(\boldsymbol{\xi}\cdot\boldsymbol{n}) - \frac{1}{2}\boldsymbol{\xi}(\boldsymbol{v}\cdot\boldsymbol{n}) \\ &= \frac{1}{2}\frac{\partial}{\partial t'}\boldsymbol{\xi}(\boldsymbol{\xi}\cdot\boldsymbol{n}) + \frac{1}{2}(\boldsymbol{\xi}\times\boldsymbol{v})\times\boldsymbol{n}, \end{aligned}\right\} \tag{39.3}$$

the vector potential may be written

$$\boldsymbol{A} = \frac{1}{R}\left[\Sigma e\boldsymbol{v} + \frac{1}{2c}\frac{\partial^2}{\partial t'^2}\Sigma e\,\boldsymbol{\xi}(\boldsymbol{\xi}\cdot\boldsymbol{n}) + \frac{1}{2c}\frac{\partial}{\partial t'}\Sigma(\boldsymbol{\xi}\times\boldsymbol{v})\times\boldsymbol{n}\right]. \tag{39.4}$$

Now $\Sigma e\boldsymbol{v} = \dot{\boldsymbol{p}}$, the time rate of change of the dipole moment of the system, and $\boldsymbol{m} = \frac{1}{2}\Sigma e(\boldsymbol{\xi}\times\boldsymbol{v})/c$ is the magnetic moment. In the second term $\Sigma e\,\boldsymbol{\xi}(\boldsymbol{\xi}\cdot\boldsymbol{n})$ is the vector whose components are $n_j p_{ij}$, where p_{ij} is the quadrupole moment of the system defined by Eq. (7.6). The form Q_{ij} of Eq. (7.15) may be substituted for p_{ij} without affecting the fields, and the vector with components $n_j\, Q_{ij}$ may be called $\boldsymbol{Q}$. Thus the whole vector potential, to terms of first order in the expansion parameter $\boldsymbol{\xi}\cdot\boldsymbol{n}/c$, is

$$\boldsymbol{A} = \frac{1}{R}\left[\dot{\boldsymbol{p}} + \frac{1}{2c}\ddot{\boldsymbol{Q}} + \dot{\boldsymbol{m}}\times\boldsymbol{n}\right]. \tag{39.5}$$

Here it is understood that the time derivatives are taken at the retarded time. For the radiation field alone $\boldsymbol{B} = \boldsymbol{A}\times\boldsymbol{n}/c$, and $\boldsymbol{E} = \boldsymbol{B}\times\boldsymbol{n}$, so that

$$\left.\begin{aligned} \boldsymbol{H} &= \frac{1}{R}\left\{\ddot{\boldsymbol{p}}\times\boldsymbol{n} + \frac{1}{2c}\dddot{\boldsymbol{Q}}\times\boldsymbol{n} + (\ddot{\boldsymbol{m}}\times\boldsymbol{n})\times\boldsymbol{m}\right\}; \\ \boldsymbol{E} &= \frac{1}{R}\left\{(\ddot{\boldsymbol{p}}\times\boldsymbol{n})\times\boldsymbol{m} + \frac{1}{2c}(\dddot{\boldsymbol{Q}}\times\boldsymbol{n})\times\boldsymbol{n} + \boldsymbol{n}\times\ddot{\boldsymbol{m}}\right\} \end{aligned}\right\} \tag{39.6}$$

represent the transverse components which contribute to the radiated energy.

The energy radiated per unit time per unit solid angle is given by the Poynting vector in the usual way. The angular distribution of radiation may be very complicated if all terms contribute, but in averaging over all directions to obtain the total radiation the cross products of the three terms give zero, so that only the mean squares of the electric dipole, quadrupole and magnetic dipole contributions remain. In the general multipole radiation formulas of Sect. 33 this result follows from the orthogonality of the spherical harmonics when integrated over a sphere; the equivalent result here is obtained by averaging over the products of the components of the unit vector $\boldsymbol{n}$ (see Ref. [*10*]). The total radiation rate from a system of charges for which the initial approximations are valid is

$$-\frac{dW}{dt'}=\frac{1}{c^3}\left[\frac{2}{3}\ddot{\boldsymbol{p}}^2+\frac{1}{20c^2}\dddot{Q}^2+\frac{2}{3}\ddot{\boldsymbol{m}}^2\right]. \tag{39.7}$$

Note that the multipole moments are here defined in the elementary way, but that harmonic time dependence is not assumed.

It is clear from the definition of the magnetic moment and from Eq. (39.7) that the ratio of magnetic dipole and electric quadrupole radiation to that of electric dipole character is in general of the order of $(v/c)^2$. For computing the radiation from charges moving with velocities approaching that of light the multipole expansion is not useful, and Eq. (36.4) must be employed. For low velocities the total radiation is adequately represented by the electric dipole term, except for systems whose electric dipole radiation vanishes. If all charges moving under their mutual interactions have the same charge to mass ratio, for example, the second time derivative of the dipole moment is proportional to the acceleration of the center of mass, and is therefore zero. Under these circumstances there is no magnetic dipole radiation either, since the magnetic moment is proportional to the angular momentum for such a system and thus $\dot{\boldsymbol{m}}=0$. For the same reason a system consisting of only two particles emits no magnetic dipole radiation, regardless of the charge-to-mass ratio.

A Fourier analysis of the motion leads to formulas for the spectral resolution of the emitted radiation, in accord with the considerations of Sect. 37. For example, if δW_ω is the total energy radiated in the frequency interval $d\omega$ during aperiodic motion such as a collision, and $\boldsymbol{p}_\omega$ is the Fourier component of $\boldsymbol{p}$,

$$\delta W_\omega=\frac{8\pi}{3c^3}(\ddot{\boldsymbol{p}}_\omega)^2\,d\omega=\frac{8\pi\omega^4}{3c^3}|\boldsymbol{p}_\omega|^2\,d\omega. \tag{39.8}$$

For periodic motion the intensity of radiation with frequency $\omega=n\omega_0$ is given by

$$-\frac{dW_n}{dt}=\frac{4\omega_0^4n^4}{3c^3}|\boldsymbol{p}_n|^2. \tag{39.9}$$

Similar considerations hold for magnetic dipole and electric quadrupole radiation.

40. Radiation during collisions (see Refs. [*10*], [*9*]). During the scattering of one charged particle by another the dipole moment is that of the system comprising the scattered particle and the scatterer relative to their common center of mass. (The center of mass is not accelerated.) If the center of mass is chosen as the origin of coordinates, with $\boldsymbol{r}_1$ the position vector of charge e_1, $\boldsymbol{r}_2$ that of charge e_2, the electric dipole moment is

$$\boldsymbol{p}=e_1\boldsymbol{r}_1+e_2\boldsymbol{r}_2, \tag{40.1}$$

or, in terms of $\boldsymbol{r}=\boldsymbol{r}_1-\boldsymbol{r}_2$, the radius vector from e_2 to e_1,

$$\boldsymbol{p}=\frac{e_1m_2-e_2m_1}{m_1+m_2}\boldsymbol{r}. \tag{40.2}$$

The instantaneous rate of dipole radiation may be written immediately as

$$-\frac{dW}{dt'}\,d\Omega = \frac{1}{4\pi c^3}\,(\ddot{\boldsymbol{p}}\times\boldsymbol{r})^2\,d\Omega \tag{40.3}$$

to show the angular distribution, and for the total radiation rate in all directions,

$$-\frac{dW}{dt'} = \frac{2}{3}\,\frac{\ddot{p}^2}{c^3}\,. \tag{40.4}$$

The total radiation emitted in a single collision may be found by integrating (40.4) over the entire trajectory, but a more useful quantity is the average radiation from a whole stream of particles of uniform current density as the stream is affected by a single scatterer.

The distance at which the colliding particles would pass if they moved in straight lines is called the impact parameter, which may be designated as b. For unit flux of particles, in which there is one incident particle per unit time per unit cross section of the beam, the total radiation is obtained by multiplying the radiation from an individual collision by the number of particles having impact parameters between b and $b+db$ and integrating over all possible impact parameters. The result has the dimensions of energy times area, and may be called the effective radiation per scatterer. Thus if ΔW is the energy radiated per collision at impact parameter b, the effective radiation χ is given by

$$\chi = \int_0^\infty \Delta W\, 2\pi\, b\, db\,. \tag{40.5}$$

It is here assumed that the scattering forces are centrally symmetric.

The angular distribution of dipole radiation emitted in the scattering of a beam of particles by a centrally symmetric field may be written quite generally. Both the scattering and the radiation have axial symmetry about the direction through the scatterer and parallel to the beam. If the dipole acceleration corresponding to scattering of a single particle is averaged over all directions at right angles to the beam, the effective radiation per unit solid angle may be written with the use of Eq. (40.3). Let the direction of the beam be x; in $(\ddot{\boldsymbol{p}}\times\boldsymbol{n})^2=\ddot{\boldsymbol{p}}^2-(\boldsymbol{n}\cdot\ddot{\boldsymbol{p}})^2$ all cross terms in two different components average out, and further,

$$\overline{\ddot{p}_y^2} = \overline{\ddot{p}_z^2} = \tfrac{1}{2}\,(\ddot{\boldsymbol{p}}^2 - \ddot{p}_x^2)\,.$$

It follows that, when averaged over the y and z directions,

$$\overline{(\ddot{\boldsymbol{p}}\times\boldsymbol{n})^2} = \tfrac{1}{2}\,(\ddot{\boldsymbol{p}}^2 + \ddot{p}_x^2) + \tfrac{1}{2}\,(\ddot{\boldsymbol{p}}^2 - 3\ddot{p}_x^2)\cos^2\vartheta\,, \tag{40.6}$$

where ϑ is the angle between the direction of the radiation and that of the beam. The effective radiation into the element of solid angle $d\Omega$ is obtained by integrating over the time and over all impact parameters. The result may be written

$$d\chi(\vartheta) = \frac{d\Omega}{4\pi c^3}\left[A + B\,\frac{3\cos^2\vartheta - 1}{2}\right], \tag{40.7}$$

where

$$A = \tfrac{2}{3}\int_0^\infty\int_{-\infty}^\infty \ddot{\boldsymbol{p}}^2\, dt\, 2\pi\, b\, db\,,\qquad B = \tfrac{1}{3}\int_0^\infty\int_{-\infty}^\infty (\ddot{\boldsymbol{p}}^2 - 3\ddot{p}_x^2)\, dt\, 2\pi\, b\, db\,. \tag{40.8}$$

The form of the second term in (40.7) is such that its integral over ϑ gives zero, and the total effective radiation is

$$\chi = \frac{1}{3c^3}\int_0^\infty\int_{-\infty}^\infty \ddot{\boldsymbol{p}}^2\, dt\, b\, db\,. \tag{40.9}$$

Formulas for the spectral resolution of the effective radiation are completely analogous to (40.7) and (40.8). That is,

$$\frac{d\chi_\omega(\vartheta)}{d\omega} = \frac{d\Omega}{c^3}\left[A(\omega) + B(\omega)\,\frac{3\cos^2\vartheta - 1}{2}\right] \tag{40.10}$$

where

$$A(\omega) = \frac{2}{3}\,\omega^4 \int\limits_0^\infty \boldsymbol{p}_\omega^2\, 2\pi\, b\, d\, b, \qquad B(\omega) = \frac{\omega^4}{3}\int\limits_0^\infty (\boldsymbol{p}_\omega^2 - p_{x\omega}^2)\, 2\pi\, b\, d\, b. \tag{40.11}$$

Similarly, the total radiation in a frequency interval $d\omega$ is given by

$$\chi_\omega = \frac{d\omega}{3c^3}\,\omega^4 \int\limits_0^\infty \boldsymbol{p}_\omega^2\, 2\pi\, b\, d\, b. \tag{40.12}$$

For frequencies such that the duration of the collision is negligible compared to the periods characterizing the radiation, the Fourier component of the field may be approximated by

$$\boldsymbol{B}_\omega = \frac{1}{2\pi}\int\limits_{-\infty}^{\infty} \boldsymbol{B}\, e^{i\omega t}\, dt = \frac{1}{2\pi}\int\limits_{-\infty}^{\infty} \boldsymbol{B}\, dt. \tag{40.13}$$

This approximation is equivalent to the assumption of zero collision time. For sufficiently low frequencies, therefore,

$$\boldsymbol{B} = \frac{1}{2\pi}(\boldsymbol{A}_2 - \boldsymbol{A}_1)\times n, \tag{40.14}$$

where $\boldsymbol{A}_2 - \boldsymbol{A}_1$ is the change in the vector potential produced during the collision. It follows that the total radiation corresponding to frequency ω emitted during collision is equal to

$$d\,W_\omega = \frac{2}{3\pi c}(\boldsymbol{A}_2 - \boldsymbol{A}_1)^2 R^2\, d\omega. \tag{40.15}$$

If $v \ll c$, Eq. (40.15) corresponds to electric dipole radiation given by

$$d\,W_\omega = \frac{2}{3\pi c^3}\left(\Sigma\, e\,(\boldsymbol{v}_2 - \boldsymbol{v}_1)\right)^2 d\omega, \tag{40.16}$$

where $\boldsymbol{v}_1$, $\boldsymbol{v}_2$ are the velocities before and after the collision. To the approximation assumed Eq. (40.16) is useful for describing the radiation from an electron which is decelerated (Bremsstrahlung). According to (40.15) and (40.16) the spectrum of radiation is constant on a frequency scale, i.e. $dW_\omega/d\omega$ is constant, although it must be borne in mind that approximation (40.13) is valid only for low frequencies. (High frequency components $\boldsymbol{B}_\omega$ will not arise if $\boldsymbol{B}$ is integrated over a finite collision time.) If the energy emitted remains constant as the frequency goes to zero, however, a paradox arises: on the basis of PLANCK's hypothesis that radiation is emitted in quanta of energy $\hbar\omega$, an infinite number of zero energy quanta would be emitted. This feature, also present in a more exact treatment, is known as the infrared problem or even as the infrared catastrophe. From an experimental viewpoint the problem is academic, but the theoretical difficulty has been the subject of extensive study[1]. In relativistic quantum field theory this difficulty is overcome by the renormalization of mass and charge.

---

[1] The fundamental paper is that of F. BLOCH and A. NORDSIECK: Phys. Rev. **52**, 54 (1937).

41. Radiation accompanying Coulomb scattering. The above considerations may be applied to find the radiation that accompanies the collision of two charged particles whose velocities are small compared with that of light (see Refs. [*9*], [*10*]).

The relative motion of two particles which attract or repel each other with a force that varies as the inverse square of the distance between them is known to be hyperbolic. (It is here assumed that the energy is positive, so that closed orbits are excluded.) If the charges attract, the equivalent one-body trajectory of a particle of reduced mass $\mu = m_1 m_2/(m_1 + m_2)$ is the hyperbola whose equation in polar coordinates is

$$1 - \epsilon \cos\varphi = \frac{a(\epsilon^2 - 1)}{r}, \tag{41.1}$$

where

$$a = \frac{|e_1 e_2|}{2W} = \frac{\alpha}{2W}, \qquad \epsilon = \sqrt{1 + \frac{2WL^2}{\mu\alpha^2}}.$$

Here W is the total energy, L is the angular momentum $\mu r^2 \dot{\varphi} = \mu v_0 b$, and the constant $|e_1 e_2|$ in the inverse square law is abbreviated to α for convenience. Since by the equation of motion

$$\mu \ddot{\mathbf{r}} = -\alpha \mathbf{r}/r^3,$$

the rate of radiation is

$$-\frac{dW}{dt} = \frac{2}{3c^3}\left(\frac{e_1 m_2 - e_2 m_1}{m_1 + m_2}\right)^2 \ddot{\mathbf{r}}^2 = \frac{2\alpha^2}{3c^3}\left(\frac{e_1}{m_1} - \frac{e_2}{m_2}\right)^2 \frac{1}{r^4}. \tag{41.2}$$

To find the total radiation during the collision we may express r in terms of φ by means of Eq. (41.1), and replace the time integration by one over the angle φ, since $dt = \mu r^2 d\varphi/L$. The integration is taken from $-\varphi_0$ to φ_0, where φ_0 is the angle between the axis of the hyperbola and its asymptote, determined by $\cos\varphi_0 = 1/\epsilon$. The result may be expressed in terms of the angle of deflection, $\Theta = \pi - 2\varphi_0$, for which

$$\cot\Theta/2 = \mu v_0^2/b\alpha$$

where b is the impact parameter. The total radiation from the collision of two charged particles which attract each other is then

$$\Delta W = \frac{\mu^3 v_0^5}{3c^3 |e_1 e_2|}\left(\frac{e_1}{m_1} - \frac{e_2}{m_2}\right)^2 \tan^3\frac{\Theta}{2}\left\{(\pi + \Theta)\left(1 + 3\tan^2\frac{\Theta}{2}\right) + 6\tan\frac{\Theta}{2}\right\}. \tag{41.3}$$

The calculation is very similar for the case of repulsion (like charges). Here the equation of the hyperbola is

$$-1 + \epsilon\cos\varphi = \frac{a(\epsilon^2 - 1)}{r}. \tag{41.4}$$

The total radiation is

$$\Delta W = \frac{\mu^3 v_0^5}{3c^3 e_1 e_2}\left(\frac{e_1}{m_1} - \frac{e_2}{m_2}\right)^2 \tan^3\frac{\Theta}{2}\left\{(\pi - \Theta)\left(1 + 3\tan^2\frac{\Theta}{2}\right) - b\tan\frac{\Theta}{2}\right\}. \tag{41.5}$$

For a head-on collision, $\Theta = \pi$,

$$\Delta W = \frac{8\mu^3 v_0^5}{45c^3 e_1 e_2}\left(\frac{e_1}{m_1} - \frac{e_2}{m_2}\right)^2 \tag{41.6}$$

gives the total radiation.

For like charges the total effective radiation may be found by integrating over all impact parameters as well as the time

$$\chi = \frac{2\alpha^2}{3c^3}\left(\frac{e_1}{m_1} - \frac{e_2}{m_2}\right)^2 2\pi \int_0^\infty \int_{-\infty}^\infty \frac{1}{r^4}\, dt\, b\, db. \tag{41.7}$$

The time integration over the trajectory may be replaced by integration over r by means of $dt = dr/v_r$, where the radial velocity v_r is

$$v_r = \sqrt{\frac{2}{\mu}\left[W - \frac{L^2}{2\mu r^2} - V(r)\right]} = \sqrt{v_0^2 - \frac{b^2 v_0^2}{r^2} - \frac{2\alpha}{\mu r}}.$$

The range of r is from infinity to the closest distance of approach r_0 to infinity again, and is thus twice the integral from r_0 to infinity. It is convenient to integrate first over b and then over r, to obtain the final result,

$$\chi = \frac{8\pi v_0}{9c^3} \frac{e_1 e_2 m_1 m_2}{m_1 + m_2} \left(\frac{e_1}{m_1} - \frac{e_2}{m_2}\right)^2, \tag{41.8}$$

for the effective radiation per unit flux of particles with initial velocity v_0.

For identical particles the dipole radiation vanishes. Moreover, there is no magnetic dipole radiation, and the radiation loss during collision is electric quadrupole in character. The quadrupole moment of the system relative to the center of mass is

$$Q_{ij} = \frac{1}{2} e \left(x_i x_j - \frac{r^2}{3}\delta_{ij}\right), \tag{41.9}$$

where the x_i are components of the vector separation of the particles. The radiation rate is obtained by substituting Eq. (41.9) in

$$-\frac{dW}{dt} = \frac{1}{20c^5} \dddot{Q}_{ij}^2. \tag{41.10}$$

The derivatives of x_i with respect to the time can be expressed in terms of the relative velocity of the particles, the components of which are in turn expressible in terms of the initial velocity and impact parameter by means of the conservation laws for energy and angular momentum. The integration over the time is replaced by integration over r, as above, and the final result of the double integral over b and r is

$$\chi = \frac{8\pi}{3} \frac{e^4 v_0^3}{m c^5} \tag{41.11}$$

for the effective radiation. We note again that the ratio of electric quadrupole to electric dipole radiation is of the order of $(v/c)^2$.

42. Spectral resolution of collision radiation (see Ref. [*10*]). The spectral resolution of radiation emitted during collisions of charged particles is obtained by computing the Fourier components of the dipole moment. In the nomenclature of Sect. 41 the time dependence of the coordinates may be expressed by parametric equations. Let us first consider the collision of unlike charges, for which these equations take the form

$$r = a(\epsilon \operatorname{Cos}\xi - 1), \quad t = \sqrt{\frac{\mu a^3}{\alpha}}(\epsilon \operatorname{Sin}\xi - \xi) = (\epsilon \operatorname{Sin}\xi - \xi)/\omega_0, \tag{42.1}$$

in which the range of ξ is from $-\infty$ to $+\infty$, and for convenience ω_0 is written for $\sqrt{\alpha/\mu a^3} = \mu v_0^3/|e_1 e_2|$. If the direction of the motion is along x, the relative coordinates in the plane of the motion are given by

$$x = a(\operatorname{Cos}\xi - \epsilon), \quad y = a\sqrt{\epsilon^2 - 1}\operatorname{Sin}\xi. \tag{42.2}$$

The Fourier components of the velocities are more readily computed than those of the coordinates themselves. From the fact that $\dot{x}_\omega = -i\omega x_\omega$ it follows that

$$\left.\begin{aligned} x_\omega &= \frac{i}{2\pi\omega}\int_{-\infty}^{\infty} \dot{x}\, e^{i\omega t}\, dt = \frac{i}{2\pi\omega}\int_{-\infty}^{\infty} e^{i\omega t}\, dx \\ &= \frac{i a}{2\pi\omega}\int_{-\infty}^{\infty} \operatorname{Sin}\xi\, e^{\frac{i\omega}{\omega_0}(\epsilon \operatorname{Sin}\xi - \xi)}\, d\xi . \end{aligned}\right\} \tag{42.3a}$$

Similarly

$$y_\omega = \frac{i a\sqrt{\epsilon^2-1}}{2\pi\omega}\int_{-\infty}^{\infty} \operatorname{Cos}\xi\, e^{\frac{i\omega}{\omega_0}(\epsilon \operatorname{Sin}\xi - \xi)}\, d\xi , \tag{42.3b}$$

which is equal to

$$y_\omega = \frac{i a\sqrt{\epsilon^2-1}}{2\pi\omega\epsilon}\int_{-\infty}^{\infty} e^{\frac{i\omega}{\omega_0}(\epsilon \operatorname{Sin}\xi - \xi)}\, d\xi , \tag{42.4}$$

as may be seen by substituting $\operatorname{Cos}\xi \equiv \operatorname{Cos}\xi - 1/\epsilon + 1/\epsilon$ in the integrand. Now from the theory of Bessel functions it is known that

$$\int_{-\infty}^{\infty} e^{\nu\xi - i x \operatorname{Sin}\xi}\, d\xi = i\pi H_\nu^{(1)}(i x), \tag{42.5}$$

where $H_\nu^{(1)}$ is the Hankel function of the first kind, which may be written without the superscript in what follows. Utilization of this formula leads to

$$\left.\begin{aligned} y_\omega &= -\frac{a\sqrt{\epsilon^2-1}}{2\omega\epsilon} H_{\frac{i\omega}{\omega_0}}\left(\frac{i\omega\epsilon}{\omega_0}\right), \\ x_\omega &= -\frac{a}{2\omega} H'_{\frac{i\omega}{\omega_0}}\left(\frac{i\omega\epsilon}{\omega_0}\right), \end{aligned}\right\} \tag{42.6}$$

where the prime on the Hankel function means differentiation with respect to its argument. The total radiation in the frequency interval $d\omega$ which is emitted in the course of a single collision is found to be

$$dW_\omega = \frac{2\pi\omega^2 a^2}{3c^3}\left(\frac{e_1 m_2 - e_2 m_1}{m_1+m_2}\right)^2 \left\{\left[H'_{\frac{i\omega}{\omega_0}}\left(\frac{i\omega}{\omega_0}\epsilon\right)\right]^2 - \frac{\epsilon^2-1}{\epsilon^2}\left[H_{\frac{i\omega}{\omega_0}}\left(\frac{i\omega}{\omega_0}\epsilon\right)\right]^2\right\}. \tag{42.7}$$

To compute the effective radiation per unit flux of particles it is necessary to integrate the radiation per collision over all values of the impact parameter. The integral over b from 0 to ∞ may be transformed to that over ϵ from 1 to ∞, by use of the fact that $b\,db = a^2\epsilon\, d\epsilon$. The resultant integral can be directly integrated with the aid of the formula

$$z\left[Z_\nu'^2 + \left(\frac{\nu^2}{z^2} - 1\right) Z_\nu^2\right] = \frac{d}{dz}(z Z_\nu Z_\nu') \tag{42.8}$$

where $Z_\nu(z)$ is any solution of the Bessel equation

$$Z_\nu'' + \frac{1}{z} Z_\nu' + \left(1 - \frac{\nu^2}{z^2}\right) Z_\nu = 0, \tag{42.9}$$

as may be seen by simple substitution of the equation in (42.8). Since $H_{i\omega/\omega_0}\left(\frac{i\omega}{\omega_0}\epsilon\right)$ goes to zero as ϵ goes to infinity, the final formula for the effective radiation of frequency ω is

$$d\chi_\omega = \frac{4\pi^2 i a^2\omega}{3c^3}\sqrt{\frac{\alpha a}{\mu}}\left(\frac{e_1 m_2 - e_2 m_1}{m_1+m_2}\right)^2 H_{\frac{i\omega}{\omega_0}}\left(\frac{i\omega}{\omega_0}\right) H'_{\frac{i\omega}{\omega_0}}\left(\frac{i\omega}{\omega_0}\right). \tag{42.10}$$

Simple approximations for Eq. (42.10) may be obtained for the limiting cases of low and high frequencies. The important contribution to the value of the defining integral (42.5) for the Hankel function comes from the range of the integration parameter for which the exponent is of order unity. Thus for low frequencies, $\omega \ll \omega_0$,

$$H_{\frac{i\omega}{\omega_0}}\left(\frac{i\omega}{\omega_0}\right) \approx -\frac{i}{\pi}\int_{-\infty}^{\infty} e^{-\frac{i\omega}{\omega_0}\operatorname{Sin}\xi}\, d\xi = H_0\left(\frac{i\omega}{\omega_0}\right), \tag{42.11}$$

and, similarly,

$$H'_{\frac{i\omega}{\omega_0}}\left(\frac{i\omega}{\omega_0}\right) \approx H'_0\left(\frac{i\omega}{\omega_0}\right). \tag{42.12}$$

For small x,

$$i\, H_0(i\, x) \approx \frac{2}{\pi}\ln\frac{2}{\gamma x}, \tag{42.13}$$

where γ is the antilogarithm of EULER'S constant, $\gamma = 1.78107\ldots$ Therefore at low frequencies,

$$d\chi_\omega = \frac{16 e_1^2 e_2^2}{3 v_0^2 c^3}\left(\frac{e_1}{m_1} - \frac{e_2}{m_2}\right)^2 \ln\left(\frac{2 v_0^3 m_1 m_2}{\gamma\,\omega\,|e_1 e_2|\,(m_1 + m_2)}\right) d\omega, \tag{42.14}$$

which depends logarithmically on the frequency.

For high frequencies the region of small ξ is important, and the integral for the Hankel function goes over to a Γ-function:

$$H_{\frac{i\omega}{\omega_0}}\left(\frac{i\omega}{\omega_0}\right) \approx -\frac{i}{\pi\sqrt{3}}\left(\frac{6\omega_0}{\omega}\right)^{\frac{1}{3}}\Gamma\left(\frac{1}{3}\right), \quad H'_{\frac{i\omega}{\omega_0}}\left(\frac{i\omega}{\omega_0}\right) \approx \frac{1}{\pi\sqrt{3}}\left(\frac{6\omega_0}{\omega}\right)^{\frac{2}{3}}\Gamma\left(\frac{2}{3}\right).$$

Since $\Gamma(x)\;\Gamma(1-x) = \pi/\sin\pi x$,

$$d\chi_\omega = \frac{16\pi\, e_1^2 e_2^2}{3\sqrt{3}\, v_0^2 c^3}\left(\frac{e_1}{m_1} - \frac{e_2}{m_2}\right)^2 d\omega, \tag{42.15}$$

which is independent of the frequency.

The computation of the radiation accompanying the collision of two charges of the same sign is very similar to that given above. The Fourier components of the coordinates are identical with Eqs. (42.6) except that they contain a factor $\exp(-\pi\omega/\omega_0)$. This factor does not affect the radiation at low frequencies, since it is then approximately equal to unity. At high frequencies, however, the effective radiation from the collision of particles under a repulsive Coulomb force is

$$d\chi_\omega = \frac{16\pi\, e_1^2 e_2^2}{3\sqrt{3}\, v_0^2 c^3}\left(\frac{e_1}{m_1} - \frac{e_2}{m_2}\right)^2 e^{-\frac{2\pi\omega e_1 e_2}{\mu v_0^3}}\, d\omega, \tag{42.16}$$

which falls off exponentially with increasing frequency.

43. Electromagnetic mass from momentum considerations (see Refs. [*3*], [*12*], [*15*]). Consider an electron moving with uniform velocity $\boldsymbol{v}$ in empty space. A virtual variation of the velocity would affect both the particle and the field, and the linear momentum of the field should contribute to the momentum balance. If the electron velocity is so low that use of the quasi-static fields $\boldsymbol{E} = e\boldsymbol{r}/r^3$ and $\boldsymbol{B} = \boldsymbol{v}\times\boldsymbol{E}/c$ is justified, the expression for the field momentum may be written down at once. The coordinate axes may be so chosen that $\boldsymbol{v}$ is in the x direction, whence there is axial symmetry about x and only $\frac{1}{4\pi c}(\boldsymbol{D}\times\boldsymbol{B})_x$ contributes to the momentum. Then

$$\left.\begin{aligned}\int\frac{\boldsymbol{E}\times(\boldsymbol{v}\times\boldsymbol{E})}{4\pi c^2}\,dV &= \int\frac{\boldsymbol{v}\,E^2 - (\boldsymbol{v}\cdot\boldsymbol{E})\,\boldsymbol{E}}{4\pi c^2}\,dV = \int\frac{\boldsymbol{v}\,(E_y^2 + E_z^2)}{4\pi c^2}\,dV\\ &= \frac{4}{3c^2}\,\boldsymbol{v}\int\frac{E^2}{8\pi}\,dV = m_{\text{electromagnetic}}\,\boldsymbol{v},\end{aligned}\right\} \tag{43.1}$$

since the field is spherically symmetric for low values of v. The integral

$$\int \frac{E^2}{8\pi} dV = U_0 \tag{43.2}$$

is the electrostatic field energy of the charge, and its value depends on the form of the charge distribution. For a point charge U_0 is infinite, since the integrand varies as $1/r^4$. If it is assumed that the charge is distributed uniformly over the surface of a sphere of radius r_0,

$$U_0 = e^2/2r_0. \tag{43.3}$$

If, on the other hand, the charge is assumed to be distributed uniformly throughout the volume of the sphere, U_0 is $\frac{6}{5}$ times the expression (43.3).

At higher velocities the computation equivalent to that above must involve the exact convection fields of (35.5) and (35.7). For what is called the *Lorentz electron*, a model which gained added significance in the light of special relativity, the exact form of the momentum expression is very easily derived. It has been noted that the convection potential of Eq. (35.14) has equipotential surfaces which, although spherical for a point charge at rest, become contracted by the factor $(1-v^2/c^2)^{\frac{1}{2}}$ along the direction of motion of the charge. This was consistent with the contraction hypothesis needed to explain the null result of the Michelson-Morley experiment (see Special Relativity) and led LORENTZ to postulate an electronic structure deformable by the same factor. If the charge density of an electron at rest is $\varrho_0(x, y, z)$, that of an electron in motion is equal to

$$\varrho(x, y, z) = \frac{\varrho_0(x/\sqrt{1-v^2/c^2}, y, z)}{\sqrt{1-v^2/c^2}} \tag{43.4}$$

where ϱ_0 is spherically symmetric in its own variables. The factor $1/(1-v^2/c^2)^{\frac{1}{2}}$ is needed to keep the total charge constant when the charge density is integrated over the volume it occupies. By a formal change in variables such as that used to obtain Eq. (30.15) it is shown that the scalar potential of the moving charge distribution (spherically symmetric when at rest and deformed as indicated when in motion) is

$$\phi(x, y, z) = \frac{1}{\sqrt{1-v^2/c^2}} \phi_0(x/\sqrt{1-v^2/c^2}, y, z), \tag{43.5}$$

where ϕ_0 is the static Coulomb potential. The electromagnetic momentum is then equal to

$$\left.\begin{aligned} &\frac{\boldsymbol{v}}{4\pi(1-v^2/c^2)c^2}\int\left[E_{0y}^2(x/(1-v^2/c^2)^{\frac{1}{2}}, y, z) + E_{0z}^2(x/(1-v^2/c^2)^{\frac{1}{2}}, y, z)\right]dV \\ &= \frac{\boldsymbol{v}}{4\pi c^2\sqrt{1-v^2/c^2}}\int\left[E_{0y}^2(x', y', z') + E_{0z}^2(x', y', z')\right]dV' \\ &= \frac{4}{3}\frac{\boldsymbol{v}}{\sqrt{1-v^2/c^2}}\frac{U_0}{c^2}, \end{aligned}\right\} \tag{43.6}$$

where U_0 is again the static field energy. Just as before, the value of U_0 depends on the details of the distribution; all that has been assumed is that this distribution is spherically symmetric when at rest and contracts by a factor $(1-v^2/c^2)^{\frac{1}{2}}$ along the direction of motion.

If the coefficient of $\boldsymbol{v}$ is interpreted as a contribution to the mass, it follows from (43.6) that

$$m_{\text{electromagnetic}} = \frac{4}{3}\frac{U_0}{c^2(1-v^2/c^2)^{\frac{1}{2}}}. \tag{43.7}$$

The prediction that electromagnetic mass varies with the velocity and becomes infinite for $v=c$ predated the theory of relativity. The experimental verification of this change of mass with velocity was at one time thought to indicate that the electronic mass is entirely electromagnetic. Actually it is unrealistic to expect that electromagnetic forces could account for the electronic mass, since there can be no stable configuration of charge under electromagnetic forces alone; for example, a static charge would "explode". Moreover, the relation between electromagnetic energy and electromagnetic mass is not that predicted by relativity theory and unambiguously confirmed by experiment, for

$$U=\frac{1+v^2/3c^2}{(1-v^2/c^2)^{\frac{1}{2}}}\,U_0 \tag{43.8}$$

if computed by the methods used above for a static electron and *not* $U_0/\sqrt{1-v^2/c^2}$. For all these reasons the electromagnetic rest mass $4\,U_0/3\,c^2$ cannot be the total mass; a "binding energy" of $-U_0/3$, having non-electromagnetic character, is needed. It is possible, for a Lorentz electron, to postulate a uniform pressure that compensates for the explosive tendency of the charge, and which may be derived from an energy potential that supplies the required correction. The nature of such a pressure, if it exists, can hardly come within the realm of electromagnetic theory.

The infinite self energy of a point charge can be avoided by assuming that an electron does have a finite extent, although its structure is not known. Since any charge e distributed symmetrically throughout a sphere of radius r_0 yields an electromagnetic mass of order e^2/r_0c^2, it is customary to define the "radius of the electron" in terms of the actual electron mass,

$$r_0=e^2/m\,c^2. \tag{43.9}$$

Results of the theory that depend explicitly on this radius, and thus on the detailed structure of the electron, cannot be interpreted literally. In fact, the finite extent of the electron produces difficulties in the theoretical considerations of energy conservation involving radiation. The resolution of these difficulties is discussed in Sects. 36 and 37 of the accompanying article on Special Relativity, but we shall here examine their roots.

44. Radiation reaction. The theory of relativity has provided a solution of the problem of high velocity sources in electromagnetic theory by showing that MAXWELL's equations are equally valid in all systems of coordinates in uniform relative motion; by so doing it sets a limit c on allowable velocities. Special relativity alone does nothing, however, to resolve the difficulties in accounting for the reaction force that must accompany radiation. The approximate calculation of this force, made initially by LORENTZ, implies a limitation on the validity of electron theory, which can be avoided only by recourse to the introduction of the empirical mass.

The energy conservation principle may be used to obtain an immediate expression for the radiation reaction if it is assumed that all fluctuations of energy may be averaged out. Consider the radiative loss at low velocities,

$$-\frac{dW}{dt}=\frac{2e^2\,\dot{v}^2}{3c^3}. \tag{44.1}$$

If energy is to be conserved, there must be a reaction $\boldsymbol{F}$ which satisfies the condition

$$\boldsymbol{F}\cdot\boldsymbol{v}+\frac{2e^2\,\dot{v}^2}{3c^3}=0. \tag{44.2}$$

But $\boldsymbol{v}$ and $\dot{\boldsymbol{v}}$ are uncorrelated, and hence (44.2) has no solution as a unique function of time. Conservation of energy on the average may be represented by the integral of (44.2) from some initial time t_1 to a final time t_2. When the radiation term of this integral is integrated by parts, we obtain

$$\int_{t_1}^{t_2}\left(\boldsymbol{F}-\frac{2e\,\ddot{\boldsymbol{v}}}{3c^3}\right)\cdot\boldsymbol{v}\,dt+\left[\frac{2e^2\boldsymbol{v}\cdot\dot{\boldsymbol{v}}}{3c^3}\right]_{t_1}^{t_2}=0. \tag{44.3}$$

If the acceleration (and hence the radiation) occurs during a limited time, or is periodic, the integrated term in (44.3) will not contribute to the average energy balance; therefore energy is conserved in the long run if the radiation reaction force is given by

$$\boldsymbol{F}_{\text{radiation}}=\frac{2e^2\ddot{\boldsymbol{v}}}{3c^3}. \tag{44.4}$$

This reaction is in addition to the negative rate of change of electromagnetic momentum computed in Sect. 43.

The total reaction follows from the LORENTZ' calculation, which also exhibits clearly the limitations of the theory. The method is to find by direct integration the effect on one part of the electron of the radiation fields set up by other parts, and then to sum such effects over the entire electron. It is impossible to carry out this program in generality. No essential limitation need be put on the velocity of the electron as a whole; it is only to simplify the computation that we may choose the element of charge de, on which the remainder of the electron acts, essentially at rest. It is, however, necessary to assume that the relative velocity of any other part of the electron, together with all time derivatives of the velocity, changes very little during the time it takes for an electromagnetic signal to cross the electron. This is equivalent to the conditions that $v\,(\text{relative})\ll c$, $\dot{v}\ll c^2/r_0$, $\ddot{v}\ll\dot{v}\,c/r_0$, etc. All velocities and their derivatives must be referred to the common time of arrival at element de of signals from de'; the change from retarded time t' to t is accomplished by expanding the functions of $t'=t-r/c$ in powers of r/c. From Eq. (35.5),

$$d\boldsymbol{E}(t)=\frac{de'}{s^3}\left\{\frac{1}{c^2}\boldsymbol{r}\times\left[\left(\boldsymbol{r}-\frac{r\,\boldsymbol{v}(t')}{c}\right)\times\dot{\boldsymbol{v}}(t')\right]+\left(1-\frac{v^2(t')}{c^2}\right)\left(\boldsymbol{r}-\frac{\boldsymbol{v}(t')\,r}{c}\right)\right\} \tag{44.5}$$

where $s=r-\boldsymbol{v}(t')\cdot\boldsymbol{r}/c$, and $v(t)$ may be put equal to zero. Then

$$\boldsymbol{v}(t')=-\dot{\boldsymbol{v}}(t)\frac{r}{c}+\frac{1}{2}\frac{r^2}{c^2}\ddot{\boldsymbol{v}}(t)-\cdots,$$

$$\dot{\boldsymbol{v}}(t')=\dot{\boldsymbol{v}}(t)-\frac{r}{c}\ddot{\boldsymbol{v}}(t)+\cdots$$

and, on collection of terms,

$$d\boldsymbol{E}(t)=de'\left[\frac{\boldsymbol{r}}{r^3}-\frac{2\boldsymbol{r}(\boldsymbol{r}\cdot\dot{\boldsymbol{v}})}{r^3c^2}+\frac{\ddot{\boldsymbol{v}}}{2c^3}+\frac{1}{2}\frac{\boldsymbol{r}(\boldsymbol{r}\cdot\ddot{\boldsymbol{v}})}{r^2c^3}+O(\dddot{\boldsymbol{v}}\,r/c^4)\right] \tag{44.6}$$

where $O(\dddot{v}\,r/c^4)$ means of the order of $v\,r/c^4$. For a spherically symmetrical charge distribution the average of the first term vanishes, and the terms proportional to $\ddot{v}$ may be combined by averaging. The average field is given by

$$\overline{d\boldsymbol{E}(t)}=de'\left[-\frac{2}{3}\frac{\dot{\boldsymbol{v}}}{c^2r}+\frac{2\ddot{\boldsymbol{v}}}{3c^3}+O(\dddot{\boldsymbol{v}}\,r/c^4)\right]. \tag{44.7}$$

The total reaction force results from integrating over de and de':

$$\left.\begin{aligned}\boldsymbol{F}=\iint de\,d\boldsymbol{E}&=-\frac{4\dot{\boldsymbol{v}}}{3c^2}\int\frac{de\,de'}{2r}+\frac{2e^2\ddot{\boldsymbol{v}}}{3c^3}+O\left(\frac{e^2\dddot{\boldsymbol{v}}\,r_0}{c^4}\right)\\&=-\frac{4}{3}\frac{\dot{\boldsymbol{v}}}{c^2}U_0+\frac{2e^2\ddot{\boldsymbol{v}}}{3c^3}+O\left(\frac{e^2\dddot{\boldsymbol{v}}\,r_0}{c^4}\right).\end{aligned}\right\} \tag{44.8}$$

Here U_0 again represents the electrostatic energy of the electron in its own field, and the coefficient of $\dot{\boldsymbol{v}}$ may be interpreted as electromagnetic mass. Only the radiation reaction term, proportional to $\ddot{\boldsymbol{v}}$, is independent of electron structure; in fact its form does not even demand spherical symmetry for the electron. The terms in $\dddot{\boldsymbol{v}}$ are presumably small; they depend explicitly on r_0 and are therefore incapable of precise physical meaning within the framework of the theory. The terms in $\dot{\boldsymbol{v}}$ and $\boldsymbol{v}$ are identical with those derived for low velocity charges by simple application of the conservation laws. The general formula for the radiation reaction, valid for velocities not small compared with that of light, is

$$\boldsymbol{F}_{\text{radiation}} = \frac{2}{3}\frac{e^2}{c^3}\frac{1}{(1-v^2/c^2)}\left\{\ddot{\boldsymbol{v}} + \frac{\boldsymbol{v}(\boldsymbol{v}\cdot\ddot{\boldsymbol{v}})}{c^2(1-v^2/c^2)} + \frac{3\dot{\boldsymbol{v}}(\boldsymbol{v}\cdot\dot{\boldsymbol{v}})}{c^2(1-v^2/c^2)} + \frac{3\boldsymbol{v}(\boldsymbol{v}\cdot\dot{\boldsymbol{v}})^2}{c^4(1-v^2/c^2)^2}\right\}. \quad (44.9)$$

The restrictions on the acceleration and higher time derivatives of the velocity cannot be circumvented, however.

Certain difficulties are inherent in these expressions. The restriction $\ddot{v} \ll \dot{v}\,c/r_0$ is equivalent to

$$|m\dot{\boldsymbol{v}}| \gg \left|\frac{2e^2\ddot{\boldsymbol{v}}}{3c^2 r_0}\frac{r_0}{c}\right| = F_{\text{radiation}}. \quad (44.10)$$

Thus in order for the theory to be valid the external force producing acceleration must be large compared with the radiation force, i.e., the behavior of the electron must be governed by external forces. Moreover, the approximation in which squared terms in the expansion are neglected predicts no radiation reaction at all for an electron whose acceleration is constant.

The contradictions that are met if the limitation (44.10) is not observed are illustrated by consideration of the motion of a free electron. If there are no external forces, the equation of motion is presumably

$$m\dot{\boldsymbol{v}} = \frac{2e^2\ddot{\boldsymbol{v}}}{3c^3}, \quad (44.11)$$

where m is the empirically determined mass. This equation has the solution

$$\dot{v} = K\,e^{3mc^3t/2e^2} = K\,e^{\frac{3}{2}\frac{t}{\tau_0}}, \quad (44.12)$$

where K is a constant and $\tau_0 = r_0/c \simeq 10^{-23}$ sec, the time characteristic of the passage of a signal across the electron. If Eq. (44.12) were justified the charge would be infinitely self-accelerated in the absence of any field, a result which is manifestly absurd. It is possible to avoid such "run-away" solutions by setting up certain boundary conditions, which correspond to averaging the driving force on an electron at time t over a time interval of order τ_0 after the time t. Ordinary causality is thus apparently violated over time intervals of order τ_0. A consistent description which yields a covariant generalization of the radiation reaction formula (44.4), and at the same time avoids singularities for a point electron by combining advanced and retarded potentials, has been given by DIRAC[1]. (See also Sect. 37, Special Relativity, and Ref. [*15*].)

VII. Classical theory of dispersion and scattering.

45. Classical role of electrons in the emission of radiation. Historically the existence of elementary atomic charge was deduced from FARADAY's laws of electrolytic conduction, and was asserted unambiguously[2] by RIECKE, HELMHOLTZ, and others as early as 1881. It was the investigation of cathode rays that led to J. J. THOMSON's measurement of the charge-to-mass ratio of free

[1] P. A. M. DIRAC: Proc. Roy. Soc. Lond. A **167**, 148 (1938).

[2] E. RIECKE: Wied. Ann. **13**, 191 (1881). — H. v. HELMHOLTZ: J. Chem. Soc. Lond. **39**, 277 (1881). — G. J. STONEY: Phil. Mag. **11**, 381 (1881).

electrons[1] in 1897, but the existence of electrons (negative particles of charge-to-mass ratio roughly a thousand times that for atomic ions) as atomic constituents was postulated by LORENTZ[2] in 1896 to account for the Zeeman effect. LORENTZ assumed that characteristic atomic radiation is due to harmonic oscillations of charged particles; each electron was assumed to have a position of equilibrium to which it was attracted by an "elastic" force of unknown origin when displaced by some excitation mechanism. In this form the hypothesis accounts for only a single spectral line, although "overtones" would be expected. The theory was thus inadequate to account for observed atomic frequencies, but the model does explain many features of atomic radiation and related effects; it has also been a useful guide to quantum theory in the sense of the correspondence principle.

Within the limitations of the applicability of the radiation reaction formula stated in Eq. (44.10), the equation of motion for a harmonically bound electron with natural frequency ω_0 is

$$\ddot{x} + \omega_0^2 x = \frac{2e^2 \dddot{x}}{3c^2 m}. \tag{45.1}$$

Here m is the empirically determined mass, which of course includes the electromagnetic mass. The term on the right represents radiative damping of the oscillator. Since this term must be small in comparison with the binding term, the approximation

$$\dddot{x} \approx -\omega_0^2 \dot{x} \tag{45.2}$$

is justifiable. If for convenience we set

$$\gamma = \frac{2e^2 \omega_0^2}{3c^3 m}, \tag{45.3}$$

the equation of motion becomes

$$\ddot{x} + \gamma \dot{x} + \omega_0^2 x = 0. \tag{45.4}$$

The solution of (45.4), valid for small γ, is

$$x = A\, e^{-i\omega_0 t} e^{-\gamma t/2}. \tag{45.5}$$

The energy of the oscillator, $\frac{1}{2}(m\dot{x}^2 + m\,\omega_0^2 x^2)$, is thus damped:

$$W \approx \frac{m A^2 \omega_0^2}{2} e^{-\gamma t}, \tag{45.6}$$

and

$$-\frac{dW}{dt} = \gamma W = \frac{e^2 \omega_0^4 A^2}{3c^3} e^{-\gamma t} \tag{45.7}$$

represents the rate of radiated energy. Eq. (45.7) corresponds to a damped wave train emitted by the oscillator after a given amplitude A has been excited by an external impulse. According to Eq. (45.7) the quantity $1/\gamma$ is the mean duration of the radiated pulse. This is the classical quantity corresponding to the quantum mechanical lifetime of an excited state produced by an external impulse.

The limitation on the frequency imposed by the condition that $\gamma \ll \omega_0$ is unimportant except for high energy gamma radiation. In terms of the quantum energy of the outgoing radiation the inequality becomes

$$\hbar\omega_0 \ll \frac{\hbar c}{e^2} m c^2 = 137\, m c^2 \approx 70 \text{ Mev} \tag{45.8}$$

for an oscillating electron.

Because of the radiative damping, the radiation emitted by an oscillator is not monochromatic. The line width can be obtained by making a Fourier analysis of the fields, for which the time variation is given by $\exp(-i\omega_0 t - \gamma t/2)$.

[1] J. J. THOMSON: Phil. Mag. **44**, 298 (1897).

[2] L. LORENTZ: Versl. Kon. Akad. Wet. Amsterd. **6**, 506, 555 (1898). But see Ref. [*18*].

If at a given point

$$\boldsymbol{E} = \boldsymbol{E}_0\, e^{-i\omega_0 t}\, e^{-\gamma t/2} = \int_{-\infty}^{\infty} \boldsymbol{E}_\omega\, e^{-i\omega t}\, d\omega \tag{45.9}$$

the field corresponding to a particular frequency ω is

$$\left.\begin{aligned} \boldsymbol{E}_\omega &= \frac{\boldsymbol{E}_0}{2\pi} \int_{-\infty}^{\infty} e^{i(\omega-\omega_0)t - \gamma t/2}\, dt \\ &= \frac{\boldsymbol{E}_0}{2\pi} \frac{1}{i(\omega-\omega_0) - \gamma/2}\,. \end{aligned}\right\} \tag{45.10}$$

Eq. (45.10) corresponds to a radiation intensity

$$I_\omega = \frac{I_0 \gamma}{2\pi} \frac{1}{(\omega-\omega_0)^2 + \gamma^2/4}, \tag{45.11}$$

which is normalized in such a way that $\int_{-\infty}^{\infty} I_\omega\, d\omega = I_0$. The width of the emitted "line" at half intensity is therefore $\Delta\omega \approx \gamma$. The corresponding width expressed in wavelength,

$$\Delta\lambda = \frac{2\pi c\, \Delta\omega}{\omega_0^2} \approx \frac{4\pi}{3} \frac{e^2}{m c^2} = \frac{4\pi}{3} r_0 \approx 10^{-12}\ \text{cm}, \tag{45.12}$$

is independent of the frequency of the oscillator. The relation between the line width and mean life, $\Delta\omega/\gamma \approx 1$, is equivalent to the quantum mechanical relation between the lifetime and the energy of a state, $\Delta E\, \Delta t \approx \hbar$, where $\Delta E = \hbar\, \Delta\omega$.

In practice radiation is not observed from a single atomic radiator, but from many. Radiation damping is therefore not the only source of width for a spectral line. Except at very low pressures the predominant effect is usually the disturbance of atomic oscillations by collisions or other interactions between atoms. LORENTZ (see Ref. [*11*]) computed the collision line breadth, on the assumption that the time intervals between collisions are distributed statistically about a mean collision time according to a simple exponential law. He found that the resulting line has the same shape as the natural line, Eq. (45.11). The breadth at half maximum is equal to twice the reciprocal of the average time between collisions. At low densities the effect of broadening by collisions becomes very small. The superposition of Doppler shifts due to thermal motions also contributes to line widths as observed. The Doppler breadth at half maximum is proportional to the average particle velocity. The shape of the line produced is not that of the natural line; the Doppler breadth may be much greater than the natural breadth, but the intensity decreases more rapidly with distance from the maximum (see Refs. [*6*], [*11*]).

46. Elementary theory of dispersion. In the phenomenological Maxwell theory the effect of a dielectric medium on the velocity with which electromagnetic fields are propagated is attributed to the volume polarization induced by the incident radiation. The elementary theory of dispersion is also based on this view, but the polarization is produced by individual displacements of elastically bound electrons, and is thus frequency dependent. For all frequencies below those of hard X-radiation it may be assumed that the electron velocities are sufficiently small that the effect of the magnetic vector may be neglected, and for wavelengths larger than atomic dimensions the electric field is to a good approximation uniform over the atomic volume. The polarizability is computed from the equation of motion of a bound electron in an external monochromatic field $\boldsymbol{E} = \boldsymbol{E}_0\, e^{-i\omega t}$, which is not, however, assumed identical with the incident field.

$$\ddot{\boldsymbol{x}} - \frac{2e^2\, \dddot{\boldsymbol{x}}}{3 m c^3} + \omega_0^2 \boldsymbol{x} = \frac{e}{m} \boldsymbol{E}_0\, e^{-i\omega t}. \tag{46.1}$$

The time dependence of the steady state solution of this equation is also given by $e^{-i\omega t}$, and therefore $\ddot{x} \approx -\omega^2 \dot{x}$. A reciprocal time γ may be defined as a function of the incident frequency,

$$\gamma = \frac{2e^2 \omega^2}{3mc^3}, \tag{46.2}$$

although the damping term is important only near resonance, and the definition (45.3) is sufficiently accurate for most purposes. The steady state solution is then

$$\boldsymbol{x} = \boldsymbol{E}_0 \frac{e}{m} \frac{1}{\omega_0^2 - \omega^2 - i\omega\gamma} e^{-i\omega t}. \tag{46.3}$$

If there are N electrons per unit volume which are bound so as to correspond to the natural frequency ω_0, the polarization is related to the individual electron displacement by

$$\boldsymbol{P} = N e \boldsymbol{x}. \tag{46.4}$$

Whether $\boldsymbol{x}$ as given by (46.3) can simply be substituted in (46.4) to yield the total polarization depends on the density of the medium.

The general effect of volume polarization is to add the polarization current $\frac{1}{c}\frac{\partial \boldsymbol{P}}{\partial t}$ to the vacuum displacement current $\frac{1}{4\pi c}\frac{\partial \boldsymbol{E}}{\partial t}$ as a circulation source of magnetic field:

$$\operatorname{curl} \boldsymbol{B} = \frac{1}{c}\left(\frac{\partial \boldsymbol{E}}{\partial t} + 4\pi \frac{\partial \boldsymbol{P}}{\partial t}\right). \tag{46.5}$$

This equation, when combined with $\operatorname{curl} \boldsymbol{E} = -\frac{1}{c}\frac{\partial \boldsymbol{B}}{\partial t}$ in the usual way, yields the homogeneous wave equations

$$\left.\begin{aligned} \nabla^2 \boldsymbol{E} - \frac{n^2}{c^2}\frac{\partial^2 \boldsymbol{E}}{\partial t^2} &= 0, \\ \nabla^2 \boldsymbol{B} - \frac{n^2}{c^2}\frac{\partial^2 \boldsymbol{B}}{\partial t^2} &= 0, \end{aligned}\right\} \tag{46.6}$$

where

$$n = \sqrt{1 + \frac{4\pi |\boldsymbol{P}|}{|\boldsymbol{E}|}} = \sqrt{\varepsilon} \tag{46.7}$$

is the index of refraction.

For dilute systems it is a good approximation to take as the local field at each atom just the field of the incoming wave. Therefore, from (46.3),

$$n^2 = \varepsilon = 1 + \frac{4\pi N e^2}{m} \frac{1}{\omega_0^2 - \omega^2 - i\omega\gamma}. \tag{46.8}$$

The damping parameter may represent collision disturbances as well as of radiative reaction, as already noted in Sect. 39; then $1/\gamma$ is to be interpreted as half the average time between collisions. If the medium is not dilute the local field cannot be equated to the external field; for homogeneous isotropic dielectrics the field at the oscillator may be approximated by Eq. (10.7),

$$\boldsymbol{E}_{\text{eff}} = \boldsymbol{E} + \frac{4\pi}{3}\boldsymbol{P}.$$

Substitution of $\boldsymbol{P} = N e \boldsymbol{x} = N \alpha \boldsymbol{E}_{\text{eff}}$, where α is the molecular polarizability, leads to the analog of the Clausius-Mossotti relation derived in Sect. 10:

$$\frac{n^2 - 1}{n^2 + 2} = \frac{4\pi N \alpha}{3} = \frac{N e^2}{m} \frac{4\pi}{3} \frac{1}{\omega_0^2 - \omega^2 - i\omega\gamma}. \tag{46.9}$$

For a gas the ratio $(n^2 - 1)/(n^2 + 2)$ at a particular frequency is thus proportional to the density of the gas. This law for density dependence was proposed independently by H. A. LORENTZ[1] of Leyden and L. LORENZ[2] of Copenhagen in 1880, and is called the Lorentz-Lorenz relation.

Eqs. (46.8) and (46.9) represent the dispersion in dielectrics whose electrons have a single binding frequency. For a non-polar substance the polarizability is a sum of contributions from the individual constituents of the atom or molecule, and these contributions may be regarded as independent of each other. Thus when there are several characteristic frequencies ω_i, their total effect is a properly weighted sum of terms having the form given by Eq. (46.9). If a fraction of the electrons f_i has a binding frequency ω_i and a damping width γ_i,

$$\frac{n^2-1}{n^2+2}=\frac{4\pi N e^2}{3m}\sum_i \frac{f_i}{\omega_i^2-\omega^2-i\omega\gamma_i}. \tag{46.10}$$

In writing (46.10) it has been assumed that the charge-to-mass ratio is the same for all oscillating particles. This is true for atoms, but in molecules contributions may also arise from the vibrations of the nuclei of constituent atoms and from molecular rotation. These contributions may be included formally by summing over e_i^2/m_i times the appropriate "strength" and function of the corresponding frequency; here m_i is of the order of the atomic mass. The observed proper frequencies of electronic oscillators ordinarily range from the visible to the far ultraviolet; molecular vibration and rotation frequencies lie in the infra-red.

Except in the neighborhood of a resonance frequency the term $i\,\omega\,\gamma$ is negligible, and the index of refraction is essentially real. If $\omega<\omega_0$ there is what is called *normal dispersion*, in which the refractive index diminishes with increasing wavelength. Dielectrics whose resonance frequencies lie in the ultra-violet exhibit normal dispersion throughout the visible spectrum. Near a resonance frequency n has an appreciable imaginary part, and the propagation of the wave in the medium is accompanied by absorption. If the index of refraction is written as

$$n=n'(1-i\beta) \tag{46.11}$$

the spatial phase factor of a wave progressing in the direction of increasing x, say, takes the form

$$e^{\frac{i n\omega}{c}x}=e^{\frac{i n'\omega x}{c}}\,e^{-\frac{\omega\beta n' x}{c}};$$

the intensity of the wave decreases exponentially, with an absorption coefficient per unit length of path equal to $2\beta\,\omega\,n'/c$. If the refractive index does not differ too much from unity even in an absorption region, the approximation

$$\frac{3(n^2-1)}{n^2+2}\approx n'^2-1+\frac{1}{3}\beta^2-2i\beta \tag{46.12}$$

may be used. The real index of refraction and the absorption coefficient 2β may then be approximated by

$$\left.\begin{aligned} n'^2-1 &\approx \frac{4\pi N e^2}{m}\sum \frac{\omega_i^2-\omega^2-\frac{1}{6}\omega\beta\gamma_i}{(\omega_i^2-\omega^2)^2+\omega^2\gamma_i^2},\\ 2\beta &\approx \frac{4\pi N e^2}{m}\sum \frac{\omega\gamma_i}{(\omega_i^2-\omega^2)^2+\omega^2\gamma_i^2}. \end{aligned}\right\} \tag{46.13}$$

In the immediate neighborhood of a resonance frequency ω_i the damping term γ_i predominates, and the effects of other resonance frequencies may be neglected. Moreover, the frequency may be replaced by the resonance frequency except in

[1] H. A. LORENTZ: Wied. Ann. **9**, 641 (1880). See also Ref. [*11*].
[2] L. LORENZ: Wied. Ann. **11**, 70 (1880).

the difference $\omega_i - \omega$, and Eqs. (46.13) reduce to

$$\left.\begin{aligned} n'^2 - 1 &\approx \frac{4\pi N e^2}{2m\,\omega_i}\,\frac{\omega_i - \omega - \frac{1}{12}\beta\,\gamma_i}{(\omega_i-\omega)^2 + \frac{1}{4}\gamma_i^2}\,, \\ 2\beta &\approx \frac{4\pi N e^2}{2m\,\omega_i}\,\frac{\frac{1}{2}\gamma_i}{(\omega_i-\omega)^2 + \frac{1}{4}\gamma_i^2}\,. \end{aligned}\right\} \tag{46.14}$$

Fig. 12 represents the reversal of the frequency variation of the refractive index in the resonance region that is called anomalous dispersion, and also the strongly selective resonance absorption. The parameter γ is a measure of the width of the absorption line and also of the region of anomalous dispersion.

On the short wavelength side of an absorption frequency the index of refraction is less than unity; this is also true for frequencies higher than any electronic resonance frequencies. For very high frequencies the electrons may be considered free, and if the system is dilute,

$$n^2 = 1 - \frac{4\pi N e^2}{m\,\omega^2}\,. \tag{46.15}$$

This formula should apply to the refraction of X-rays by dielectrics of low atomic weight. A volume distribution of free electrons is also encountered in the ionosphere, and in discharge tubes. In terms of the reduced wavelength, $\lambdabar = c/\omega$ and the classical electron radius, $r_0 = e^2/m\,c^2$, Eq. (46.15) becomes

$$n^2 = 1 - 4\pi N r_0 \lambdabar^2. \tag{46.16}$$

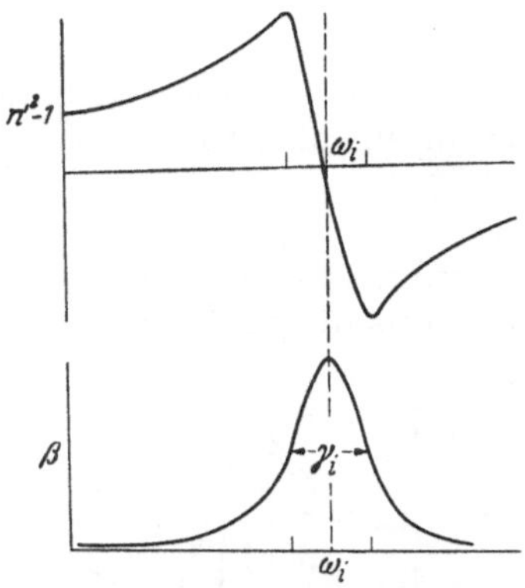

Fig. 12. Showing dependence of refractive index and absorption coefficient on ω near ω_i.

It is of interest to note that for wavelengths longer than a certain value, $\lambdabar > 1/\sqrt{4\pi N r_0}$, n^2 is negative and the refractive index is pure imaginary. For such frequencies the boundary of a free electron distribution will exhibit total reflection.

47. Velocities of wave propagation. What is ordinarily meant by the velocity of light is the speed with which the phase of a monochromatic component is propagated. It is the phase velocity that is related to the dielectric constant and the index of refraction in Sect. 40: $v = c/n$ is the phase velocity. In a dispersive medium, in which the phase velocity depends on the frequency, the passage of radiation through the medium is modified by the differences in phase velocity among its component frequencies.

To describe the transmission of a pulse of waves, or wave packet, the concept of group velocity is introduced. The spectrum of a pulse is continuous, but it is confined to a narrow band of frequencies. Let $\psi(x, t)$ represent a field corresponding to such a group of plane waves moving in the x direction, which can be represented by a Fourier integral,

$$\psi(x, t) = \int a(k)\, e^{i(kx - \omega t)}\, dk\,, \tag{47.1}$$

where $k = \omega/v = 2\pi/\lambda$. If the amplitude $a(k)$ differs from zero only in the neighborhood of a particular frequency designated by ω_0 and the corresponding k_0, we may write

$$\omega(k) = \omega_0 + \Delta\omega \approx \omega_0 + \left(\frac{d\omega}{dk}\right)_{\omega=\omega_0} \Delta k\,, \tag{47.2}$$

where $\Delta k = k - k_0$. Then

$$\left.\begin{aligned}\psi(x,t) &= e^{i(k_0 x - \omega_0 t)} \int a(k_0 + \Delta k)\, e^{i\left(x - \left(\frac{d\omega}{dk}\right)_0 t\right)\Delta k}\, d\,\Delta k \\ &= A\left(x - \left(\frac{d\omega}{dk}\right)_0 t\right) e^{i(k_0 x - \omega_0 t)}.\end{aligned}\right\} \tag{47.3}$$

If A is slowly varying, and the time t is not permitted to become too large, the groups may be considered as a wave of angular frequency ω_0 propagated with amplitude A. But $A\left(x - \left(\frac{d\omega}{dk}\right)_0 t\right)$ is clearly propagated with speed $\left(\frac{d\omega}{dk}\right)_0$ which is called the *group velocity:*

$$v_g(\omega_0) = \left(\frac{d\omega}{dk}\right)_{\omega=\omega_0}. \tag{47.4}$$

The expression for the group velocity may be written in terms of the phase velocity:

$$v_g = \frac{d\omega}{dk} = v - \lambda \frac{dv}{d\lambda}. \tag{47.5}$$

The reciprocal of v_g is readily determined from a graph of the index of refraction by means of the relation

$$\frac{c}{v_g} = n + \omega \frac{dn}{d\omega} \quad \left(n = \frac{c}{v}\right). \tag{47.6}$$

Moreover, the ratio of the phase velocity to the group velocity is simply expressed in terms of the dielectric constant ε,

$$\frac{v}{v_g} = 1 + \frac{\omega}{2\varepsilon}\frac{d\varepsilon}{d\omega} \tag{47.7}$$

Fig. 13. Relation of v (phase velocity), v_g (group velocity) and v_s (signal velocity) to ω in the neighborhood of ω_0.

For a medium whose index of refraction may be represented by Eq. (46.8) the ratio of the phase velocity to the group velocity is

$$\frac{v}{v_g} = 1 + \frac{4\pi N e^2}{m} \frac{\omega^2}{(\omega_0^2 - \omega^2)^2}, \tag{47.8}$$

except in the region of anomalous dispersion where the effect of the damping term cannot be neglected.

With phase velocity defined as c over the real part of the square root of the dielectric constant it is formally possible to compute the group velocity at all frequencies by the formulas given above. This calculation is valid for the "wings" of a resonance line; there the group velocity is less than c, although the phase velocity is greater than c on the short wavelength side of the resonance frequency. Within the region of anomalous dispersion, however, $dv/d\lambda$ is negative, and the group velocity as given formally by (47.5) becomes very large. Actually a group would become so distorted in this region that the concept of group velocity cannot be applied.

For practical purposes it is often desirable to know the rate of propagation for a *signal*; the arrival of a signal may be defined as the onset of waves of appreciable amplitude at some point inside the medium. The problem corresponding to a plane wave that starts through the medium at a particular time has been treated by SOMMERFELD and by BRILLOUIN[1]. The method is to make a Fourier analysis of the incident pulse, let each component be propagated with the phase velocity $v(\omega)$, and then resynthesize the pulse. It is found that the signal velocity

---

[1] A. SOMMERFELD: Ann. Physik **44**, 117—202 (1914). — L. BRILLOUIN: Ann. Physik **44**, 203—240. A more complete account of this work by BRILLOUIN may be found in Volume II of Congrès International d'Electricité, Paris, 1932.

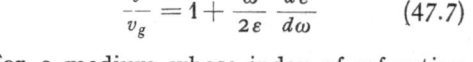

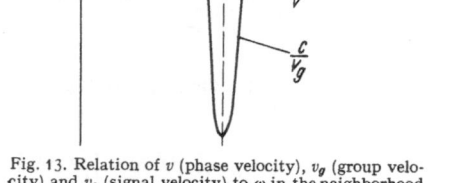

coincides with the group velocity except in the neighborhood of a resonance frequency; it is never greater than c. The reciprocal of the signal velocity, as determined by BRILLOUIN, is shown in Fig. 13.

A more general proof that no signal travels faster than c even if there are frequencies for which both the phase velocity and the group velocity (as formally defined) exceed c follows from the relation between dispersion and absorption—the so-called *dispersion relation*. This relation is an extension of the relation between dispersion and absorption as computed on the classical electron model (see Refs. [*10b*] and [*15*]).

48. Absorption of radiation by a harmonically bound electron (see Refs. [*11*], [*6*]). The total absorption coefficient for an electron for which the binding corresponds to a natural frequency ω_0 may be found by considering its behavior in the presence of radiation of density $I(\omega)\,d\omega$. If the field is represented by $\boldsymbol{E}(\omega)\cos(\omega t - \vartheta_\omega)$, where ϑ_ω is the phase of the Fourier component of frequency ω,

$$\left.\begin{aligned} I(\omega)\,d\omega &= \frac{c}{4\pi} E^2(\omega)\,\overline{\cos^2(\omega t + \vartheta_\omega)}\,d\omega \\ &= \frac{c}{8\pi} E^2(\omega)\,d\omega. \end{aligned}\right\} \tag{48.1}$$

It is assumed that the phases ϑ_ω are distributed at random. For the total absorption the radiation reaction may be omitted, and the equation of motion is

$$\ddot{\boldsymbol{x}} + \omega_0^2 \boldsymbol{x} = \frac{e}{m}\boldsymbol{E}(\omega)\cos(\omega t + \vartheta_\omega). \tag{48.2}$$

Since the energy transfer at frequency ω depends on the relative phases of oscillator and field, the free vibration of the oscillator must be taken into account. The expression

$$\boldsymbol{x} = \frac{e}{m}\boldsymbol{E}(\omega)\frac{1}{\omega_0^2 - \omega^2}\left[\cos(\omega t + \vartheta_\omega) - \cos(\omega_0 t + \vartheta_\omega)\right] + b\sin(\omega_0 t + \vartheta) \tag{48.3}$$

is a solution of (48.2) such that at $t=0$ only the free vibration is excited. Here b is the amplitude and ϑ the phase of the oscillator at time $t=0$.

The work performed on the oscillator by the light wave per unit time per unit frequency interval is

$$W_\omega = e\dot{\boldsymbol{x}}\cdot\boldsymbol{E}(\omega)\cos(\omega t + \vartheta_\omega), \tag{48.4}$$

and the total energy absorbed in time τ is

$$\int_0^\tau dt \int_0^\infty d\omega\, W_\omega.$$

The steady state term of (48.3) does not contribute to the total absorption; that is, the integral of the term involving only $(\omega t + \vartheta_\omega)$ over an integer number of periods vanishes. In general, then,

$$\left.\begin{aligned} \int_0^\tau W_\omega\,dt = \frac{e^2 E^2(\omega)}{m}\,\frac{\omega_0}{\omega_0^2 - \omega^2}\int_0^\tau dt\sin(\omega_0 t + \vartheta_\omega)\cos(\omega t + \vartheta_\omega) \\ + e\,E(\omega)\cdot b\int_0^\tau dt\cos(\omega_0 t + \vartheta)\cos(\omega t + \vartheta_\omega), \end{aligned}\right\} \tag{48.5}$$

where τ is an integer number of periods. The second integral of (48.5) depends on the relation between ϑ_ω and ϑ, and may even be negative, in which case it represents induced emission of light. Since the distribution of radiation phases is assumed to be random, however, we may average over ϑ_ω. The second integral of (48.5) then vanishes, and the first becomes

$$\int_0^\tau W_\omega\, dt = \frac{e^2 E^2(\omega)}{2m} \frac{\omega_0}{\omega_0^2 - \omega^2} \frac{1 - \cos(\omega_0 - \omega)\,\tau}{\omega_0 - \omega}. \tag{48.6}$$

This function falls off rapidly from a sharp maximum at $\omega = \omega_0$. It is therefore permissible to assume $\omega \approx \omega_0$, and if $\tau\,\omega_0 \gg 2\pi$,

$$\int_0^\tau dt \int_0^\omega d\omega\, W_\omega = \frac{\tau\,\pi\, e^2}{4m} E^2(\omega_0) = \frac{\tau\, 2\pi^2 e^2}{m\,c} I(\omega_0), \tag{48.7}$$

since the integral has the form

$$\int_{-\infty}^{\infty} \frac{1 - \cos x}{x^2}\, dx = \pi.$$

The average energy absorbed per unit time per unit intensity of incoming radiation at the resonance frequency is the absorption coefficient per oscillator,

$$a = \frac{2\pi^2 e^2}{m\,c}, \tag{48.8}$$

which is independent of the resonance frequency.

The classical dispersion formula (46.3) may be written in terms of the absorption coefficient per oscillator,

$$\frac{n^2 - 1}{n^2 + 2} = \frac{N\,c\,a}{3} \frac{2}{\pi} \frac{1}{\omega_0^2 - \omega^2 - i\,\omega\,\gamma}, \tag{48.9}$$

or, for a dilute system in which $n \approx 1$,

$$n - 1 = \frac{N\,c\,a}{\pi} \frac{1}{\omega_0^2 - \omega^2 - i\,\omega\,\gamma} \tag{48.10}$$

If there are several electrons per atom (or molecule), each of oscillator strength f_i,

$$f_i = \frac{m\,c\,a_i}{2\pi^2 e^2} \tag{48.11}$$

where a_i is the absorption coefficient corresponding to the binding frequency ω_i.

The generalization of (48.10) to include various allowed frequencies for a single oscillator (as permitted first by BOHR's theory) is accomplished by introducing

$$f(\omega')\, d\omega' = \frac{c\,m\,a(\omega')\, d\omega'}{2\pi^2 e^2}, \tag{48.12}$$

where ω' includes both the discrete and the continuous spectrum. Then Eq. (48.10) may be replaced by

$$n - 1 = \frac{c}{\pi} \int_0^\infty \frac{a(\omega')\, d\omega'}{\omega'^2 - \omega^2}, \tag{48.13}$$

if ω is not too near a resonance frequency. KRAMERS[1] showed that the validity of (48.13) is unlimited if evaluated in the complex plane, with principal values taken at the singularities. Thus

$$n(\omega) = 1 + \frac{c}{\pi} \lim_{\varepsilon \to 0_+} \int_0^\infty \frac{a(\omega')\, d\omega'}{\omega'^2 + (\omega + i\,\varepsilon)^2}. \tag{48.14}$$

---

[1] H. A. KRAMERS: Estratto dagli Atti del Congresso Internazionale dei Fisici, Como (Nicolo Zanichelli, Bologna, 1927), also Ref. [*9*]. We have here traced the historical development of (48.14) as suggested by the quantum theory of dispersion. KRAMERS' final proof is essentially classical, however, and contains no assumption except the physical condition that the polarization of an atom (or other particle) cannot precede the arrival of the field that produces it. (See Refs. [*10b*] and [*15*].)

By examining a disturbance which begins at time t KRAMERS proved that a signal cannot travel faster than c for a medium for which (48.14) is satisfied. It was shown by DE KRONIG[1] that the dispersion relation is a necessary as well as a sufficient condition for this result. More recently[2] it has been proved that the validity of a dispersion relation is logically equivalent to "strict causality" in many fields of physics.

49. Scattering of light. In Sect. 46 loss of energy from a beam of light transmitted through a dielectric has been attributed to radiation resistance. Now a fraction of the incident energy may be scattered in all directions and thus lost to the original beam; we shall see that this is equivalent to radiation damping for normally dispersive dilute systems. Treatment of scattering is somewhat more complicated than the elementary view of the coherent forward beam, and makes explicit some of the assumptions inherent in the elementary theory of dispersion.

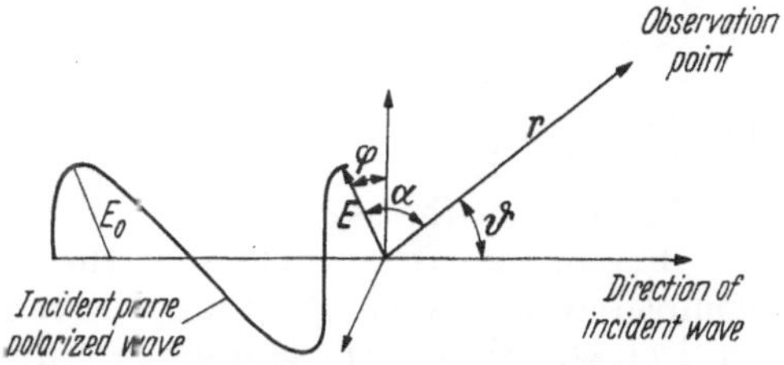

Fig. 14. Showing angles used to describe scattering.

Consider first an individual scatterer, elastically bound with a force that corresponds to a single frequency. In a wave whose electric field is given by $E = E_0\, e^{-i\omega t}$ the steady state displacement x corresponds to an acceleration

$$\ddot{x} = \frac{e}{m} \frac{-\omega^2}{\omega_0^2 - \omega^2 - i\,\omega\,\gamma} E. \tag{49.1}$$

This acceleration gives rise to a dipole radiation field of magnitude

$$E(r) = \frac{e \sin\alpha\, \ddot{x}}{r\, c^2} \tag{49.2}$$

where α is the angle between $\boldsymbol{E}$ and $\boldsymbol{r}$, as shown in Fig. 14. The rate at which the re-radiated energy crosses unit area at this angle, i.e., the magnitude of the Poynting vector in the direction of r, is

$$|\boldsymbol{N}| = \frac{c}{4\pi} \left| \frac{e \sin\alpha\, \ddot{x}}{r\, c^2} \right|^2. \tag{49.3}$$

In terms of the primary intensity

$$I_0 = \frac{c\,\overline{E^2}}{4\pi} \tag{49.4}$$

Eq. (49.3) becomes

$$|\boldsymbol{N}| = \frac{I_0\, r_0^2 \sin^2\alpha}{r^2} \frac{\omega^4}{(\omega_0^2 - \omega^2)^2 + (\gamma\,\omega)^2}, \tag{49.5}$$

where $r_0 = e^2/m c^2$ is the classical electron radius. The total rate of re-radiated energy is

$$-\frac{dW}{dt} = \int r^2 |\boldsymbol{N}|\, d\Omega = \frac{8\pi}{3} r_0^2 I_0 \frac{\omega^4}{(\omega_0^2 - \omega^2) + (\gamma\,\omega)^2}. \tag{49.6}$$

If (49.6) is divided by the intensity of the incoming radiation the resulting expression gives the effective scattering cross section per electron,

$$\sigma = \frac{8\pi r_0^2}{3} \frac{\omega^4}{(\omega_0^2 - \omega^2)^2 + (\gamma\,\omega)^2}. \tag{49.7}$$

Special cases of Eq. (49.7) were originally derived for free electrons (high frequencies) and tightly bound electrons ($\omega \ll \omega_0$). The first case is known as *Thomson*

[1] R. DE L. KRONIG: Nederl. Tijdschr. Natuurk. 9, 402 (1942).
[2] E. g., J. S. TOLL: Phys. Rev. 104, 1760 (1956).

scattering[1], and yields a cross section which is independent of frequency:

$$\sigma(\omega \gg \omega_0) = \frac{8\pi}{3} r_0^2 = \sigma_0, \tag{49.8}$$

which is of the order of magnitude of the geometrical cross section of the electron. At resonance the cross section as given by Eq. (49.7) may become very large. Since $\gamma = 2r_0^2\,\omega^2/3c$,

$$\sigma(\omega = \omega_0) = \sigma_0 \left(\frac{\omega_0}{\gamma}\right)^2 = 6\pi \lambda\!\!\!^{-}{}_0^2 \tag{49.9}$$

for $\omega = \omega_0$, which is greatly in excess of the classical area of the electron. For strong binding, $\omega \ll \omega_0$, $\gamma \ll \omega_0$, Eq. (49.7) becomes

$$\sigma(\omega) = \sigma_0 \left(\frac{\omega}{\omega_0}\right)^4, \tag{49.10}$$

giving a cross section that is proportional to the fourth power of the incident wavelength. This law was first proposed by Lord RAYLEIGH[2] in 1899 to account for the blueness of the sky.

It is of interest to note that the Rayleigh scattering of a dilute system is completely equivalent to the dissipative absorption for normal dispersion discussed in Sect. 46. With a single resonance frequency $\omega_0 \gg \omega$ the absorption coefficient $2\beta\,\omega\,n'/c$ becomes, by Eqs. (46.13) and (46.2),

$$2\beta\,\omega\,n'/c = \frac{8\pi r_0^2}{3}\,\frac{N\omega^4}{\omega_0^4} = N\,\sigma_0 \left(\frac{\omega}{\omega_0}\right)^4, \tag{49.11}$$

the loss to be expected per unit length by virtue of N uncorrelated scatterers per unit volume. It is also clear from the same equations that to first order in γ the absorption coefficient satisfies the relation

$$2\beta\,\omega\,n'/c = \frac{2}{3N}\left(\frac{\omega}{c}\right)^4 \frac{(n'^2-1)^2}{n'}, \tag{49.12}$$

which takes account of the normal dispersion in the frequency dependence of the real index of refraction n'. Eq. (49.12) represents RAYLEIGH's original formulation of the law.

The angular distribution of scattered radiation is most conveniently expressed in terms of the scattering angle ϑ and the polarization angle φ between $\boldsymbol{E}$ and the plane of the incident and scattered radiation (see Fig. 14). The angles α, ϑ, and φ are related by $\cos\alpha = \cos\varphi \sin\vartheta$, and thus

$$\sin^2\alpha = 1 - \cos^2\varphi\,(1 - \cos^2\vartheta)$$

may be substituted in Eq. (49.5). If the primary radiation is unpolarized we must average over φ; since the average of $\cos^2\varphi$ is $\frac{1}{2}$,

$$\overline{\sin^2\alpha} = \tfrac{1}{2}\,(1 + \cos^2\vartheta). \tag{49.13}$$

The differential scattering cross section per unit solid angle for randomly polarized incident radiation is therefore

$$\frac{d\sigma}{d\Omega} = \frac{r_0^2\,(1+\cos^2\vartheta)}{2}\,\frac{\omega^4}{(\omega_0^2-\omega^2)^2 + (\gamma\,\omega)^2}. \tag{49.14}$$

The dependence of Eq. (49.12) on wavelength was used by Lord RAYLEIGH to account for the blueness of the sky and the reddening effect of the atmosphere at sunset. With the assumption that the individual molecules scatter in a wholly

[1] The formula was developed in connection with the scattering of x-rays. For an account see, e.g., J. J. THOMSON and G. P. THOMSON: Conduction of Electricity through Gases, Vol. 2. Cambridge 1933.

[2] RAYLEIGH: Phil. Mag. **47**, 375 (1899).

random fashion, so that the phases are random and the intensities add, together with average empirical values of ω_0 in the expression for n', Eq. (49.12) leads quantitatively to generally good agreement with observed atmospheric scattering[1]. The assumption that each molecule of the atmosphere can be considered to scatter separately, with no net interference from other molecular scatterers, must be examined carefully in view of the law of superposition of electromagnetic fields[2].

Let us assume that the gas is sufficiently dilute that the local field at each oscillator may be taken equal to the primary field, but that the wavelength of the radiation is large compared with the average distance between molecules—conditions that are satisfied in the atmospheric scattering of visible light. Consider the gas as made up of small "cells", elementary volumes of equal size large enough to contain many particles but sufficiently small that all the particles in the j-th cell contribute a scattered field at a distant point P with a particular phase φ_j. Then the total electric field at P due to scattering contains a factor

$$\Phi = \sum_j N_j \, e^{i\varphi_j} \tag{49.15}$$

where N_j is the number of particles in the j-th cell. The quantity (49.15) vanishes if N_j is the same for all cells, since φ_j takes on all values. All N_j are equal if the atoms or molecules are arranged in a regular pattern in space, and thus a perfect crystal does not scatter light at all. More generally, the same conclusion applies to any perfectly homogeneous distribution of matter. For wavelengths long compared with the average molecular spacing, scattering can only be due to deviations from uniformity, such as would arise in a fluid from thermal fluctuations. If δN_j is the difference of the actual number of scattering electrons in the j-th cell from the mean number per cell, the scattering intensity is proportional to

$$|\Phi|^2 = \Big|\sum_j \delta N_j \, e^{i\varphi_j}\Big|^2 = \sum_j (\delta N_j)^2 + \sum_{j \neq k} e^{i(\varphi_j - \varphi_k)} \, \delta N_j \, \delta N_k \, , \tag{49.16}$$

which must be averaged over all cells. In dilute systems the fluctuations of different cells are independent, and the average of the second term on the right vanishes. If the fluctuations are small, the dispersion is given by

$$\overline{(\delta N_j)^2} = N_j \tag{49.17}$$

and therefore the total scattering is proportional to $\sum_j N_j = N$, the total number of scatterers, in agreement with RAYLEIGH'S assumption.

The polarization of skylight[3], discovered by ARAGO in 1811, can be easily understood. The incident unpolarized light may be represented by the superposition of two incoherent waves of equal amplitude polarized in planes at right angles to each other. The directions of these two electric vectors may be taken in the plane of scattering and perpendicular to this plane. The observer will receive dipole radiation from the vibrations induced by the perpendicular vector which is independent of the scattering angle, whereas the scattered electric field in the plane formed by the directions of incidence and scattering will be proportional to $\cos\vartheta$, the cosine of the scattering angle. The ratio of the intensities of the two incoherent components is then $\cos^2\vartheta$; polarization of the scattered light is thus complete at right angles to the incident beam.

---

[1] See "Scattering in Atmospheres" in: The Atmospheres of the Earth and Planets, G. P. KUIPER, ed. Chicago: Chicago University Press 1952.

[2] M. SMOLUCHOWSKI: Ann. d. Physik **25**, 205 (1908). — A. EINSTEIN: Ann. d. Physik **33**, 1275 (1910).

[3] See Z. SEKERA: Polarization of Skylight. This Encyclopedia, Vol. 48, p. 288.

To describe the scattering of light by non-ideal gases the complete expression (49.16) must be used, since the fluctuations are correlated. The marked scattering of a gas near its critical point known as critical opalescence was investigated in this way, primarily by Ornstein and Zernike[1] (see Ref. [*18*]). Essentially the method consists of evaluating a correlation function, which in general depends on the scattering angle. To a first approximation, however, the phenomenon can be described by an average correlation function G which is related to the coefficient of isothermal compressibility,

$$1 + G = \frac{\overline{\Delta N_v^2}}{\overline{N_v}} = -\frac{NkT}{V}\left(\frac{\partial V}{\partial p}\right)_T, \tag{49.18}$$

on the assumption that the elementary volume is sufficiently large that the use of macroscopic parameters is justified. This simple expression becomes infinite at the critical point, but is in accord with observation in the neighborhood of the critical temperature. For detailed and complete treatment of critical opalescence see Refs. [*18*].

50. The rigorous theory of dispersion (see Refs. [*18*] and [*4b*]). The elementary theory of dispersion takes for granted the existence of an average polarization, which is then inserted in the Maxwell wave equation and propagated as a transverse wave. The justification for this procedure is not obvious from the electron theory viewpoint, in which any wave must result from the superposition of the incident radiation and secondary wavelets from electronic oscillators. A complete treatment includes the correlation between scatterers, whether in non-ideal gases or in crystals, in the latter case to account for double refraction and dichroïsm if the symmetry is lower than cubic. The fundamental problem, however, is that of simple transmission through an isotropic medium, which we may take to be sufficiently dilute that the fluctuations of density are uncorrelated; if the macroscopic theory is justified for this case it can be applied with confidence to the more complicated problems of the transmission of light through non-ideal gases and crystalline solids.

The first step in a rigorous theory is to show that the interference of incident and secondary radiation leads to an average polarization of the type

$$\boldsymbol{P} = \chi(\boldsymbol{E}_i + \boldsymbol{E}_d) \tag{50.1}$$

in which $\boldsymbol{E}_i$ denotes the incident wave and $\boldsymbol{E}_d$ is the field exerted on a molecule of the medium by the dipoles induced in all the other molecules; $\boldsymbol{E}_d$ is thus related to the polarization $\boldsymbol{P}$ by an integral equation. It is then required to prove that $\boldsymbol{P}$, as determined by Eq. (50.1), does satisfy a wave equation

$$\nabla^2 \boldsymbol{P} + n^2\left(\frac{\omega}{c}\right)^2 \boldsymbol{P} = 0, \tag{50.2}$$

where n is the (as yet undetermined) index of refraction of the medium, and the transversality condition,

$$\operatorname{div} \boldsymbol{P} = 0. \tag{50.3}$$

The proof consists of showing the compatibility of these three equations, with n properly chosen.

The dipole moment $\boldsymbol{p}\,e^{-i\omega t}$ induced in a single molecule by any electric field $\boldsymbol{E}\,e^{-i\omega t}$ is, as in the static case treated in Sect. 10,

$$\boldsymbol{p} = \alpha \boldsymbol{E} \tag{50.4}$$

[1] L. S. Ornstein and F. Zernike: Proc. Kon. Ned. Acad. Wet. Amsterd. **17**, 793 (1914).

where α is now the molecular polarizability

$$\alpha = \frac{e^2}{m}\left(\frac{1}{\omega_0^2 - \omega^2 - i\omega\gamma}\right) \tag{50.5}$$

from Eq. (46.3), or is given by the appropriate sum of such expressions over all the natural frequencies of the molecular oscillators. The radiative reaction of such a dipole on itself can be interpreted as the work done on a system of vibrating charges by a force equivalent to the field

$$\boldsymbol{E}_r = \frac{2}{3c^3}\dddot{\boldsymbol{p}} = \frac{2i}{3c^3}\omega^3 \boldsymbol{p}. \tag{50.6}$$

The total field acting on any molecule is therefore

$$\boldsymbol{E} = \boldsymbol{E}_i + \boldsymbol{E}_r + \boldsymbol{E}_d. \tag{50.7}$$

The partial field produced on a molecule at point P by a dipole situated at point P' is a linear function of $\boldsymbol{p}(P')$, and may therefore be written $F(P, P')\cdot\boldsymbol{p}(P')$ by means of a symmetrical tensor $F(P, P')$. For an assembly of molecules at $P_0, P_1, P_2, \ldots$ the dipole field at P_0 is given formally by

$$\boldsymbol{E}_d = \sum{}' F(P_0, P_i)\cdot\boldsymbol{p}(P_i), \tag{50.8}$$

the prime on the summation sign indicating that the point P_0 must be omitted from the summation. Thus

$$\boldsymbol{p}(P_0) = \alpha\left[\boldsymbol{E}_i + \frac{2i}{3}\left(\frac{\omega}{c}\right)^3\boldsymbol{p}(P_0) + \sum_i{}' F(P_0, P_i)\cdot\boldsymbol{p}(P_i)\right]. \tag{50.9}$$

In view of the smallness of the parameter α this equation may be solved by successive approximation,

$$\left.\begin{aligned}
\boldsymbol{p}^{(0)}(P_0) &= \alpha\,\boldsymbol{E}_i(P_0);\\
\boldsymbol{p}^{(n)}(P_0) &= \alpha\left[\sum_i{}' F(P_0, P_i)\cdot\boldsymbol{p}^{(n-1)}(P_i) + \frac{2i}{3}\left(\frac{\omega}{c}\right)^3\boldsymbol{p}^{(n-1)}(P_0)\right];\\
\boldsymbol{p}(P_0) &= \sum_{n=0}^{\infty}\boldsymbol{p}^{(n)}(P_0).
\end{aligned}\right\} \tag{50.10}$$

It can be proved that the average of (50.10) over a physically infinitesimal volume element yields a well defined value of the average polarization $\boldsymbol{P}$ at any point in the medium, provided the wavelength of the radiation is sufficiently large compared with atomic dimensions that the field may be assumed uniform over an infinitesimal volume. (See Refs. [*18*] and [*4a*] for details of this proof.) For a gaseous medium, with neglect of density fluctuation correlations,

$$\left.\begin{aligned}
\boldsymbol{P}^{(0)}(P) &= N\alpha\,\boldsymbol{E}_i(P);\\
\boldsymbol{P}^{(n)}(P) &= N\alpha\int_{s(P)}^{S} F(P, P')\cdot\boldsymbol{P}^{(n-1)}(P')\,dV' + \alpha\,\frac{2i}{3}\left(\frac{\omega}{c}\right)^3\boldsymbol{P}^{(n-1)}(P);\\
\boldsymbol{P}(P) &= \sum_{n=0}^{\infty}\boldsymbol{P}^{(n)}(P),
\end{aligned}\right\} \tag{50.11}$$

where the integration excludes a small sphere $s(P)$ containing at most one dipole but otherwise extends to the boundary $\boldsymbol{S}$ of the medium. The corresponding integral equation relating the average polarization to the incident field is

$$\boldsymbol{P}(P) = N\alpha\left[\boldsymbol{E}_i(P) + \int\limits_{s(P)}^{S} F(P, P')\cdot\boldsymbol{P}(P')\,dV'\right] + \frac{2i\alpha}{3}\left(\frac{\omega}{c}\right)^3\boldsymbol{P}(P). \quad (50.12)$$

This relation may be put in the form

$$\boldsymbol{P}(P) = \chi\left[\boldsymbol{E}_i + \int\limits_{s(P)}^{S} F(P, P')\cdot\boldsymbol{P}(P')\,dV'\right] = \chi\left[\boldsymbol{E}_i + \boldsymbol{E}_d\right] \quad (50.13)$$

where

$$\chi = N\left(\frac{1}{\alpha} - \frac{2i}{3}\left(\frac{\omega}{c}\right)^3\right)^{-1} \quad (50.14)$$

includes the effect of radiation damping.

According to electron theory the elementary fields are transmitted as vacuum fields, so that each may be derived from a retarded Hertz potential; that is, the effect of $\boldsymbol{P}(P')$ at point P is

$$F(P, P')\cdot\boldsymbol{P}(P') = \operatorname{curl}\operatorname{curl}\boldsymbol{P}(P')\,\frac{e^{i\omega r/c}}{r} \quad (50.15)$$

where r is the distance from P' to P, and the derivatives are to be taken at point P. Therefore

$$\boldsymbol{E}_d = \int\limits_{s(P)}^{S} \operatorname{curl}\operatorname{curl}\boldsymbol{P}(P')\,\frac{e^{i\omega r/c}}{r}\,dV'. \quad (50.16)$$

In transforming the dipole field to the curl curl of a total Hertz potential, i.e., in taking the differential operator outside the integral, the small sphere $s(P)$ undergoes the same infinitesimal displacement as its center, thus giving rise to a surface integral. This surface integral is finite for a second derivative, and explicit calculation yields

$$\boldsymbol{E}_d = \operatorname{curl}\operatorname{curl}\int\limits_{s(P)}^{S}\boldsymbol{P}(P')\,\frac{e^{i\omega r/c}}{r}\,dV' - \frac{8\pi}{3}\boldsymbol{P}(P). \quad (50.17)$$

Therefore Eq. (50.13) may be written

$$\left(1 + \frac{8\pi}{3}\chi\right)\boldsymbol{P} = \chi\left[\boldsymbol{E}_i + \operatorname{curl}\operatorname{curl}\int\limits_{s(P)}^{S}\boldsymbol{P}\,\frac{e^{i\omega r/c}}{r}\,dV'\right]. \quad (50.18)$$

Since $\operatorname{div}\boldsymbol{E}_i = 0$ it is evident that $\boldsymbol{P}$ satisfies the transversality condition $\operatorname{div}\boldsymbol{P} = 0$.

It remains to show that Eq. (50.18) is satisfied by a polarization vector that is propagated according to Eq. (50.2) with $n^2 \neq 1$, despite the fact that the incident wave and the elementary wavelets are propagated with velocity c. This is done by assuming Eq. (50.2), and finding that the last term in Eq. (50.18) consists of two parts, one of which exactly cancels the incident wave, while the other obeys Eq. (50.2). Thus the incident wave is "extinguished" by interference with the dipole field, and replaced by one which travels with a velocity characteristic of the assemblage of dipoles constituting the medium, as is assumed in the elementary theory. The proof results from an application of GREEN's theorem in the form

$$\int\limits_{s(P)}^{S}(\boldsymbol{P}\nabla^2 G - G\nabla^2\boldsymbol{P})\,dV' = \int\limits_{S}\left(\boldsymbol{P}\frac{\partial G}{\partial n} - \frac{\partial\boldsymbol{P}}{\partial n}\right)dS' - \int\limits_{S(P)}\left(\boldsymbol{P}\frac{dG}{dn} - G\frac{d\boldsymbol{P}}{dr}\right)dS', \quad (50.19)$$

where $G=(1/r)\,e^{i\omega r/c}$ and $\partial/\partial n$ is the normal derivative in the outward direction. Now $\nabla^2 G=-(\omega^2/c^2)\,G$, and by assumption $\nabla^2 \boldsymbol{P}=-n^2(\omega^2/c^2)\,\boldsymbol{P}$, so that the left side of Eq. (50.19) is just

$$(n^2-1)\frac{\omega^2}{c^2}\int\limits_{s(P)}^{S} G\,\boldsymbol{P}(P')\,dV'.$$

The surface integral over the small sphere $s(P)$ has the limiting value $-4\pi\boldsymbol{P}$. By applying the operator curl curl to both sides of Eq. (50.19), noting that $\operatorname{curl}\operatorname{curl}\boldsymbol{P}=n^2(\omega^2/c^2)\,\boldsymbol{P}$ by virtue of the assumed wave equation and $\operatorname{div}\boldsymbol{P}=0$, we obtain

$$\left.\begin{aligned}\operatorname{curl}\operatorname{curl}\int\limits_{s(P)}^{S} G\,\boldsymbol{P}\,dV' &= \frac{4\pi n^2}{n^2-1}\boldsymbol{P}+\\ &+\frac{c^2}{(n^2-1)\,\omega^2}\operatorname{curl}\operatorname{curl}\int\limits_{S}\left(\boldsymbol{P}\frac{\partial G}{\partial n}-G\frac{\partial\boldsymbol{P}}{\partial n}\right)dS',\end{aligned}\right\}\qquad(50.20)$$

which exhibits the separation into the two parts mentioned above. Substitution in Eq. (50.18) yields

$$\left[1-\frac{4\pi\chi}{3}\left(\frac{n^2+2}{n^2-1}\right)\right]\boldsymbol{P}=\chi\left[\boldsymbol{E}_i+\frac{c^2}{\omega^2(n^2-1)}\operatorname{curl}\operatorname{curl}\int\limits_{S}\left(\boldsymbol{P}\frac{\partial G}{\partial n}-G\frac{\partial\boldsymbol{P}}{\partial n}\right)dS'\right],\qquad(50.21)$$

an equality between two sets of terms that obey different wave equations. This is possible only if the two sides of Eq. (50.21) vanish separately. The condition

$$\boldsymbol{E}_i+\frac{c^2}{\omega^2(n^2-1)}\operatorname{curl}\operatorname{curl}\int\limits_{S}\left(\boldsymbol{P}\frac{\partial G}{\partial n}-G\frac{\partial\boldsymbol{P}}{\partial n}\right)dS'=0\qquad(50.22)$$

expresses the *extinction theorem*, due to EWALD[1] (crystalline media) and OSEEN[2] (isotropic media). Clearly the "extinction" of the incident wave is a contribution from the dipoles on the boundary of the medium.

The second condition implied by Eq. (50.21) relates the hitherto undetermined index of refraction n to the average polarizability χ:

$$\chi=\frac{3}{4\pi}\,\frac{n^2-1}{n^2+2}\,.\qquad(50.23)$$

The average field at a random point in the medium differs from the field $\boldsymbol{E}_i+\boldsymbol{E}_d$ on a dipole by the contribution from $s(P)$. On the average this is just the additional field that exists within a homogeneous polarized sphere, i.e., $-4\pi\boldsymbol{P}/3$.

$$\boldsymbol{E}_{\text{av}}=\boldsymbol{E}_i+\boldsymbol{E}_d+\frac{4\pi\boldsymbol{P}}{3}=\frac{4\pi\boldsymbol{P}}{n^2-1}\qquad(50.24)$$

and thus

$$\boldsymbol{D}=\boldsymbol{E}+4\pi\boldsymbol{P}=n^2\boldsymbol{E},\qquad(50.25)$$

which completes the justification for the elementary theory of dispersion.

The entire arguments goes through as above for non-ideal gases with the introduction of a correlation function for the average phases of the elementary wavelets, except for an extra term in the reciprocal of the average polarizability,

[1] P. P. EWALD: Diss. München 1912. Ann. d. Physik **49**, 1 (1916).
[2] C. W. OSEEN: Ann. d. Physik **48**, 1 (1915).

Eq. (50.14) (see Ref. [*18*]). The chief contribution of correlation at moderate densities is an imaginary term which varies as (ω^3/c^3) and may be combined with the radiation damping to represent a net dissipative force. The calculations are entirely analogous to those for the scattering from a non-ideal gas. Indeed, the dissipation of energy from the beam is physically due to scattering under the conditions of a steady state, and the two points of view are entirely equivalent. For a perfect crystal, in which there is no scattering and no dissipation provided that the wavelength is long compared with the separation of the atomic or molecular constituents of the crystal, the correlation exactly cancels the radiation damping. For crystals of lower symmetry than cubic it is of course necessary to take account of the fact that $\boldsymbol{P}$ is not parallel to $\boldsymbol{E}$, which gives rise to a separation into two waves, but the treatment is not qualitatively different from that given here; in particular the extinction theorem remains valid, as the keystone of justification for the elementary view of assuming an average polarization which is propagated in accord with the macroscopic theory.

VIII. Hamiltonian formulation of the field equations.

51. The radiation field. The classical theory of electromagnetic radiation may be put into the form of classical mechanics, i.e., into canonical form. No new physical information is obtained in this way, and for radiation alone the results are equivalent to those obtained by considering the allowed solutions of the wave equation in a cavity with perfectly conducting walls (Sect. 23). Historically, however, it was the correlation of the field with a mechanical system, together with the application of energy equipartition as a condition for equilibrium, which most fully exhibited the classical difficulties in describing blackbody radiation. Subsequently a canonical formulation of the field equations was necessary for the transition to a quantum description of the field and its interaction with particles.

The canonical formulation of the electromagnetic field equations may start from the construction of an equivalent Lagrangian function[1]. This Lagrangian is usually expressed as the integral over all space of a Lagrangian density, with the wave field amplitudes at all points as coordinates corresponding to particle coordinates in classical mechanics. Alternatively, it is possible to expand the field in a set of orthonormal functions, with the expansion coefficients regarded as field coordinates[2]. The second alternative will be followed here, essentially as in Ref. [*6*].

For the radiation field alone, which satisfies the homogeneous wave equation (i.e., div $\boldsymbol{E}=0$), one may readily choose the gauge so that the scalar potential vanishes. Then

$$\boldsymbol{E} = -\frac{1}{c}\dot{\boldsymbol{A}}, \qquad \boldsymbol{H} = \operatorname{curl}\boldsymbol{A}, \tag{51.1}$$

$$\nabla^2\boldsymbol{A} - \frac{1}{c^2}\ddot{\boldsymbol{A}} = 0, \quad \operatorname{div}\boldsymbol{A} = 0. \tag{51.2}$$

In unlimited space there are no limitations on the radiation frequencies, and the spectrum is continuous. It is possible to introduce a denumerably infinite set of frequencies by confining the field to a box of dimensions L, and considering the behavior of the field for wavelengths small in comparison with L.

[1] W. HEISENBERG and W. PAULI: Z. Physik **56**, 1 (1929); **59**, 168 (1930).
[2] E. FERMI: Rev. Mod. Phys. **4**, 87 (1932).

If the fields (and therefore the vector potential $\boldsymbol{A}$) are assumed to be periodic on the surface of the enclosure, $\boldsymbol{A}$ can be represented by

$$\boldsymbol{A} = \sum_\lambda q_\lambda(t)\, \boldsymbol{A}_\lambda(\boldsymbol{r}), \tag{51.3}$$

where $q_\lambda(t)$ is a periodic function of the time and $\boldsymbol{A}_\lambda(\boldsymbol{r})$ is periodic in the dimensions of the box. The equations satisfied by $q_\lambda(t)$ and $\boldsymbol{A}_\lambda(\boldsymbol{r})$ are, from (51.2),

$$\ddot{q}_\lambda + \omega_\lambda^2 q_\lambda^2 = 0 \tag{51.4}$$

and

$$\nabla^2 \boldsymbol{A}_\lambda + \frac{\omega_\lambda^2}{c^2}\, \boldsymbol{A}_\lambda = \nabla^2 \boldsymbol{A}_\lambda + k_\lambda^2\, \boldsymbol{A}_\lambda = 0. \tag{51.5}$$

The values of $\boldsymbol{k}_\lambda$ are limited to the discrete set characterized by three integers $n_{\lambda x}$, $n_{\lambda y}$, $n_{\lambda z}$, such that

$$|\boldsymbol{k}_\lambda| = \frac{2\pi}{L} \sqrt{n_{\lambda x}^2 + n_{\lambda y}^2 + n_{\lambda z}^2} = \frac{\omega_\lambda}{c} \tag{51.6}$$

The index λ must allow for two independent polarizations for each set of integers, as well as all possible combinations of integers. The periodic functions may, for example, involve sines and cosines, of $(\boldsymbol{k}_\lambda \cdot \boldsymbol{r})$, much as in Sect. 23, but the two polarizations are denoted by different values of λ. That is, we may choose

$$\boldsymbol{A}_\lambda = \sqrt{\frac{8\pi c^2}{L^3}}\, \boldsymbol{\epsilon}_\lambda \cos(\boldsymbol{k}_\lambda \cdot \boldsymbol{r}), \tag{51.7}$$

where $\boldsymbol{\epsilon}_\lambda$ is a unit vector representing the direction of polarization, and takes two values for each value of $(\boldsymbol{k}_\lambda \cdot \boldsymbol{r})$. Here positive integers $n_{\lambda x}$, etc., give the complete spectrum, and the functions are so normalized that

$$\int \boldsymbol{A}_\lambda \cdot \boldsymbol{A}_\mu\, dV = 4\pi c^2 \delta_{\lambda\mu}. \tag{51.8}$$

In terms of the normalized orthogonal functions $\boldsymbol{A}_\lambda$ any particular field is represented by the amplitudes $q_\lambda(t)$. The Eq. (51.4) satisfied by q_λ is that of a harmonic oscillator, which is described in Hamiltonian formulation by

$$\left.\begin{aligned} H_\lambda &= \frac{1}{2}(p_\lambda^2 + \omega_\lambda^2 q_\lambda^2), \\ \frac{\partial H_\lambda}{\partial q_\lambda} &= -\dot{p}_\lambda, \qquad \frac{\partial H_\lambda}{\partial p_\lambda} = \dot{q}_\lambda = p_\lambda. \end{aligned}\right\} \tag{51.9}$$

The entire field is then represented by the set of variables p_λ, q_λ, with the Hamiltonian function

$$H = \sum_\lambda H_\lambda. \tag{51.10}$$

By direct substitution of the fields and utilization of the orthogonality condition on the functions $\boldsymbol{A}_\lambda$ it is straightforward to show that

$$U = \int \frac{E^2 + H^2}{8\pi}\, dV = \sum_\lambda H_\lambda = \sum_\lambda \frac{1}{2}(\dot{q}_\lambda^2 + \omega_\lambda^2 q_\lambda^2). \tag{51.11}$$

The radiation field is thus represented as a set of harmonic oscillators, each of frequency ω_λ. This is identical to the standing wave representation of Sect. 23.

The field may also be represented by the amplitudes of progressive waves. If complex exponential functions are used instead of sines and cosines a real vector potential may be written as

$$A = \sum_{y} [q_\lambda(t)\, A_\lambda(r) + q_\lambda^*(t)\, A_\lambda^*(r)], \tag{51.12}$$

where

$$A_\lambda = \varepsilon_\lambda \sqrt{\frac{4\pi c^2}{L^3}}\, e^{i(k_\lambda \cdot r)}, \qquad q_\lambda = |q_\lambda|\, e^{-i\omega_\lambda t}. \tag{51.13}$$

Each A_λ now represents a progressive wave, with a propagation vector k_λ. The integers related to k_λ by (51.6) may now be either positive or negative, and $A_{\ \lambda}$ has the propagation vector $-k_\lambda$. The polarization vector $\varepsilon_{-\lambda}$ is equal to ε_λ, however. Orthogonaltiy and normalization of the A_λ is expressed by

$$\int A_\lambda \cdot A_\mu^*\, dV = \int A_\lambda \cdot A_{-\mu}\, dV = 4\pi c^2\, \delta_{\mu\lambda}. \tag{51.14}$$

The Hamiltonian corresponding to the energy density associated with a single oscillator is $2\omega_\lambda^2\, q_\lambda\, q_\lambda^*$, but q_λ and q_λ^* are not canonical variables in this representation. Real canonical variables are given in terms of the q_λ and q_λ^* by

$$\left.\begin{aligned} Q_\lambda &= q_\lambda + q_\lambda^*, \\ P_\lambda &= -i\,\omega_\lambda (q_\lambda - q_\lambda^*) = \dot{Q}_\lambda. \end{aligned}\right\} \tag{51.15}$$

In terms of P_λ and Q_λ the Hamiltonian takes the familiar form

$$H_\lambda = \frac{1}{2}\,(P_\lambda^2 + \omega_\lambda^2\, Q_\lambda^2), \tag{51.16}$$

and the Hamiltonian equations in these variables lead to the Eq. (51.4) for q_λ and q_λ^*. The P_λ and Q_λ are of course linear combinations of the p_λ and q_λ in the standing wave representation. As may be readily verified by means of the field equations the A_λ representing the radiation field are transverse to the vector k_λ.

52. Hamiltonian for particles in a field. The equations of motion for a charged particle in an electromagnetic field must correspond to the Lorentz force

$$F = e\left(E + \frac{1}{c}\, v \times B\right). \tag{52.1}$$

If the fields are expressed in terms of the vector and scalar potentials in the usual way,

$$E = -\operatorname{grad}\phi - \frac{1}{c}\dot{A}, \qquad B = \operatorname{curl} A, \tag{52.2}$$

and if we confine our attention to velocities for which $v \ll c$, the Hamiltonian for a single particle of charge e_k which yields canonical equations of motion corresponding to (52.1) is

$$H_k = e_k\,\phi + \frac{\left(p_k - \frac{e}{c}\, A_k\right)^2}{2m_k}. \tag{52.3}$$

With a system of particles the field must include the interaction of the particles with each other as well as the external field. The externally imposed field ϕ^e, A^e can be taken into account implicitly by interpreting p_k as $p_k - (e_k/c)\, A^e$, H_k as $H_k - e_k\,\phi^e$, so that the field which appears explicitly is that due to the charges.

This field then satisfies the equations

$$\left.\begin{aligned} \nabla^2 \boldsymbol{A} - \frac{1}{c^2}\ddot{\boldsymbol{A}} &= -\frac{4\pi}{c}\varrho\,\boldsymbol{v}, \\ \nabla^2 \phi - \frac{1}{c^2}\ddot{\phi} &= -4\pi\varrho \end{aligned}\right\} \tag{52.4}$$

and for Lorentz gauge,

$$\operatorname{div}\boldsymbol{A} + \frac{1}{c}\dot{\phi} = 0. \tag{52.5}$$

A general vector field $\boldsymbol{A}$ may be divided into two parts, one of which is solenoidal and the other irrotational: $\boldsymbol{A} = \boldsymbol{A}_1 + \boldsymbol{A}_2$, $\operatorname{div}\boldsymbol{A}_1 = 0$, $\boldsymbol{A}_2 = \operatorname{grad}\psi$. If the fields are again inclosed in a box of volume L^3 with periodic boundary conditions, the expansion of $\boldsymbol{A}_1$ is identical to that of $\boldsymbol{A}$ given in the preceding section. Similarly, $\boldsymbol{A}_2$ may be expanded:

$$\boldsymbol{A}_2 = \sum_\sigma q_\sigma(t)\,\boldsymbol{A}_\sigma(r), \tag{52.6}$$

where the $\boldsymbol{A}_\sigma$ are periodic solutions of the equation

$$\nabla^2 \boldsymbol{A}_\sigma + \frac{\omega_\sigma^2}{c^2}\boldsymbol{A}_\sigma = 0, \tag{52.7}$$

with ω_σ limited to discrete values as in Eq. (51.6). By means of general vector relations it can be proved that the $\boldsymbol{A}_\sigma$ are orthogonal to the $\boldsymbol{A}_\lambda$, and they may be normalized so that

$$\int \boldsymbol{A}_\sigma \cdot \boldsymbol{A}_\varrho\, dV = 4\pi c^2\,\delta_{\sigma\varrho}. \tag{52.8}$$

The partial field $\boldsymbol{A}_2$ represented in terms of $\boldsymbol{A}_\sigma$ consists of longitudinal waves, which do not arise in a pure radiation field.

Similarly the scalar potential may be developed in terms of periodic functions:

$$\phi = \sum_\sigma q_{0\sigma}(t)\,\phi_\sigma, \tag{52.9}$$

with ϕ_σ satisfying the equation

$$\nabla^2\phi_\sigma + \frac{\omega_\sigma^2}{c^2}\phi_\sigma = 0. \tag{52.10}$$

The scalar functions ϕ_σ are related to the $\boldsymbol{A}_\sigma$, since the irrotational $\boldsymbol{A}_\sigma$ may be expressed as the gradient of a scalar, which then satisfies Eq. (52.10) and the same boundary conditions as ϕ_σ. If ϕ_σ is normalized so that

$$\int \phi_\sigma\,\phi_\varrho\, dV = 4\pi c^2\,\delta_{\sigma\varrho}, \tag{52.11}$$

this relation is

$$\boldsymbol{A}_\sigma = \frac{c}{\omega_\sigma}\operatorname{grad}\phi_\sigma. \tag{52.12}$$

The coefficients of ϕ_σ and $\boldsymbol{A}_\sigma$ must satisfy the Lorentz condition; substitution of the expanded potentials in (52.5) leads to

$$\dot{q}_{0\sigma}(t) = \omega_\sigma\, q_\sigma(t). \tag{52.13}$$

The differential equations for the wave amplitudes, like those of the potentials themselves, are now inhomogeneous; they are found by substituting the series expansions into (52.4) and taking advantage of the orthogonality of the expansion functions to integrate over space. The results are, on the assumption of point charges,

$$\left.\begin{aligned} \ddot{q}_\lambda + \omega_\lambda^2 q_\lambda &= \frac{1}{c}\sum_k e_k \boldsymbol{v}_k \cdot \boldsymbol{A}_\lambda(k), \\ \ddot{q}_\sigma + \omega_\sigma^2 q_\sigma &= \frac{1}{c}\sum_k e_k \boldsymbol{v}_k \cdot \boldsymbol{A}_\sigma(k), \\ \ddot{q}_{0\sigma} + \omega_\sigma^2 q_{0\sigma} &= \sum_k e_k \phi_\sigma(k), \end{aligned}\right\} \tag{52.14}$$

where $\boldsymbol{A}_\lambda(k)$, for example, signifies the value of $\boldsymbol{A}_\lambda$ at the k-th particle. The Lorentz condition can be stated as an initial condition on the solution for q_σ and $q_{0\sigma}$: if, at time $t=0$, $\ddot{q}_{0\sigma} = \omega_\tau \dot{q}_\sigma$ in addition to $\dot{q}_{0\sigma} = \omega_\sigma q_\sigma$, then the latter equation, (52.13), is satisfied by solutions of (52.14) for all times t.

The Hamiltonian for the entire system is a sum of terms corresponding to particles and field, but the field enters the particle Hamiltonian in such a way as to express the interaction of the particles with the field. For each of the three types of oscillators representing the amplitudes of $\boldsymbol{A}_\lambda$, $\boldsymbol{A}_\sigma$, and $\boldsymbol{\phi}_\sigma$ there is simply a harmonic oscillator Hamiltonian, but in order that the canonical equations of motion be identical with (52.14) it is necessary that $\frac{1}{2}(p_{0\sigma}^2 + \omega_\sigma^2 q_{0\sigma}^2)$ be introduced with a negative sign. Thus

$$H = \sum_k H_k + \sum_\lambda H_\lambda + \sum_\sigma H_\sigma \tag{52.15}$$

where

$$H_k = e_k \sum_\sigma q_{0\sigma} \phi_\sigma(k) + \frac{1}{2m}\left[p_k + \frac{e_k}{c}\left(\sum_\lambda q_\lambda \boldsymbol{A}_\lambda(k) + \sum_\sigma q_\sigma \boldsymbol{A}_\sigma(k)\right)\right]^2 \tag{52.16}$$

represents the energy of the k-th particle,

$$H_\lambda = \frac{1}{2}(p_\lambda^2 + \omega_\lambda^2 q_\lambda^2) \tag{52.17}$$

represents a transverse radiation wave of frequency ω_λ, while a sum of terms H_σ, with

$$H_\sigma = \frac{1}{2}(p_\sigma^2 + \omega_\sigma^2 q_\sigma^2) - \frac{1}{2}(p_{0\sigma}^2 + \omega_\sigma^2 q_{0\sigma}^2), \tag{52.18}$$

represents the energy of longitudinal and scalar waves. As was shown by FERMI, these terms express the Coulomb interaction between the particles. The four sets of canonical equations corresponding to the total Hamiltonian (52.15) are entirely equivalent to the equations of motion for the particles and the differential equations (52.14).

Appendices.

53. Note on vector spherical harmonics. The functions encountered in the solution of the wave equation in polar coordinates include spherical harmonics, conveniently expressed as vector spherical harmonics for the description of the fields, and spherical Bessel functions. For convenience we define vector spherical harmonics, and summarize some of their properties and uses.

As is well known, the associated Legendre functions may be defined for positive integers $m \leqq l$ and for $-1 \leqq x \leqq 1$ in the notation of HOBSON[1] by

$$P_l^m(x) = (1-x^2)^{m/2} \frac{d^m}{dx^m} P_l(x), \tag{1}$$

where

$$P_l(x) = \frac{1}{2^l l!} \frac{d^l}{dx^l} (x^2-1)^l. \tag{2}$$

These functions satisfy the condition that

$$\int_{-1}^{1} [P_l^m(x)]^2 \, dx = \frac{2}{2l+1} \frac{(l+m)!}{(l-m)!}. \tag{3}$$

Spherical harmonics are conveniently defined for the description of the angular dependence of the Debye potentials by

$$Y_l^m(\vartheta, \varphi) = \left[(2l+1) \frac{(l-m)!}{(l+m)!}\right]^{\frac{1}{2}} P_l^m(\cos\vartheta) \, e^{i m \varphi}, \tag{4}$$

so normalized that the mean value of $Y_l^m(\vartheta, \varphi)$ over a sphere is unity for all l and m. Now the differential equation satisfied by P_l^m,

$$(1-x^2)(P_l^m)'' - 2x(P_l^m)' + \left(l(l+1) - \frac{m^2}{1-x^2}\right) P_l^m = 0, \tag{5}$$

involves only m^2, and negative values of m have physical significance in the dependence on azimuthal angle, $e^{im\varphi}$. By writing the series obtained for $P_l^m(x)$ by differentiation [Eq. (2)], and including the normalizing factor, it may be seen that substitution of $-m$ for m leaves the value of $P_l^m(x)$ unchanged if m is even, and simply changes the sign of the series if m is odd. Therefore we may write

$$Y_l^{-m}(\vartheta, \varphi) = (-1)^m [Y_l^m(\vartheta, \varphi)]^* \tag{6}$$

where Y_l^{m*} is the complex conjugate of Y_l^m.

Vector spherical harmonics may be defined by means of the operator introduced in Sect. 24:

$$\boldsymbol{L} Y_l^m = -i\boldsymbol{r} \times \operatorname{grad} Y_l^m = i \operatorname{curl}(\boldsymbol{r} Y_l^m) \equiv \boldsymbol{X}_l^m(\vartheta, \varphi). \tag{7}$$

These vector spherical harmonics are orthogonal over a sphere, and so normalized that

$$\int \boldsymbol{X}_l^m \cdot \boldsymbol{X}_l^{m*} \, d\Omega = 4\pi l(l+1). \tag{8}$$

They have no radial component, and describe fully the angular dependence of the wholly transverse multipole fields, i.e., the magnetic fields of electric multipoles, and the electric fields of magnetic multipoles. Thus for electric multipoles a "unit" field may be written

$$\left.\begin{aligned} \boldsymbol{H}_E(l, m; \boldsymbol{r}) &= k\, z_l(k r)\, \boldsymbol{X}_l^m(\vartheta, \varphi), \\ \boldsymbol{E}_E(l, m; \boldsymbol{r}) &= i \operatorname{curl}[z_l(k r)\, \boldsymbol{X}^m(\vartheta, \varphi)], \end{aligned}\right\} \tag{9}$$

and for magnetic multipoles,

$$\left.\begin{aligned} \boldsymbol{E}_M(l, m; \boldsymbol{r}) &= k\, z_l(k r)\, \boldsymbol{X}_l^m(\vartheta, \varphi), \\ \boldsymbol{H}_M(l, m; \boldsymbol{r}) &= -i \operatorname{curl}[z_l(k r)\, \boldsymbol{X}_l^m(\vartheta, \varphi)], \end{aligned}\right\} \tag{10}$$

[1] E. W. HOBSON: The Theory of Spherical and Ellipsoidal Harmonics. Cambridge 1931.

where $z_l(k r)$ is the spherical Bessel or Hankel function appropriate to the boundary conditions, to be specified as $h_l(k r)$ for the representation of the radial dependence of an outgoing wave from a source at the origin. [Definitions of the $z_l(k r)$ in terms of ordinary cylinder functions as defined by WATSON are given in Sect. 24.] We note that for a given multipole in the elementary sense a combination of $\boldsymbol{X}_l^m$ with $-l \leqq m \leqq l$ is implied; e.g., for a linear dipole of arbitrary orientation a combination of $\boldsymbol{X}_1^{-1}$, $\boldsymbol{X}_1^0$, and $\boldsymbol{X}_1^1$ is required. In fact, the angular dependence of the magnetic field of an electric dipole along the z axis is given by $\boldsymbol{X}_1^0$, and dipoles in the $x-y$ plane produce fields which are represented by linear combinations of $\boldsymbol{X}_1^1$ and $\boldsymbol{X}_1^{-1}$. The dependence of the amplitude of each multipole on the distribution of current density in a localized source is given in Sect. 33.

A plane wave may be represented in terms of vector spherical harmonics in a series which constitutes an expansion in terms of multipole fields. The basic formula is the scalar expansion of a plane wave propagated along the polar axis in spherical coordinates:

$$e^{ikz} = e^{ikr\cos\vartheta} = \sum_{l=1}^{\infty} i^l (2l+1)\, j_l(k r)\, P_l(\cos\vartheta), \tag{11}$$

where $j_l(kr)$ is the spherical Bessel function $j_l(kr) = \sqrt{\frac{\pi}{2kr}}\, J_{l+\frac{1}{2}}(kr)$ which is regular at the origin and finite everywhere. The elementary method for finding the coefficients in the corresponding series of vector functions is to identify the radial component of the desired vector in a formal expansion, since

$$\left.\begin{aligned} \boldsymbol{r}\cdot \operatorname{curl}\operatorname{curl}\,[\boldsymbol{r}\, j_l(kr)\, Y_l^m(\vartheta,\varphi)] &= l(l+1)\, j_l(k r)\, Y_l^m(\vartheta,\varphi) \\ &= -\, i\, \boldsymbol{r}\cdot \operatorname{curl}\,[j_l(kr)\, \boldsymbol{X}_l^m(\vartheta,\varphi)], \end{aligned}\right\} \tag{12}$$

and $\boldsymbol{X}_l^m$ has no component along $\boldsymbol{r}$. Identification of the radial components of both $\boldsymbol{E}$ and $\boldsymbol{H}$ of the plane wave, together with utilization of the reciprocal curl relations between $\boldsymbol{E}$ and $\boldsymbol{H}$, makes the identification of the entire series complete.

Let us consider a circularly polarized wave, for which the vector character is given by $(\hat{\boldsymbol{x}} \pm i\,\hat{\boldsymbol{y}})/\sqrt{2}$, where $\hat{\boldsymbol{x}}$ and $\hat{\boldsymbol{y}}$ are unit vectors in the direction of x and y respectively. The polar coordinates of a vector of constant amplitude with circular polarization are

$$\frac{\hat{\boldsymbol{x}} + i\,\hat{\boldsymbol{y}}}{\sqrt{2}}\, e^{ikz} = \frac{e^{ikr\cos\vartheta}}{\sqrt{2}} \left(\sin\vartheta\, e^{i\varphi}\, \hat{\boldsymbol{r}} + \cos\vartheta\, e^{i\varphi}\, \hat{\boldsymbol{\vartheta}} + i\, e^{i\varphi}\, \hat{\boldsymbol{\varphi}}\right), \tag{13}$$

where $\hat{\boldsymbol{r}}$, $\hat{\boldsymbol{\vartheta}}$ and $\hat{\boldsymbol{\varphi}}$ are unit vectors in the direction of increasing r, ϑ and φ, respectively. Examination of the radial component leads us to consider that

$$\sin\vartheta\, e^{ikr\cos\vartheta} = -\frac{1}{i k r}\frac{\partial}{\partial\vartheta}\left(e^{ikr\cos\vartheta}\right), \tag{14}$$

and

$$\frac{d\, P_l(\cos\vartheta)}{d\vartheta} = -\, P_l^1(\cos\vartheta), \tag{15}$$

which, together with the dependence on azimuthal angle, show that the value of m is restricted to 1 (or to -1, for light polarized in the opposite sense). Thus with the substitution of the series (11) in Eq. (13) the coefficient of $\boldsymbol{E}_E$ as defined in Eq. (9) may be identified at once. Similar considerations lead to the identification of the coefficient of $\boldsymbol{H}_M$, as defined in Eq. (10), in the corresponding series expansion of the magnetic field of the plane wave. When the relation between the electric and magnetic vectors is utilized, the final result for a plane wave with polarization $(\hat{\boldsymbol{x}} \pm i\,\hat{\boldsymbol{y}})/\sqrt{2}$ is, in terms of the vector spherical harmonics,

$$\left.\begin{aligned} \boldsymbol{E} &= \sum_{l=1}^{\infty} i^l \sqrt{\frac{2l+1}{2l(l+1)}} \left\{\frac{\pm 1}{k} \operatorname{curl}\left[j_l(k\,r)\,\boldsymbol{X}_l^{\pm 1}\right] + j_l(k\,r)\,\boldsymbol{X}_l^{\pm 1}\right\}, \\ \boldsymbol{H} &= \sum_{l=1}^{\infty} i^{l+1} \sqrt{\frac{2l+1}{2l(l+1)}} \left\{\mp j_l(k\,r)\,\boldsymbol{X}_l^{\pm 1} - \frac{1}{k} \operatorname{curl}\left[j_l(k\,r)\,\boldsymbol{X}_l^{\pm 1}\right]\right\}. \end{aligned}\right\} \tag{16}$$

A linearly polarized plane wave propagated along the polar axis may be readily represented as a linear combination of the expressions for right and left circularly polarized light, although the multipole expansion of linearly polarized light is often carried through with real functions of the azimuthal angle, i.e., $\cos m\,\varphi$ and $\sin m\,\varphi$ instead of $e^{\pm i m \varphi}$ (e.g., see Ref. [*15*]).

Vector spherical harmonics have been used by many authors, as noted in Ref. [*3*], but there has been no general uniformity in normalization.

54. Conversion factors.

Multiply the number of Gaussian units below	by	to obtain the number of mks units in
Current in abamperes	10	amperes
Current density in abamperes/cm²	10^5	amperes/meter²
Charge in esu	$\frac{1}{3} \times 10^{-9}$ *	coulombs
Charge density in esu/cm³	$\frac{1}{3} \times 10^{-3}$ *	coulombs/meter³
Capacitance in cm	$\frac{1}{3} \times 10^{11}$ *	farads
Inductance in emu	10^{-9}	henrys
Resistance in esu of potential per abampere	30 *	ohms
Potential in esu	300 *	volts
Field intensity E in esu	3×10^4 *	volts/meter
Displacement D in esu	$\frac{1}{12\pi} \times 10^{-5}$ *	coulombs/meter²
Magnetic flux in maxwells	10^{-8}	webers
Flux density B in gauss	10^{-4}	webers/meter²
H in oersteds	$\frac{1}{4\pi} \times 10^3$	ampere-turns/meter
Conductivity in abamperes per cm² per esu of field intensity	$\frac{10\,*}{3}$	ohms/meter
Magnetomotive force in gilberts	$\frac{10}{4\pi}$	ampere turns

* In all conversion factors marked with an asterisk 3 is used for $c/10^{10}$, where c is measured in cm/sec. If higher accuracy is desired, a more precise value of c must be substituted. J. W. M. DuMond and E. R. Cohen [(1961 adjustment of Natural constants. Ann. of Physics (1962) and Nuovo Cim. (1962), to be published] give $c = (299792.5)$ km/sec.

General references.

[*1a*] Abraham, M., and R. Becker: The Classical Theory of Electricity and Magnetism, 2nd ed. London and Glasgow: Blackie & Son 1950.
[*1a*] Becker, R.: Theorie der Elektrizität, Bd. II. Leipzig: Teubner 1933.
[*1c*] Sauter, F.: Einführung in die Maxwellsche Theorie. Elektronentheorie und Relativitätstheorie, 16. Aufl. von [*1a*]. Stuttgart: Teubner 1957.
[*2*] Bates, L. F.: Modern Magnetism. Cambridge 1951.
[*3*] Blatt, J., and V. F. Weisskopf: Theoretical Nuclear Physics. New York: Wiley 1952.
[*4a*] Born, M.: Optik. Berlin: Springer 1933.
[*4b*] Born, M., and E. Wolf: Principles of Optics. London: Pergamon Press 1959.

[5] Encyclopedia of Physics (Handbuch der Physik). Berlin-Göttingen-Heidelberg: Springer. In references throughout this text the volume number is given, e.g., [5—16] for Vol. 16 of this Encyclopedia.
[6] HEITLER, W.: The Quantum Theory of Radiation, 3rd ed. Oxford 1954.
[7] HERTZ, H.: Electric Waves. London: Macmillan 1893.
[8] JEANS, J.H.: The Mathematical Theory of Electricity and Magnetism, 5th ed. Cambridge 1925.
[9] KRAMERS, H.A.: Collected Scientific Papers. Amsterdam 1956.
[10a] LANDAU, L., and E. LIFSCHITZ: The Classical Theory of Fields. Moscow 1948. English trans., Reading, Mass. Addison-Wesley 1951.
[10b] LANDAU, L., and E. LIFSCHITZ: Electrodynamics of Continuous Media. Moscow 1959. English trans. London: Pergamon Press 1960.
[11] LORENTZ, H.A.: Collected Papers. The Hague: M. Nijhoff 1934—1939.
[12] LORENTZ, H.A.: The Theory of Electrons, 2nd ed. Leipzig: Teubner 1909 (reprinted by Dover, 1952).
[13] MAXWELL, J.C.: A Treatise on Electricity and Magnetism, 3rd ed. Oxford 1904.
[14] NISBET, A.: Proc. Roy. Soc. Lond. A **231**, 250 (1955). — Physica, Haag **21**, 799 (1955).
[15] PANOFSKY, W.K.H., and M. PHILLIPS: Classical Electricity and Magnetism, 2nd ed. Reading, Mass.: Addison-Wesley 1962.
[16] PLANCK, M.: Theorie der Elektrizität und des Magnetismus, 1. Aufl. Leipzig: Hirzel 1922. English translation: Theory of Electricity and Magnetism. London: Macmillan 1932.
[17] ROSE, M.E.: Multipole Fields. New York: Wiley 1955.
[18] ROSENFELD, L.: Theory of Electrons. Amsterdam: North Holland Publishing Co. 1951.
[19] SOMMERFELD, A.: Vorlesungen über theoretische Physik, Bd. III, Elektrodynamik. Wiesbaden: Dietrich 1948. — English edition: Lectures on Theoretical Physics, Vol. III, Electrodynamics. New York: Academic Press 1952.
[20] SOMMERFELD, A.: Vorlesungen über theoretische Physik, Bd. IV, Optik. Wiesbaden: Dietrich 1950. — English edition: Lectures on Theoretical Physics, Vol. IV, Optics. New York: Academic Press 1954.
[21] STRATTON, J.A.: Electromagnetic Theory. New York-Toronto-London: McGraw-Hill 1941.
[22] VAN VLECK, J.H.: The Theory of Electric and Magnetic Susceptibilities. Oxford 1932.
[23] WHITTAKER, E.: A History of the Theories of Aether and Electricity, Vol. I (1951), Vol. II (1953). London and New York: T. Nelson.

The Special Theory of Relativity[1].

By

PETER G. BERGMANN.

A. Foundations.

With 1 Figure.

I. Preliminaries.

1. Classical relativity. The *special* (or *restricted*) *theory of relativity* was proposed by ALBERT EINSTEIN[2] in response to the failure of classical relativity to explain the propagation characteristics of electromagnetic waves. To permit the reader to appreciate the significance of that failure we shall start off with a brief resume of classical relativity (as we would call it today) and its bearing on the propagation of light and other electromagnetic waves.

By definition any theory of relativity is concerned with the possible choice of *frame of reference* for the description of the physical universe and for the formulation of the laws of nature. Frame of reference is the technical term for the combination of a spatial coordinate system and a time scale. Because the coordinate system must be defined for all times, its choice also involves a determination of the state of motion of the observer who is thought of as being connected with the frame of reference. In discussing frames of reference, we are then concerned with the origin of the spatial coordinates, the unit of length, the directions of the coordinate axes, with the zero of the time scale, the unit of time, and with the state of motion of the coordinate system. For all of these aspects we must attempt to answer the question whether among all the conceptually possible choices some appear particularly appropriate for the description of nature. In classical antiquity not only was the Earth assumed to be the center of the Universe (at least by PTOLEMY and his followers) but Hellas was the center of the Earth, or even a particular Greek city, hence the origin of the spatial coordinate system was fixed in that location. Likewise in a cosmogonic theory that fixes the origin of the universe at some particular instant of time in the past, the zero of the time scale is fixed. Similarly, at various times it has been thought that there are natural units of length and time, or that there are preferential states of motion. Today, in special relativity, we tend to relate the units of length and time to each other by means of the (universal) speed of light c. Any given theory of relativity will contend that some of the possible choices indicated are completely arbitrary and without effect on the form of the laws of nature, whereas others among the parameters may be determined by considerations of simplicity. In any event, according to the point of view of this article, such theoretical claims are concerned not with logical possibilities but with properties of nature; according to a given theory, such-and-such a choice will render the description of nature essentially more simple than any other choice, or conversely, all choices lead to

[1] Bibliography at the end of the contribution on General Relativity, p. 272.

[2] A. EINSTEIN: Ann. Physik **17**, 891 (1905). Reprinted in H. A. LORENTZ, Das Relativitätsprinzip [7].

the same form. Clearly, such assertions are bound up with the form of the laws of nature. As a matter of historical fact, the special theory of relativity was made necessary by MAXWELL'S laws of the electromagnetic field. Once the theory had been constructed, it had, and continues to have, a profound influence on the theoretical work on new dynamic laws.

The classical theory of relativity is connected with the work of COPERNICUS, GALILEO, and NEWTON. They recognized, first of all, that the Earth was not the center of the universe. Accordingly, the choice of origin of the coordinate system was arbitrary, and so was the choice of directions of the axes. Likewise, the new laws of mechanics were presumably eternal. The zero of time was also to be chosen at will. There were no restrictions on the units of length and time. NEWTON even determined the exact degree of arbitrariness left in the choice of frame of reference. His laws of mechanics deal with the accelerations that material bodies suffer under the influence of their mutual interaction. Hence, they are the same in any two frames of reference that are moving relative to each other in such a manner that the acceleration of a material body with respect to both is the same. That is the case if the two frames are unaccelerated with respect to each other. Moreover, in Newtonian mechanics the strength of interaction between two bodies is determined by their distance. This distance must appear the same to all legitimate observers. Hence, in any two admissible frames of reference the distance between two material objects at the same time must appear to be the same.

An *admissible* frame of reference is one in which the First Law of NEWTON is satisfied: In the absence of external forces (i.e. interaction with other bodies) a body will persist in its state of rest or uniform linear motion. Such frames of reference are called *inertial frames*. For any two inertial frames, time interval measurements and distance measurements between material events must yield the same values, the coordinate origin of one must be in uniform motion (or rest) with respect to the other. The angles between the axes of one and the axes of the other must not change in the course of time. Altogether, if we restrict ourselves to Cartesian coordinate systems, the equations relating the space and time coordinates of some particular event in one inertial frame to those in another take the following form:

$$\left.\begin{aligned} x^{k\prime} &= \sum_{l=1}^{3} c^k{}_l\, x^l + u^k t + d^k, \quad k = 1, 2, 3, \\ t' &= t + d^0, \qquad \sum_{l=1}^{3} c^i{}_l\, c^j{}_l = \delta^{ij} = \begin{cases} 1 & \text{if } i = j, \\ 0 & \text{if } i \neq j. \end{cases} \end{aligned}\right\} \tag{1.1}$$

In these equations, we have labeled the coordinates in each frame x^1, x^2, x^3, respectively, instead of the more common x, y, z. We shall follow this notation consistently throughout this article and also the one on general relativity. Moreover, in all that follows we shall omit summation symbols relating to indices that appear twice in a term (EINSTEIN'S *summation convention*). Eqs. (1.1) contain a total of 10 parameters that characterize the freedom of choice of a new inertial frame of reference. They are called the equations of a *Galilean transformation* (i.e. transition to a new frame of reference). These equations characterize completely and concisely the *classical principle of relativity*.

2. Non-relativistic Maxwell-Lorentz theory. Whereas the basic equations of Newtonian mechanics are covariant with respect to the Galilean transformation equations (1.1) (i.e. they reproduce without change in the primed coordinate system or frame of reference if postulated in the unprimed frame), MAXWELL'S equations of the electromagnetic field and LORENTZ's ponderomotive equations for a point charge do not. If we assume this system of laws to be valid in some particular frame of reference and if we then go over to a new frame by means of Eqs. (1.1), we find that we cannot choose field strengths and other quantities in a manner so as to avoid the appearance of new terms in the equations, which depend on the relative velocity $\boldsymbol{u}$ of the two frames of reference with respect to each other.

For the sake of simplicity we shall not bother with the angles that the new coordinates may have with respect to the original coordinates. These angles contribute nothing of interest. Instead we shall simply work with the following version of the transformation equations:

$$\boldsymbol{x}' = \boldsymbol{x} + \boldsymbol{u}\,t, \qquad t' = t. \tag{2.1}$$

In that case, the partial derivatives occurring in the Maxwell-Lorentz equations expressed in terms of the new coordinates become:

$$\nabla' = \nabla, \qquad \frac{\partial}{\partial t} = \frac{\partial}{\partial t'} - \boldsymbol{u}\cdot\nabla'. \tag{2.2}$$

If, then, we write the basic equations in the original frame of reference in the form

$$\left.\begin{aligned} &\operatorname{div}\boldsymbol{D} = 4\pi\sigma, \qquad \operatorname{curl}\boldsymbol{E} + \frac{1}{c}\frac{\partial \boldsymbol{B}}{\partial t} = 0,\\ &\operatorname{div}\boldsymbol{B} = 0, \qquad \operatorname{curl}\boldsymbol{H} - \frac{1}{c}\frac{\partial \boldsymbol{D}}{\partial t} = \frac{4\pi}{c}\boldsymbol{j},\\ &\boldsymbol{f} = e\left(\boldsymbol{E} + \frac{\boldsymbol{u}}{c}\times\boldsymbol{B}\right), \end{aligned}\right\} \tag{2.3}$$

we can introduce the primed frame of reference and obtain the following set of equations:

$$\left.\begin{aligned} &\operatorname{div}'\boldsymbol{D} = 4\pi\sigma, \qquad \operatorname{curl}'\left(\boldsymbol{E} - \frac{\boldsymbol{u}}{c}\times\boldsymbol{B}\right) + \frac{1}{c}\frac{\partial \boldsymbol{B}}{\partial t'} = 0,\\ &\operatorname{div}'\boldsymbol{B} = 0, \qquad \operatorname{curl}'\left(\boldsymbol{H} + \frac{\boldsymbol{u}}{c}\times\boldsymbol{D}\right) - \frac{1}{c}\frac{\partial \boldsymbol{D}}{\partial t'} = \frac{4\pi}{c}(\boldsymbol{j} + \boldsymbol{u}\,\sigma),\\ &\boldsymbol{f}' = e\left(\boldsymbol{E} - \frac{\boldsymbol{u}}{c}\times\boldsymbol{B} + \frac{\boldsymbol{u}'}{c}\times\boldsymbol{B}\right). \end{aligned}\right\} \tag{2.4}$$

These equations deal with the original field variables but in terms of the new space and time coordinates. There are several possibilities how we may introduce new field variables. One possibility that suggests itself is this:

$$\left.\begin{aligned} &\boldsymbol{E}' = \boldsymbol{E} - \frac{\boldsymbol{u}}{c}\times\boldsymbol{B}, \qquad &&\boldsymbol{H}' = \boldsymbol{H} + \frac{\boldsymbol{u}}{c}\times\boldsymbol{D},\\ &\boldsymbol{D}' = \boldsymbol{D}, &&\boldsymbol{B}' = \boldsymbol{B},\\ &\boldsymbol{j}' = \boldsymbol{j} + \boldsymbol{u}\,\sigma, &&\sigma' = \sigma. \end{aligned}\right\} \tag{2.5}$$

With these new variables, Eqs. (2.3) reproduce themselves completely in the new frame of reference; but even in the absence of dielectrics and magnetic materials there is no simple relationship between electric field strength and electric displacement, nor between magnetic field strength and magnetic induction, in the primed frame of reference, whereas in the original (unprimed) frame of reference $\boldsymbol{D}=\boldsymbol{E}$, and $\boldsymbol{B}=\boldsymbol{H}$.

Scalar and vector potentials are determined by the electric field strength and the magnetic induction. In the new frame of reference the potentials turn out to be:

$$\boldsymbol{A}' = \boldsymbol{A}, \qquad \varphi' = \varphi + \frac{\boldsymbol{u}}{c}\cdot\boldsymbol{A}. \tag{2.6}$$

With the help of the transformation laws (2.5), (2.6) we may now undertake to investigate the possibilities of observing experimentally the motion of our earth-bound frame of reference, i.e. the value of the vector $\boldsymbol{u}$ for terrestrial observations. Such observations may be of two kinds: (a) The electromagnetic field corresponding to a given charge-current distribution will show some signs of our motion relative to the frame in which $\boldsymbol{E}=\boldsymbol{D}$ and $\boldsymbol{H}=\boldsymbol{B}$. We shall examine the theory of such effects, and treat at least one experiment actually performed (Trouton and Noble[1]). (b) Even in the absence of any charges and currents, the propagation of electromagnetic waves will be anisotropic (Michelson and Morley[2]).

For the electromagnetic potentials we can derive a set of fully separated "wave equations",

$$\left.\begin{aligned}\left[\left(\frac{1}{c}\frac{\partial}{\partial t}+\frac{\boldsymbol{u}}{c}\cdot\nabla\right)^2-\nabla^2\right]\left(\varphi-\frac{\boldsymbol{u}}{c}\cdot\boldsymbol{A}\right)&=4\pi\sigma,\\ \left[\left(\frac{1}{c}\frac{\partial}{\partial t}+\frac{\boldsymbol{u}}{c}\cdot\nabla\right)^2-\nabla^2\right]\boldsymbol{A}&=\frac{4\pi}{c}(\boldsymbol{j}-\boldsymbol{u}\sigma),\end{aligned}\right\}\tag{2.7}$$

provided we adopt the gauge condition

$$\nabla\cdot\boldsymbol{A}+\left(\frac{1}{c}\frac{\partial}{\partial t}+\frac{\boldsymbol{u}}{c}\cdot\nabla\right)\left(\varphi-\frac{\boldsymbol{u}}{c}\cdot\boldsymbol{A}\right)=0.\tag{2.8}$$

Although Eqs. (2.7), (2.8) are valid in the new frame of reference, all primes have been omitted, for ease of reading. The ponderomotive equation, in terms of the potentials, is, as usual,

$$\boldsymbol{f}=-e\left[\operatorname{grad}\left(\varphi-\frac{\boldsymbol{v}}{c}\cdot\boldsymbol{A}\right)+\frac{1}{c}\frac{d\boldsymbol{A}}{dt}\right].\tag{2.9}$$

We shall now consider the field surrounding a stationary charge distribution. By assumption this field is independent of the time t. Accordingly, the equations of electrostatics reduce to the following system:

$$\left.\begin{aligned}\Delta^*\varphi&=-4\pi\left(1-\frac{u^2}{c^2}\right)\sigma,\\ \Delta^*\boldsymbol{A}&=\frac{4\pi}{c}\boldsymbol{u}\sigma,\\ \Delta^*&\equiv\nabla^2-\left(\frac{\boldsymbol{u}}{c}\cdot\nabla\right)^2.\end{aligned}\right\}\tag{2.10}$$

For a point charge placed at the origin, the solutions of this system of differential equations are:

$$\left.\begin{aligned}\varphi&=\left(1-\frac{u^2}{c^2}\right)\frac{e}{r^*},\quad \boldsymbol{A}=-\frac{\boldsymbol{u}}{c}\frac{e}{r^*},\\ r^{*2}&\equiv\left(1-\frac{u^2}{c^2}\right)r^2+\left(\frac{\boldsymbol{u}}{c}\cdot\boldsymbol{r}\right)^2.\end{aligned}\right\}\tag{2.11}$$

If we are interested in the interactions between charges at rest (with respect to the new frame, in which all our expressions are valid), we may disregard the expression for $\boldsymbol{A}$ and work just with the static potential of a point charge, φ. The equipotential surfaces are, as we can see from the expression for r^* in Eq. (2.11), not spherical but ellipsoidal. Hence the electrostatic force between two charges is not in the connecting straight line unless this connecting straight line is either parallel or perpendicular to the direction of $\boldsymbol{u}$. Exactly, the force between two

[1] Trouton and Noble: Phil. Trans. Roy. Soc. Lond. A **202**, 165 (1903). — Proc. Roy. Soc., Lond. **72**, 132 (1903).

[2] Michelson and Morley: Amer. J. Sci. **34**, 333 (1887). — Phil. Mag. **24**, 449 (1887).

charges e_1 and e_2, with the connecting straight line $\boldsymbol{r}$, comes out as

$$\boldsymbol{f} = \frac{e_1 e_2}{r^{*3}} \left[\left(1 - \frac{u^2}{c^2}\right) \boldsymbol{r} + \left(\frac{\boldsymbol{u}}{c} \cdot \boldsymbol{r}\right) \frac{\boldsymbol{u}}{c} \right]. \tag{2.12}$$

Suppose, then, that the two charges form a dipole. In that event, we can easily determine the torque acting on the dipole because of the internal forces of the system:

$$\boldsymbol{T} \equiv \boldsymbol{r} \times \boldsymbol{f} = \frac{e^2}{r^{*3}} \left(\frac{\boldsymbol{u}}{c} \cdot \boldsymbol{r}\right) \left(\frac{\boldsymbol{u}}{c} \times \boldsymbol{r}\right). \tag{2.13}$$

This torque will act about an axis at right angles both to the connecting straight line and to $\boldsymbol{u}$. Its magnitude will be

$$T = \frac{1}{2} e_1 e_2 \frac{r^2}{r^{*3}} \frac{u^2}{c^2} \sin 2\vartheta, \qquad \vartheta = \sphericalangle (\boldsymbol{u}, \boldsymbol{r}), \tag{2.14}$$

and it will tend to turn the dipole into the direction of $\boldsymbol{u}$.

Trouton and Noble attempted to determine this torque by suspending a charged condenser from a torsion balance. The result was negative; no torque was discovered.

Eqs. (2.7) are also convenient if we wish to study the propagation of plane waves in a frame of reference that leads to "transport terms". We set the right-hand sides zero and obtain expressions for the electromagnetic potentials whose dependence on space and time coordinates is imaginary-exponential, of the form $\exp[i(\boldsymbol{k}\cdot\boldsymbol{x} - \omega t)]$. To satisfy Eqs. (2.7), ω must obey the relationship:

$$\omega = c k + \boldsymbol{u} \cdot \boldsymbol{k}. \tag{2.15}$$

Because the propagation is anisotropic, we cannot simply assume that the direction of propagation of a light ray is parallel to the vector $\boldsymbol{k}$. That vector is perpendicular to the wave fronts, but with an anisotropic law of propagation rays and wave fronts are not normal to each other.

To obtain an effective "index of refraction" we may proceed as follows: In Eq. (2.15) we replace the wave propagation vector $\boldsymbol{k}$, which has the dimension L^{-1}, by the dimensionless vector $\boldsymbol{p} = \frac{c}{\omega} \boldsymbol{k}$, which satisfies the equation

$$p^2 - \left(1 - \frac{\boldsymbol{u}}{c} \cdot \boldsymbol{p}\right)^2 = 0. \tag{2.16}$$

This new vector $\boldsymbol{p}$ is the gradient of the eikonal of ray optics. Eq. (2.16) represents the eikonal equation (Hamilton-Jacobi equation) of geometrical optics (also known as the Huygens-Fresnel principle, cf. this Encyclopedia, Vol. XXIV, p. 48ff.). To obtain the time of transit along a ray trajectory, or the optical path length (which, by definition, is c times the time of transit), we may treat the expression (2.16) as the Hamiltonian and find the direction of the rays by differentiation of H with respect to $\boldsymbol{p}$:

$$\frac{d\boldsymbol{x}}{d\vartheta} = \alpha \left[\boldsymbol{p} + \left(1 - \frac{\boldsymbol{u}}{c} \cdot \boldsymbol{p}\right) \frac{\boldsymbol{u}}{c}\right]. \tag{2.17}$$

The constant factor α depends on the choice of the parameter ϑ along the ray trajectory and is to this extent arbitrary. But as the canonical momentum $\boldsymbol{p}$ must in turn be the derivative of the Lagrangian L with respect to $(d\boldsymbol{x}/d\vartheta)$ we get for the Lagrangian itself the expression

$$L = \left(1 - \frac{u^2}{c^2}\right)^{-1} \left[\sqrt{\left(\frac{d\boldsymbol{x}}{d\vartheta}\right)^2 - \left(\frac{\boldsymbol{u}}{c} \times \frac{d\boldsymbol{x}}{d\vartheta}\right)^2} - \frac{\boldsymbol{u}}{c} \cdot \frac{d\boldsymbol{x}}{d\vartheta}\right]. \tag{2.18}$$

This expression, multiplied by $d\vartheta$, represents directly the optical path length of an infinitesimal segment of the ray trajectory. It is clearly independent of the choice of the parameter ϑ.

Suppose we consider a straight-line trajectory of length A. Its optical path length will be given by the expression

$$A^* = \left(1 - \frac{u^2}{c^2}\right)^{-1} \left[\sqrt{1 - \frac{u^2}{c^2}\sin^2\gamma} - \frac{u}{c}\cos\gamma\right] A, \tag{2.19}$$

where γ is the angle between the ray path and the direction of the velocity of the frame of reference $\boldsymbol{u}$. The optical path length is not the same for the two directions in which the path A may be traversed.

For purposes of an experimental determination of the vector $\boldsymbol{u}$ in the laboratory frame of reference, the dependence of the optical path length on the direction of propagation apparently could be observed. Considering the fact that the speed of the Earth relative to the Sun is about 30 km sec^{-1}, or $10^{-4}\,c$, it was to be expected that (u/c) would be at least of the order of magnitude of 10^{-4}. The dependence of A^* on (u/c) is linear because of the last term in the square bracket of Eq. (2.19) but otherwise quadratic. Actually an observation of the optical path length itself is impractical. We do not observe the absolute phase of electromagnetic waves but the relative phase of two alternative paths, ordinarily through interferometric observations. To observe the phase of a travelling wave itself would involve the observation of times of arrival, e.g. of a pulse of electromagnetic energy, at various points along the path of transmission. Such a determination would involve the synchronization of clocks stationed at these points; and this synchronization can be accomplished only with the help of timing signals passed from one clock to another. Presumably these time signals travel at a rate that is also affected by $\boldsymbol{u}$. Accordingly, in a first-order [in (u/c)] experiment the necessary calibrations would be affected by the anisotropy that forms the subject matter of the experiment itself.

The actual experiment involves the transmission of a light signal over a closed path, which in the Michelson-Morley experiment was chosen to be a straight path of transmission traversed forth and back. For the two trips in opposite directions the total optical path length equals

$$\left.\begin{aligned} A^{**} &= 2\left(1 - \frac{u^2}{c^2}\right)^{-1} \sqrt{1 - \frac{u^2}{c^2}\sin^2\gamma}\, A \\ &= 2\left(1 - \frac{u^2}{c^2}\right)^{-1} \sqrt{1 - \frac{1}{2}\frac{u^2}{c^2} + \frac{1}{2}\frac{u^2}{c^2}\cos 2\gamma}\, A. \end{aligned}\right\} \tag{2.20}$$

It is largest when the path A is parallel to $\boldsymbol{u}$, smallest when A and $\boldsymbol{u}$ are at right angles. MICHELSON and MORLEY constructed an interferometer with two arms at right angles to each other. The whole interferometer could be turned about an axis. If the two effective arms had lengths A and B, respectively, and if the light beam would travel along either A or B back and forth, the waves propagated along these two paths being finally permitted to interfere with each other, then the difference in optical path length would be

$$A^{**} - B^{**} = 2\left[\left(1 + \frac{3}{4}\frac{u^2}{c^2}\right)(A - B) + \frac{1}{4}\frac{u^2}{c^2}(A + B)\cos 2\gamma\right]. \tag{2.21}$$

This difference in optical path lengths determines the location of the interference fringes. If the whole instrument is turned about its axis, the interference fringes should be displaced back and forth, with one period completed every 180° of arc, with an amplitude that depends on the effective lengths of the arms of the interferometer and should equal (u^2/c^2) times the average of the two arms, or at least about 100 Å ($10^{-2}\,\mu$) for each m of arm length. Through multiple reflections, effective arm lengths of about 10 m have been produced, leading to a possible Michelson-Morley effect of about 0.2 wavelengths of visible light. Such a shift is, of course, easily observable. The principal sources of error of the experiment are mechanical distortions of the geometry of the instrument during a run, caused by stresses or by differential warming. Many repeated performances by various investigators have led to the conclusion that there is no effect of anywhere near the expected minimum magnitude.

II. Einstein's special theory of relativity.

3. The concept of simultaneity. Classical mechanics obeys a principle of relativity, which establishes the equivalence of all inertial frames of reference. Maxwellian electrodynamics, as we have seen, is inconsistent with this classical or Galilean principle of relativity; that theory appears to require a single frame corresponding to absolute rest. However, when it comes to determining that frame experimentally (or rather our state of motion with respect to it), it proves singularly elusive: It turns out that the largest observable effects are of the second order in (u/c), and all attempts to observe these second-order effects have been unsuccessful, even when the anticipated effect would have been well within the limits of error. All the experiments carried out suggested that Maxwell's and Lorentz's equations are valid rigorously with respect to our laboratory frame of reference.

In this situation a number of workers attempted to develop new approaches that would explain the uniform failure of all experiments to show transport effects. H.A. Lorentz[1] conjectured that the motion of any physical object through the "ether" (a conjectural medium of transmission, which represents the frame of absolute rest) would modify its internal cohesive properties so that it would contract its spatial dimensions in the direction of motion and that its dynamics would slow down. With appropriate choice of these coefficients of contraction and time dilation it will become impossible to determine by any conceivable experiment the motion of an observer through the ether, as scales and clocks used for a determination of the relative speed of propagation of electromagnetic waves, and in fact the electromagnetic fields themselves, become distorted in such a manner as to lead to the *apparent* validity of Maxwell's equations with respect to any inertial frame of reference. According to Lorentz, then, there is a frame of absolute rest, but its identification is vitiated once and for all by the systematic errors of all measurements carried out with the help of instruments that move through the ether along with the observer.

Poincaré[2] discovered that Maxwell's field equations are covariant with respect to a group of coordinate transformations that deviate from those of Eq. (1.1) or (2.1); he discovered the transformations that we know today as Lorentz's transformation equations. These are, in fact, the same equations that Lorentz used to go over from "true" to "local" (apparent) space and time coordinates. But Poincaré, being a mathematician, was unimpressed by the need for maintaining Newton's absolute time scale. He considered coordinates, both of space

[1] Cf. Lorentz, Einstein, Minkowski: Das Relativitätsprinzip (editor O. Blumenthal). Leipzig: Teubner 1913, 3rd edit. 1920. English translation. London: Methuen 1923; New York: Dover, undated [7].

[2] H. Poincaré: International Congress of Physics, Paris, 1900; La science et l'hypothèse, Paris, 1903; International Congress of Arts and Science, St. Louis, 1904.

and of time, as a means for identifying space-time points. Each frame of reference constituted a possible coordinate system; if the Maxwell equations reproduced themselves under a certain transformation group, then that meant that all these frames of reference were equivalent. Accordingly, POINCARÉ accepted a new principle of relativity, with which the laws of electrodynamics were compatible. However, POINCARÉ treated space and time coordinates purely formally, without a serious attempt to relate them to physical observations. It remained for EINSTEIN to combine the new principle of relativity (which was identical with POINCARÉ's) with a physical interpretation of the transformation equations.

At the time of his own work EINSTEIN was probably unaware of LORENTZ's and POINCARÉ's work. His own first paper[1] is devoid of any references; later he called the transformation equations after LORENTZ. This conflict of priorities, which appears to have caused no trouble between the protagonists at the time (1905), has recently been discussed at some length though obliquely by Sir EDMUND WHITTAKER[2], who apparently felt that POINCARÉ's contributions had not received proper credit. With due regard for WHITTAKER's historical researches and penetrating analyses, it appears reasonable that the contemporaries, at least the physicists, were convinced of the validity of the new principle of relativity by EINSTEIN's arguments rather than by POINCARÉ's, even though POINCARÉ was a famous professor and EINSTEIN an unknown young civil servant in the Swiss Federal Patent Office.

Let us consider two frames of reference as exemplified by two observers moving relative to each other, and each equipped with basic physical instruments, such as scales, clocks, and possibly other devices. It is to be understood that these instruments are at rest relative to their respective observers. Suppose the two observers wish to calibrate their instruments relative to each other. Such a calibration involves that the two observers can agree to consider two distances marked on their respective scales as equal, and likewise to consider periods of their respective clocks as equal. We shall now see that such a calibration involves a postulate concerning the nature of simultaneity.

Two scales will be considered to have equal lengths if it is possible to place them side by side in such a manner that their respective end points coincide. If the two scales are moving relatively to each other, we might compare their respective lengths by either of two procedures. We may change the state of motion of one so that it will be at rest relative to the other; the other procedure is to let the two scales glide past each other and to ascertain whether their respective endpoints coincide *simultaneously*. The first procedure imposes on our scale or rod an acceleration, which may well affect its mechanical properties permanently. Whether or not such a permanent change occurs, the scale that we have accelerated is no longer useable for laying out a Cartesian coordinate system in the frame of reference for which it was originally intended. For the second procedure it is not necessary to change the state of motion of either rod, but we have to judge the simultaneous occurrence of two events that occur at points in space separated by the length of either scale. (This is at least true for scales parallel to the direction of relative motion; for scales at right angles this problem does not arise.)

Likewise, in order to compare the rates of two clocks that are moving relative to each other, we must compare with each other the lengths of their respective periods. But at least for one of the two endpoints of these periods the clocks are spatially removed from each other. Again we must determine the simultaneous occurrence of spatially distant events. To the extent that the feasibility of such a determination was at all discussed in Newtonian mechanics, it was assumed that simultaneity was unrelated to the state of motion of the observer. Such an assump-

[1] A. EINSTEIN: Ann. Phys., Leipzig **17**, 891 (1905). See also LORENTZ et al., The Principle of Relativity [7].

[2] E. WHITTAKER: History of the Theori^es of Aether and Electricity. New York: Philosophical Library 1951 [*14*].

tion would be justified as a matter of course if we could transmit information about events with arbitrarily large signal velocity ("infinite speed of transmission"). Practically, this would mean that to send a signal from some point in space A to another point B and back again to A (where the identity of the space point A at the beginning and at the end of this transmission is to be established in relation to some observer) would take arbitrarily little time.

Actually no known signal travels at a speed greater than that of electromagnetic radiation in empty space, c. We could ignore this fact and continue to hope that such signals may yet be discovered in the future. To the contrary, the special theory of relativity is based on the assumption that there are in fact no signals that travel at a speed exceeding c. With this assumption EINSTEIN was able to modify the then current ideas of space and time so that he could explain naturally the negative outcome of all experiments designed to observe transport effects in electrodynamics.

If the speed of transmission of light signals represents an upper limit to the transmission of signals generally, then it is impossible to devise an experiment by which to determine the simultaneity of distant events independently of the state of motion of the observer. In most discussions the standard definition adopted is that two events are to be considered simultaneous if an observer stationed midway between them "sees" them simultaneously. (It is, of course, immaterial whether this "seeing" involves visible light, radio waves, or some other means of signal transmission that propagates at the rate c.) Other definitions of simultaneity are possible. For instance, let the observer be stationed at a point A and have a mirror stationed at a distant point B. Let A also possess a clock, that is a periodic device with constant period. If A now sends out a light flash at some time t_1, which is reflected at B, so that the flash is observed by A in the mirror at B at some time t_2, A will presumably conclude that the reflection at B took place at the time $\frac{1}{2}(t_1+t_2)$. Conversely, if A has a mirror as well and if B can observe both the original flash at A and also the reflection of its own reflection, and finally A's clock as well, then B may conclude that the reflection at its own (B's) mirror took place midway between the readings of A's clock at the flash and the (second) reflection, but this conclusion would be reasonable only if the distance between A and B remains unchanged all the time.

If B moves relative to A, then the two observers might argue as follows: "A" would take the attitude that as his observations on B involve only one instant in time (the reflection in B's mirror), the going and the returning signal travel the same distance regardless of B's state of motion; hence no correction is to be applied. "B", on the other hand, must make two observations on A, at the two times when A's clock reads t_1 and t_2, respectively. If A and B are moving apart, the second time of transmission must be greater than the first; hence the time of reflection at its own (B's) mirror must precede the instant at which A's clock reads $\frac{1}{2}(t_1+t_2)$. If the two observers approach each other, the reverse conclusion holds.

4. The Lorentz transformation. Different observers will disagree as to the simultaneity of distant events if both of them consider that light signals propagate at the speed c with respect to themselves. In other words, the notion of universal time, which implies the assumption of absolute simultaneity, is incompatible with a universal law of propagation for electromagnetic waves in empty space. The question arises whether some different ideas concerning the nature of space and time intervals may not be consistent with a universal law of propagation of light signals. This, it was found, is indeed the case. We shall now derive the exact form of the new transformation equations, which relate space and time coordinates in one frame of reference to those in another. Having determined the form of these transformation equations, we shall examine the question of their internal consistency.

We shall consider two observers with coordinate systems $\boldsymbol{x}', t'$ and $\boldsymbol{x}, t$, respectively, moving relatively to each other with the velocity $\boldsymbol{v}$. Both frames of reference are to be *inertial frames*. If uniform motion in one frame of reference is to be observed as uniform motion in the other, the relationship between the primed and the unprimed coordinates must be linear. To avoid trivial additive constants, we shall assume, moreover, that at the time $t=0$ and at the origin $\boldsymbol{x}=0$ the time and space coordinates of the primed frame of reference, t' and $\boldsymbol{x}'$, also vanish. With these assumptions, we may write the transformation equations with five parameters, α through η, as yet to be determined:

$$\left.\begin{aligned} \boldsymbol{x}' &= \alpha\boldsymbol{x} + \boldsymbol{v}(\beta\boldsymbol{v}\cdot\boldsymbol{x} + \gamma t), \\ t' &= \varepsilon\boldsymbol{v}\cdot\boldsymbol{x} + \eta t. \end{aligned}\right\} \tag{4.1}$$

The prerelativistic transformation equations (2.1) must hold at least approximately, as long as $v \ll c$. Accordingly, α, γ, and η must approximate 1, whereas β and ε must approximate 0.

To determine these five parameters, we shall first introduce the requirements that the primed origin has the velocity $-\boldsymbol{v}$ in the unprimed frame, and that the unprimed origin has the velocity $+\boldsymbol{v}$ in the primed frame of reference. In other words, for $\boldsymbol{x}'=0$ we must have $\boldsymbol{x}=-\boldsymbol{v}t$, and for $\boldsymbol{x}=0$ we must have $\boldsymbol{x}'=\boldsymbol{v}t'$. The two requirements yield the relations

$$\gamma - v^2\beta = \alpha, \qquad \gamma = \eta. \tag{4.2}$$

Next we formulate the universality of the speed of light signals. Suppose at the time $t=0$ a flash occurs at the origin, and the light spreads in all directions. At any given point with the coordinates $\boldsymbol{x}$ the signal will arrive at a time t given by $c^2t^2=\boldsymbol{x}^2$, $t\geqq 0$. But this same relationship will hold, identically in $\boldsymbol{x}$, with respect to t' and $\boldsymbol{x}'$. Thus we obtain three more relations between the parameters:

$$\left.\begin{aligned} \sqrt{1-\frac{v^2}{c^2}}\,\gamma &= \alpha, \\ c^2\varepsilon^2 &= (2\alpha + v^2\beta)\beta, \\ c^2\varepsilon - v^2\beta &= \alpha. \end{aligned}\right\} \tag{4.3}$$

The five conditions (4.2), (4.3) are homogeneous in the five unknowns, hence four are algebraically dependent on the remaining one. The solutions may be written in the form

$$\beta = \left[\left(1-\frac{v^2}{c^2}\right)^{-\frac{1}{2}} - 1\right]\frac{\alpha}{v^2}, \qquad \gamma = \eta = c^2\varepsilon = \left(1-\frac{v^2}{c^2}\right)^{-\frac{1}{2}}\alpha. \tag{4.4}$$

The remaining constant α may be fixed by the further requirement that the transformation inverse to (4.1), which leads from the primed back to the unprimed frame of reference, is to have the same form as the transformation (4.1) except that $\boldsymbol{v}$ is to be replaced by $-\boldsymbol{v}$. This requirement is based on the principle of relativity in its most general form, that the two frames of reference $(\boldsymbol{x}, t)$ and $(\boldsymbol{x}', t')$ have equal status; either one, looked at from the point of view of the other, appears similar. This reciprocity requirement can be satisfied only if we set α either equal to 1 or to -1. If for small values of (v/c) our transformation law is to go over into the law (2.1), we must choose the positive sign. Substituting our values into the expression (4.1) we finally obtain the transformation law

$$\left.\begin{aligned} \boldsymbol{x}' &= \boldsymbol{x} - \frac{\boldsymbol{v}}{v^2}(\boldsymbol{v}\cdot\boldsymbol{x}) + \left(1-\frac{v^2}{c^2}\right)^{-\frac{1}{2}}\left(\frac{\boldsymbol{v}\cdot\boldsymbol{x}}{v^2}+t\right)\boldsymbol{v},\\ t' &= \left(1-\frac{v^2}{c^2}\right)^{-\frac{1}{2}}\left(t+\frac{\boldsymbol{v}\cdot\boldsymbol{x}}{c^2}\right), \end{aligned}\right\} \tag{4.5}$$

which is one of the forms of the so-called Lorentz transformation.

If the vector $\boldsymbol{v}$ is parallel to the x-axis, and hence also parallel to the x'-axis, these expressions reduce to

$$x' = \frac{x+vt}{\sqrt{1-\frac{v^2}{c^2}}}, \quad y' = y, \quad z' = z, \quad t' = \frac{t+\frac{v}{c^2}x}{\sqrt{1-\frac{v^2}{c^2}}}. \tag{4.6}$$

5. The space-time interval. The linear transformation (4.5) connects with each other two frames of reference moving, relative to each other, at the uniform velocity $\boldsymbol{v}$, in such a manner that a linear motion that has the speed c in one frame of reference has that same speed in the other frame, regardless of the direction of propagation. Accordingly, the vanishing of the quadratic form of time and coordinate differentials $(d\boldsymbol{x}^2 - c^2\,dt^2)$ in one frame of reference implies the vanishing of the same form $(d\boldsymbol{x}'^2 - c^2\,dt'^2)$ in the other. It can be shown, by straightforward computation, that these two forms are equal to each other in general if the transformation equations (4.5) are satisfied. That is to say, we have, in full generality,

$$\left.\begin{aligned} d\tau^2 &= dt^2 - \frac{1}{c^2}\,d\boldsymbol{x}^2 = dt'^2 - \frac{1}{c^2}\,d\boldsymbol{x}'^2,\\ d\sigma^2 &= d\boldsymbol{x}^2 - c^2\,dt^2 = d\boldsymbol{x}'^2 - c^2\,dt'^2. \end{aligned}\right\} \tag{5.1}$$

Depending on the relative magnitude of dt and $c^{-1}\,d|\boldsymbol{x}|$, either $d\tau$ or $d\sigma$ will be real. In the former case, we call the relationship between the two neighboring points in space and time "*time-like*", in the latter case "*space-like*". If the relationship is time-like, then one can find a frame of reference in which $d\boldsymbol{x}$ vanishes and in which $dt = d\tau$; for a space-like relationship, on the other hand, one can find a frame in which dt is zero and $d|\boldsymbol{x}| = d\sigma$. Either $d\tau$ or $d\sigma$ is often referred to as the (invariant) *space-time interval*.

If the differentials $d\boldsymbol{x}$, dt are associated with the motion of a material point (mass point), the interval $d\tau$ represents the lapse of time in a frame of reference associated (at that instant) with the moving particle. In that case we call the interval the *proper time* or *eigen-time* of the mass point. The same designation applies to the integral $\tau \equiv \int d\tau$. This integral may be given the form

$$\tau_{21} = \int\limits_{t=t_1}^{t_2} \sqrt{dt^2 - \frac{1}{c^2}\,d\boldsymbol{x}^2} = \int\limits_{t_1}^{t_2} \sqrt{1-\frac{\dot{\boldsymbol{x}}^2}{c^2}}\,dt. \tag{5.2}$$

In contrast to the elapsed time $(t_2 - t_1)$ the proper time integral (5.2) will have the same numerical value in all frames of reference connected with each other through a Lorentz transformation.

Given two points in space and time, relativity denies the absolute significance of the (coordinate) time interval or of the spatial distance between them but substitutes instead the (time-like or space-like) interval. For a moving particle, if we consider two points on its trajectory, the interval will always be time-like[1]. The proper time (5.2) taken along the trajectory will be smaller than the interval $\Delta\tau^2 = \Delta t^2 - \frac{\Delta x^2}{c^2}$ if the motion is accelerated. Both of these two quantities have an absolute significance in that their numerical values are independent of the frame of reference in which they are computed.

6. Self-consistency of the Lorentz transformation. Whether the relativistic space-time concepts are self-consistent must be investigated in some detail. Such an inquiry will pass through two distinct stages. First we shall look into the qualitative implications of a transformation law which leads to different determinations of time intervals by different observers and connects these differences with spatial relationships. Afterwards, we shall proceed to a second, and more precise, examination, which will establish the so-called group property of the Lorentz transformations.

In Sect. 3 we studied the relative nature of simultaneity of two spatially distant events. Given two relatively moving frames of reference, S and S', it is quite conceivable that for some two events $\Delta t = 0$, wereas, according to Eq. (4.5) or (4.6), $\Delta t' \neq 0$. Moreover, for a given point in space, which we might mark by stationing there a material body, t and t' will increase at different rates, in accordance with the same transformation equations. This is not to say that if we station at that point a physical clock that it will not show a well-defined rate; in fact, relativity assumes that this rate is the same for different clock mechanisms provided they are all competently designed and built. This common rate is identical with the rate of growth of the time coordinate of a co-moving frame of reference. For all other observers (or frames of reference) the rate of growth of the time coordinate will be greater (cf. Sect. 8); that is to say if an observer who is in unaccelerated relative motion to the marked point observes the clock stationed there and corrects his observations for the finite speed of transmission of light signals, he will find that clock to be slow. Likewise, a solid body, ideally a coordinate scaffold, will have some definite dimensions in a co-moving frame of reference; but it will have different dimensions as determined by an observer in a different state of motion. If again allowance is made for the transit time of the light signals, the body will be foreshortened in the direction of relative motion, but unchanged at right angles to that direction[2].

These so-called kinematic effects of the Lorentz transformation, which will be treated more fully in Sect. 8, neither lead to self-contradiction, nor are they contrary to the principle of relativity. For instance, our theory predicts that the rate of a clock appears greatest in the co-moving frame of reference. Let us consider two clocks of similar design, which presumably run at the same proper rate; let one be at rest in the frame S, the other in the frame S'. An observer, placing himself at rest in either frame, will find that the rate of "his" clock is greater. But in order to compare the two clock rates, he will have to judge the simultaneity of at least two sets of "events", for instance when the hands of the two clocks move through the marks "0" and "1", respectively. At least one of these two sets of events will be spatially distant from each other; hence it is quite consistent with our original analysis of Sect. 3 if the two observers arrive at seemingly contradictory conclusions.

In analyzing this, or any of the other relativistic effects, we must keep in mind that the theory of relativity removes the characteristic of "physical reality" (intrinsic property) from some types of observations. In the case of the physical

[1] Otherwise, signals could travel with speeds exceeding c.

[2] TERRELL has shown that a body is not contracted by relative motion unless this correction is made. Phys. Rev. **116**, 1041 (1959).

clock, for instance, the theory of relativity tells us that a clock will possess a *proper rate,* whose magnitude is independent of the observer, but that its *coordinate rate* possesses no more physical reality than, for instance, the angle under which a line segment appears to an observer. Both of these pieces of observational data, the (coordinate) rate of a clock, and the angle of view, depend on combinations of intrinsic ("real") properties of the physical object and its relationship to the observer. In both cases, the "observer" need not be an intelligent being, it may be some automatic piece of equipment that makes a record of its "observations", which may be inspected by the human operator at any later time.

After these somewhat qualitative and intuitive comments we are ready to examine the mathematical self-consistency of the Lorentz transformations. Given a transformation law between two frames of reference S and S' of the form (4.5), and a similar relationship between the frame S' and a third frame S'', does there exist a Lorentz transformation that leads directly from S to S''. Only if the Lorentz transformation is "transitive" in this sense, does it have any claim to physical significance. After all, the physical motivation behind the construction of the Lorentz transformation law is the asserted equivalence of all inertial frames of reference, along with the universality of the law of propagation of light signals. If the Lorentz transformation is the proper law that leads from one inertial frame to another, then a Lorentz transformation must certainly lead from S to S''.

The transitivity of the Lorentz transformation is a mathematical question; it must be answered as such. It turns out that in order to establish transitivity we must generalize the concept of the Lorentz transformation slightly. We shall call any linear transformation of the four space-time coordinates a Lorentz transformation if it reproduces the quadratic form $d\tau^2$ of Eq. (5.1). In addition to the transformations (4.5) this definition includes the combination of a transformation of that type with a three-dimensional orthogonal coordinate transformation, which involves only the space but not the time coordinates (and with a time reversal). With this extension, the Lorentz transformation law indeed assumes *group character*[1].

To prove transitivity of the extended transformation law we need only point out that the successive application of two linear transformations represents a linear transformation, and that the preservation of any mathematical structure, such as the quadratic form (5.1) is transitive as well. Hence, the successive application of two (or more) linear transformations each of which reproduces the form (5.1) results in a total transformation having the same properties. A set of transformations whose combinations are transformations of the same type and each of which possesses an inverse are called a *transformation group. The Lorentz transformations form such a group.*

In the next section we shall show that the most general set of Lorentz transformation coefficients can indeed be described as a function of two vectors

---

[1] The orthogonal group, consisting of the linear transformations that lead from one Cartesian coordinate system to another, may be divided into the *proper* orthogonal group, which involves only rotation of the coordinate directions without reflection and which forms a group of its own, and the *improper* orthogonal transformations which do involve reflection; together they form the *full* orthogonal group. Likewise there is the *proper* Lorentz group, in which neither space is reflected nor time reversed, then the *orthochronous* Lorentz group in which time is not reversed, the group of transformations which do not reflect space directions, and the *full* Lorentz group, which includes both time-reversing and space-reflecting transformations.

(relative velocity, and angle-plus-axis of spatial rotation) and of two signs (which refer to space reflection and to time reversal).

7. The most general Lorentz transformation. In what follows we shall use matrix notation in order to reduce the visible bulk of the notation. For details, cf. the article on *Algebra* by G. FALK in Vol. II of this Encyclopedia.

We shall introduce two three-dimensional vectors $\boldsymbol{v}$ and $\boldsymbol{w}$ and a three-by-three matrix Γ as well as an ordinary numeric η to write down the most general linear transformation equations between two sets of four coordinates each,

$$x' = \Gamma x + v t, \qquad t' = w^T x + \eta t, \qquad (w^T x \equiv \boldsymbol{w} \cdot \boldsymbol{x}). \tag{7.1}$$

Whereas v and x are to be considered column vectors, the superscript T (transpose) denotes a row vector. We now require that $(x'^T x' - c^2 t'^2)$ equal $(x^T x - c^2 t^2)$, identically in x and t. Accordingly we obtain the following set of equalities for the coefficients:

$$\Gamma^T \Gamma - c^2 w w^T = I, \qquad \Gamma^T v - c^2 \eta w = 0, \qquad c^2 \eta^2 - v^2 = c^2 \qquad (v^2 \equiv v^T v). \tag{7.2}$$

These conditions can be solved immediately with respect to the numeric η and the vector w,

$$\eta = \sqrt{1 + \frac{v^2}{c^2}}, \qquad w = \frac{\Gamma^T v}{c^2 \sqrt{1 + \frac{v^2}{c^2}}}. \tag{7.3}$$

There is a choice in the sign of η (a negative sign would imply reversal of time); the vector v is unrestricted, and w is uniquely determined by the remaining quantities.

The matrix Γ satisfies the matrix equation

$$\Gamma^T \left(I - \frac{v v^T}{c^2 + v^2}\right) \Gamma = I, \tag{7.4}$$

which we shall reduce to another similar equation. We note that

$$I - \frac{v v^T}{c^2 + v^2} = (I - \alpha v v^T)^2, \qquad \alpha = \frac{1}{v^2}\left(1 - \frac{c}{\sqrt{c^2 + v^2}}\right), \tag{7.5}$$

so that

$$C^T C = I, \qquad C = (I - \alpha v v^T) \Gamma. \tag{7.6}$$

The matrix Γ is completely determined by the new matrix C, as

$$\Gamma = (I + \beta v v^T) C, \qquad \beta = \frac{1}{v^2}\left(\sqrt{1 + \frac{v^2}{c^2}} - 1\right). \tag{7.7}$$

Hence, once we have solved the equation $C^T C = I$, we have determined the matrix Γ as well. This last equation, however, is the defining equation for orthogonal transformations. The solutions are well known. They can be described conveniently in terms of a choice of sign (which identifies proper and improper orthogonal transformations) and a vector, which is parallel to the axis of rotation and whose magnitude equals the angle through which one (three-dimensional) coordinate system is turned relative to the other. If this vector is denoted by the symbol ϑ and if we form a three-by-three matrix Θ, which is anti-symmetric and whose components are

$$\Theta = \begin{pmatrix} 0, & \vartheta_3, & -\vartheta_2 \\ -\vartheta_3, & 0, & \vartheta_1 \\ \vartheta_2, & -\vartheta_1, & 0 \end{pmatrix}, \qquad \Theta^T = -\Theta, \tag{7.8}$$

then the expression for C is:

$$C = \pm e^{\Theta}, \tag{7.9}$$

where the exponential function is symbolic for the usual power series. Hence C, and with it Γ, depends on a choice of one sign and one arbitrary vector. Herewith our assertion at the end of Sect. 6 is proved.

8. The kinematic effects. Even without a complete investigation of relativistic electrodynamics and relativistic mechanics it is possible to predict certain effects solely on the basis of the Lorentz transformation (4.5) and the principle of relativity, according to which a physical system will exhibit comparable behavior in any frame of reference with respect to which it is in a specified state of motion. Two identical clocks, such as the spectral lines of two identical atoms, will possess the same frequency in the two frames of reference with respect to which they are, respectively, at rest. Similarly, two material scales, such as the radii of the first Bohr orbits of two hydrogen atoms, will have the same dimensions in their two respective frames of reference. Granted these very general assumptions, we can predict the spatial dimensions of an object in any frame of reference if they are known in the rest frame, and so forth.

Lorentz-Fitzgerald contraction. We shall determine the primed coordinate differences between two material points at rest in the unprimed coordinate system of Eq. (4.5), at a fixed instant in (primed) time, $\Delta t' = 0$. By straightforward computation we find

$$\Delta \boldsymbol{x}' = \Delta \boldsymbol{x} - \frac{\boldsymbol{v}}{v^2} (\boldsymbol{v} \cdot \Delta \boldsymbol{x}) \left(1 - \sqrt{1 - \frac{v^2}{c^2}}\right). \tag{8.1}$$

There is no change in the length of a measuring rod at right angles to the relative motion $\boldsymbol{v}$, but in the direction of $\boldsymbol{v}$ there is a contraction by a factor $\sqrt{1 - v^2/c^2}$; in all other directions there is a change in direction of the rod as well as a contraction (less than the factor $\sqrt{1 - v^2/c^2}$).

The rate of clocks. Given a clock at rest in the frame of reference S ($\boldsymbol{x} = 0$), its proper time will be identical with the coordinate time t. With respect to the frame S', we find

$$t = \sqrt{1 - \frac{v^2}{c^2}}\, t'. \tag{8.2}$$

The clock appears to be slow compared to the time of the frame S' by a factor $\sqrt{1 - v^2/c^2}$.

This apparent slowing down of clocks was first demonstrated experimentally by Ives[1], who observed spectral lines emitted by canal rays under conditions where he was able to compensate for first-order Doppler effect. More recently, the Mössbauer effect[2] was used in an ultracentrifuge to show that an "observer" moving at right angles to the source (where Doppler effect is excluded) will see the frequency of monochromatic γ-rays shifted in accordance with Eq. (8.2)[3].

The addition of velocities. A particle with the velocity $\boldsymbol{u}$ with respect to the frame S, $\boldsymbol{x} = \boldsymbol{u}t$, will have a velocity $\boldsymbol{u}'$ with respect to the frame S' which will

[1] H. Ives: J. Opt. Soc. Amer. **28**, 215 (1938).
[2] R. Mössbauer: Z. Physik **151**, 124 (1958).
[3] H. Hay, J. Schiffer, T. Cranshaw and P. Egelstaff: Phys. Rev. Letters **4**, 165 (1960).

not be simply the vector sum of $\boldsymbol{u}$ and $\boldsymbol{v}$, as it is in prerelativistic physics. Instead we have:

$$\left.\begin{aligned} \boldsymbol{x}' &= \left[\boldsymbol{u} + \frac{\boldsymbol{v}}{\sqrt{1-\frac{v^2}{c^2}}} + \boldsymbol{v}\frac{\boldsymbol{v}\cdot\boldsymbol{u}}{v^2}\left(\frac{1}{\sqrt{1-\frac{v^2}{c^2}}} - 1\right)\right] t, \\ t' &= \frac{1}{\sqrt{1-\frac{v^2}{c^2}}}\left(1 + \frac{\boldsymbol{v}\cdot\boldsymbol{u}}{c^2}\right) t, \end{aligned}\right\} \tag{8.3}$$

and hence

$$\boldsymbol{u}' = \frac{\boldsymbol{v} + \boldsymbol{u}\sqrt{1-\frac{v^2}{c^2}} + \frac{\boldsymbol{v}}{v^2}(\boldsymbol{v}\cdot\boldsymbol{u})\left(1-\sqrt{1-\frac{v^2}{c^2}}\right)}{1 + \frac{\boldsymbol{v}\cdot\boldsymbol{u}}{c^2}}. \tag{8.4}$$

If, in particular, $\boldsymbol{u}$ and $\boldsymbol{v}$ happen to be parallel, then the expression (8.4) reduces to

$$u' = \frac{u+v}{1+\frac{v\,u}{c^2}}. \tag{8.5}$$

Eqs. (8.4), (8.5) are particularly applicable if the speed of light is measured in a dielectric which is moving relative to the observer. If the propagation of the light and the motion of the transmitting medium are parallel, then the apparent index of refraction comes out as

$$n' \approx n \pm \frac{v}{c}\left(1 - \frac{1}{n^2}\right).$$

With the help of Eq. (8.4) one can show that if the relative velocity of the two frames, $\boldsymbol{v}$, is less than c [and if it were not, the coefficients of the Lorentz transformation (4.5) would become complex], then $\boldsymbol{u}'$ will be less than c as long as $\boldsymbol{u}$ is. For proof we calculate the square of $\boldsymbol{u}'$, which turns out to be

$$\left.\begin{aligned} u'^2 &= \left(1 + \frac{\boldsymbol{v}\cdot\boldsymbol{u}}{c^2}\right)^{-2}\left[(\boldsymbol{v}+\boldsymbol{u})^2 - \frac{1}{c^2}(\boldsymbol{v}\times\boldsymbol{u})^2\right] \\ &= c^2\left[1 - \frac{\left(1-\frac{v^2}{c^2}\right)\left(1-\frac{u^2}{c^2}\right)}{\left(1+\frac{\boldsymbol{v}\cdot\boldsymbol{u}}{c^2}\right)^2}\right] < c^2, \end{aligned}\right\} \tag{8.6}$$

under the assumptions made above.

The twin "paradox". Given two clocks of identical construction, let one remain at rest in some (inertial) frame of reference. Let the other clock travel at a uniform rate in some direction for a certain length of time and then return. At the time the two clocks are again close together, will they show the same elapsed time, or will they be different? This question, frequently phrased in terms of two twins, one of whom stays home while the other travels ("twin paradox"), can be answered unambiguously on the assumption that both clocks show proper time. Under this assumption the clock that was permanently at rest in an inertial frame will show a greater amount of time elapsed. The supposed paradox results from the apparent symmetry between the two clocks, which suggests that in a different frame of reference the answer would turn out differently.

The apparent paradox is based on the acceptance of the two clocks and their respective states of motion as "equivalent". Clearly it is impossible to have two points move uniformly in such a manner that their trajectories intersect twice. Hence for the two clocks to be spatially adjacent both at the beginning and at the end of the conceptual experiment, it is unavoidable that at least one of them be accelerated at some time during the experiment. In our case the second clock undergoes this acceleration at the midpoint of the experiment, when it reverses its direction of travel, whereas by assumption the first clock never experiences any acceleration.

We may carry out the calculation of the difference in proper time between the two clocks in any frame of reference. Let us first work in the frame in which the first clock is permanently at rest. Let the second clock travel at a speed v for a time t_0, traversing a distance $v t_0$ and then return at the same speed until it meets again the first clock. The first clock will then show a reading of elapsed time equal to $2t_0$, whereas the second clock will show a reading, in accordance with Eq. (8.2), of $2t_0 \sqrt{1-v^2/c^2}$. We can, instead, do our calculation in the frame of reference in which the second clock is at rest during the first part of the experiment. In that frame of reference the first clock is now moving at the rate v throughout the experiment, which lasts for a time $2t_0(1-v^2/c^2)^{-\frac{1}{2}}$ (new coordinate time). For an initial time $t_0 \sqrt{1-v^2/c^2}$ the second clock is at rest; then it changes its speed to a value

$$v' = \frac{2v}{1+\frac{v^2}{c^2}}, \tag{8.7}$$

in accordance with Eq. (8.5), and continues at that rate for a length of time which equals the difference between the total duration of the experiment and the initial period,

$$\left.\begin{aligned} t' &= \frac{2t_0}{\sqrt{1-\frac{v^2}{c^2}}} - t_0\sqrt{1-\frac{v^2}{c^2}} \\ &= \frac{1+\frac{v^2}{c^2}}{\sqrt{1-\frac{v^2}{c^2}}}\, t_0 . \end{aligned}\right\} \tag{8.8}$$

After this length of time, the second clock has caught up with the first one. If we now compute the proper times of both clocks along their trajectories, we shall find exactly the same values as before.

It is, of course, possible to change frames of reference at the midpoint and to keep the second clock at rest permanently. In this type of calculation we must use the Lorentz equations (4.6) to redetermine the locations as well as the time coordinates of the two clocks at the instant at which we go over from one frame to the other. Again, if the calculation is done correctly, the same values for proper time will result.

All of the foregoing discussion is based on the assumption that both clocks will show proper time. This assumption though permissible is by no means basic to the theory of relativity. All that the theory of relativity claims is that a clock that is free of acceleration and free of external stresses will possess a proper rate independent of its velocity. Its behavior under acceleration will depend both on its internal structure and on the cause of the acceleration.

III. The Minkowski universe.

9. Minkowski geometry. A physically motivated program in special relativity will take as its point of departure the equations of the Lorentz transformation and then proceed to investigate the formal possibilities for laws of nature that take the same form in every inertial frame of reference. In the course of such a program we shall find in subsequent sections that the laws of the electromagnetic field require hardly any modification at all, whereas the laws of mechanics call for some revision of the Newtonian scheme. However, the pursuit of this program becomes extremely awkward without the help of the four-dimensional formalism usually associated with the name of H. MINKOWSKI[1].

We start with the invariance of the proper interval, Eq. (5.1), and reinterpret this interval as the invariant metric of a four-dimensional space whose coordinates are the usual space *and* time coordinates. This metric is *indefinite*, i.e. the quadratic form $dt^2 - c^{-2}\,d\boldsymbol{x}^2$ may take either sign (and may also be zero for non-vanishing line elements of the four-dimensional space. Instead of considering space and time as physically disparate elements, we shall call a point in space at an instant of time a *world point* (sometimes also referred to as an "event"), and x, y, z, and t uniformly its coordinates. The world points are the structural elements of the *Minkowski space*. In that space we can also construct one-dimensional *world lines* and *curves*, two-dimensional surfaces, and three-dimensional hypersurfaces. Volume elements in Minkowski space are naturally four-dimensional.

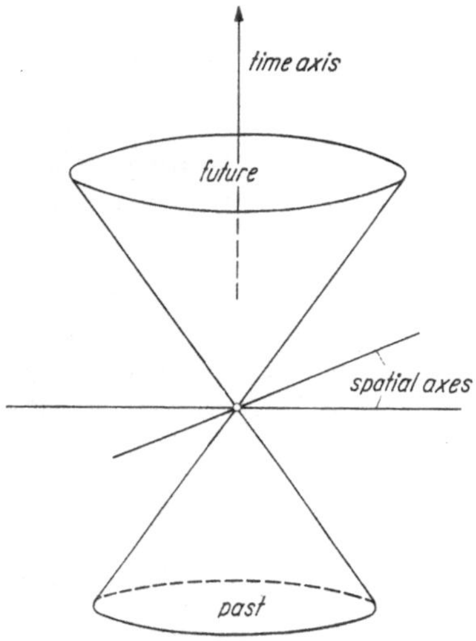

Fig. 1. Light cone.

Because its metric is indefinite, the Minkowski space differs in its geometric properties from ordinary Euclidean spaces. For instance, the so-called triangle inequality (or SCHWARZ's inequality) does not hold; in other words, the "distance" between two world points A and C can be greater than the distances AB and BC of A and C from a third world point B. In fact, it is possible to have the intervals AB and BC both equal to zero and the interval AC either time-like or spacelike.

Through any world point there passes a (three-dimensional) double cone of null directions (Fig. 1), called the *light cone*. The interiors are called *future* and *past* directions, respectively. The exterior, which is connected (in the sense in which this term is used in topology), is formed by the space-like directions. The light cone is mapped on itself by any Lorentz transformation. The *proper* Lorentz transformations preserve the identity of the future and the past light cones, and the determinant of the 3×3 transformation matrix of the spatial coordinates [the matrix Γ of Eq. (7.1) ff.] is positive. The proper Lorentz transformations form a *group*, that is, the succession of two proper Lorentz transformations is again proper, and the inverse of a proper Lorentz transformation is itself one. There are three kinds of improper Lorentz transformations, those in which the past light cone is mapped on the future, and vice versa, those in which Det Γ is negative, and those in which both of these things happen. The proper and the improper Lorentz transformations together form the *full Lorentz group*.

In all that follows, we shall designate the four coordinates x, y, z, ct by the new symbols x^1, x^2, x^3, x^0 or, collectively, by x^α. Generally, Greek indices (whether

[1] H. MINKOWSKI: Phys. Z. **10**, 104 (1909). Cf. also H. POINCARÉ: Rend. Palermo **21**, 129 (1906).

subscripts or superscripts) are to take all values from 0 to 3, whereas Latin indices run from 1 to 3[1]. We shall designate the fundamental quadratic form (5.1) in this manner:

$$\left.\begin{aligned} d\tau^2 &= \eta_{\mu\nu}\, dx^\mu\, dx^\nu \\ &= dx^T\, \eta\, dx, \end{aligned}\right\} \tag{9.1}$$

using either tensor notation with Einstein summation convention[2] or matrix notation, depending on our convenience[3]. If a general Lorentz transformation matrix is denoted by A, so that

$$dx' = A\, dx, \qquad dx'^T = dx^T A^T, \tag{9.2}$$

represents the infinitesimal line segment dx but in terms of the new coordinate system x', then the condition that the proper interval is preserved leads to

$$\left.\begin{aligned} dx'^T\, \eta\, dx &= dx^T A^T\, \eta\, A\, dx \\ &\equiv dx^T\, \eta\, dx \end{aligned}\right\} \tag{9.3}$$

and hence to

$$A^T\, \eta\, A = \eta. \tag{9.4}$$

Eq. (9.4) represents a concise formulation of the conditions obeyed by all Lorentz transformations.

10. Some properties of the Lorentz group. Because the matrix η,

$$\eta = \begin{pmatrix} -1 & 0 & 0 & 0 \\ 0 & -1 & 0 & 0 \\ 0 & 0 & -1 & 0 \\ 0 & 0 & 0 & 1 \end{pmatrix} \tag{10.1}$$

is its own inverse, $\eta^2 = I$, Eq. (9.4) may also be written in the form

$$A^T = \eta\, A^{-1}\, \eta. \tag{10.2}$$

The Lorentz group has some formal similarities with the orthogonal group, whose members obey the first Eq. (7.6). It differs significantly in a number of ways.

Just as the orthogonal transformations, the Lorentz transformations can be written as exponential functions of matrices. Let a be an otherwise arbitrary skewsymmetric 4×4 matrix, $a^T + a = 0$. Then

$$A = e^{a\eta}, \qquad A^T = e^{-\eta a}, \tag{10.3}$$

is a solution of Eq. (9.4) and represents a proper Lorentz transformation. The three kinds of improper Lorentz transformations can be written in the forms:

$$A = -\, e^{a\eta}, \qquad A = \pm\, \eta\, e^{a\eta}. \tag{10.4}$$

---

[1] This convention differs from that adopted by H. TIETZ, Article "Geometrie" in Vol. 2 of this Encyclopedia, but corresponds to the prevailing usage in the physical literature.

[2] H. TIETZ, previous footnote. Each index occurring twice in a product is to be considered a summation index without further ado.

[3] Matrix notation is described in both the articles "Geometrie" by H. TIETZ and "Algebra" by G. FALK in Vol. 2 of this Encyclopedia. For vectors, such as dx, which without additional symbol are always to be understood as column vectors, the superscript T ("transpose") identifies a row vector. For square and rectangular matrices, T simply denotes the transposed matrix.

With the help of the expressions (10.3) one can show that for any proper Lorentz transformation there are always at least two directions that are invariant under the transformation. Even under a non-identical Lorentz transformation there may be four such directions. For proof, we note first that any eigenvector of the exponent $a\eta$ will also be an eigenvector of A itself. Now we solve the secular equation for the exponent by setting

$$a = \begin{pmatrix} 0 & \alpha_3 & -\alpha_2 & \beta_1 \\ -\alpha_3 & 0 & \alpha_1 & \beta_2 \\ \alpha_2 & -\alpha_1 & 0 & \beta_3 \\ -\beta_1 & -\beta_2 & -\beta_3 & 0 \end{pmatrix}, \quad a\,\eta = \begin{pmatrix} 0 & -\alpha_3 & -\alpha_2 & \beta_1 \\ \alpha_3 & 0 & -\alpha_1 & \beta_2 \\ -\alpha_2 & \alpha_1 & 0 & \beta_3 \\ \beta_1 & \beta_2 & \beta_3 & 0 \end{pmatrix}, \tag{10.5}$$

and obtaining directly

$$\mathrm{Det}\,(a\,\eta - \lambda\,I) = \lambda^4 + (\boldsymbol{\alpha}^2 - \boldsymbol{\beta}^2)\,\lambda^2 - (\boldsymbol{\alpha}\cdot\boldsymbol{\beta})^2 = 0, \tag{10.6}$$

where $\boldsymbol{\alpha}$ and $\boldsymbol{\beta}$ are understood to be ordinary three-dimensional vectors. The solution of the secular equation (10.6) is

$$\lambda^2 = |\boldsymbol{B}| - |\boldsymbol{A}| \pm |\boldsymbol{A} + \boldsymbol{B}|, \tag{10.7}$$

where

$$|\boldsymbol{A}| = \boldsymbol{\alpha}^2, \quad |\boldsymbol{B}| = \boldsymbol{\beta}^2, \quad \sphericalangle\,\boldsymbol{A}, \boldsymbol{B} = 2 \sphericalangle\,\boldsymbol{\alpha}, \boldsymbol{\beta}. \tag{10.8}$$

As in any equation with real coefficients, the solutions of Eq. (10.6) are either real or occur in conjugate complex pairs. But the form of the solution (10.7) shows that all the eigenvalues λ are either real or imaginary, depending on whether λ^2 is positive or negative. Now the three terms on the right of Eq. (10.7) may be interpreted as the lengths of three sides of an ordinary plane triangle. Hence if the sign of the last term is chosen positive, then, because of the triangle inequality, λ^2 cannot be negative (though it may be zero); if the same sign is chosen negative, then λ^2 is non-positive (again the value zero is not excluded). Accordingly in general two roots of the secular equation will be positive, two negative. In the special case that the latter two roots are zero (which implies that $|\boldsymbol{\beta}| \geqq |\boldsymbol{\alpha}|$ and that $\boldsymbol{\alpha}\cdot\boldsymbol{\beta} = 0$), all four roots are real.

The Lorentz transformation preserves the norm of a four-vector, including that of an eigenvector of the transformation. For the transformation matrix A to have an eigenvalue 1, the matrix $a\eta$ must have an eigenvalue 0. All other eigenvalues must be associated with directions whose norm vanishes. Hence the invariant directions of Lorentz transformations are usually null directions, i.e. they lie on the light cone. In those cases in which zero is a double root of the secular equation, the two associated invariant directions generate a whole invariant plane, which might be considered the "axis" of the transformation. Our argument shows that the existence of such an axis is the exception rather than the rule.

However every proper Lorentz transformation possesses two planes that are mapped on themselves (though not every direction in that plane is invariant under the transformation). These are the planes that belong to the two pairs of eigenvectors, with real eigenvalues and with imaginary eigenvalues. That a linear combination of two eigenvectors goes over into a linear combination of the same two eigenvectors is obvious, whether or not the eigenvalues are equal. Hence the plane containing the two eigenvectors with real values of λ is certainly mapped on itself; the eigenvectors belonging to the imaginary values of λ do not have real components, but are each other's conjugate complexes. Hence these two complex eigenvectors possess two real linear combinations, their sum and their difference divided by i. These two vectors again determine a plane that is mapped on itself. In the special case that all four roots of the secular equation are real (in which case at least two are 0), there is a whole plane of invariant directions; any plane containing two invariant directions will be mapped on itself.

As we shall explain in greater detail in Sect. 12, the invariant product of two four-vectors U and V is defined as $U^T \eta V$. If it vanishes, the two vectors are said to be *normal* to each other (null vectors are normal to themselves). It is easy to show that any two eigenvectors of $a\eta$ are normal to each other unless the sum of the corresponding eigenvalues vanishes. Furthermore, two time-like vectors cannot be normal to each other. Inasmuch as the two planes that are mapped on themselves are normal to each other, it follows that only one of them can contain a time-like direction. If the two real eigenvalues are non-zero, their eigenvectors are null vectors; in that case there exist linear combinations that are time-like. Hence the plane that belongs to the imaginary eigenvalues is completely space-like. This result may be shown to hold even when the pair of real eigenvalues happens to be zero.

11. Infinitesimal Lorentz transformations. Commutators. We form an infinitesimal Lorentz transformation by considering a one-parametric set of continuous, differentiable Lorentz transformations, $A(s)$, so that $A(0)=I$, and by taking the derivative $\partial A/\partial s$ for $s=0$:

$$\left.\begin{aligned} &\delta x = P x, \\ &\delta x \equiv \left(\frac{\partial x'}{\partial s}\right)_{s=0}, \qquad P = \left(\frac{\partial A}{\partial s}\right)_{s=0}. \end{aligned}\right\} \tag{11.1}$$

The matrix P satisfies the condition

$$P^T \eta + \eta P = 0, \tag{11.2}$$

which may be verified by differentiation of the condition (9.4) for finite Lorentz transformations. We can satisfy the condition for infinitesimal Lorentz transformations (11.2) by setting

$$P = a\eta, \qquad a + a^T = 0, \tag{11.3}$$

emphasizing thereby the connection between infinitesimal Lorentz transformations and the exponential representation (10.3) of finite Lorentz transformations.

The infinitesimal Lorentz transformations permit the investigation of certain structural features of the group with greater ease than the finite Lorentz transformations. We have stated previously that any proper Lorentz transformation may be characterized by two three-dimensional vectors. However, if we wish to express the "product" of two Lorentz transformations, that is the transformation that results if we carry out the two "factor" transformations in succession, then it is difficult to find the two vectors describing the "product" in terms of the two pairs of vectors characterizing the original transformations. For infinitesimal transformations the product reduces to the sum of the original transformation matrices, say a' and a''. Accordingly, the characteristic vectors also simply add.

For finite transformations, say A' and A'', the product depends in general on which of the two transformations we carry out first, that is

$$\left.\begin{aligned} A'(A'' x) &\neq A''(A' x), \\ A' A'' &\neq A'' A'. \end{aligned}\right\} \tag{11.4}$$

Lorentz transformations do *not* form a *commutative group*. To examine this noncommutativity, one usually forms the so-called commutator, i.e. the quadruple product $A' A'' A'^{-1} A''^{-1}$. For the infinitesimal transformations this commutator reduces to an expression that is bilinear in the two "factors". In terms of the skewsymmetric matrices a' and a'', defined in terms of Eq. (11.3), the infinitesimal commutator is characterized by the skewsymmetric matrix C:

$$C = a' \eta a'' - a'' \eta a'. \tag{11.5}$$

The two vectors describing C, $\boldsymbol{\mu}$ and $\boldsymbol{\nu}$, are determined by the two vector pairs $\boldsymbol{\alpha}'$, $\boldsymbol{\beta}'$ and $\boldsymbol{\alpha}''$, $\boldsymbol{\beta}''$ as follows:

$$\left.\begin{aligned} \boldsymbol{\mu} &= \boldsymbol{\alpha}' \times \boldsymbol{\alpha}'' - \boldsymbol{\beta}' \times \boldsymbol{\beta}'', \\ \boldsymbol{\nu} &= \boldsymbol{\alpha}' \times \boldsymbol{\beta}'' + \boldsymbol{\beta}' \times \boldsymbol{\alpha}''. \end{aligned}\right\} \tag{11.6}$$

The commutator (11.5) or (11.6) may be interpreted in several distinct ways. The obvious interpretation is that of the (infinitesimal) change in x if two infinitesimal Lorentz transformations are applied in succession and then again reversed but in the wrong order. In view of the fact that the vector $\boldsymbol{\alpha}$ describes a spatial rotation, with $\boldsymbol{\alpha}$ as the axis, and $\boldsymbol{\beta}$ a true Lorentz transformation, with $\boldsymbol{\beta}$ being the relative velocity of the two frames (in units of c), we can read the following facts directly off Eqs. (11.6):

1. The commutator of two true Lorentz transformations not involving spatial rotations is a pure spatial rotation.

2. The commutator of two spatial rotations is a spatial rotation.

3. The commutator of a pure Lorentz transformation and a pure spatial rotation is a pure Lorentz transformation.

4. Two spatial rotations will commute if they have the same axis of rotation, and two pure Lorentz transformations if the two relative velocities are parallel.

5. A pure Lorentz transformation will commute with a spatial rotation if the relative velocity of the former is parallel to the axis of rotation of the latter.

Another possible interpretation is the following. Suppose the infinitesimal oransformation a'' leads from a frame P to a frame P''. Now if we subject both af these frames to the transformation a', which transforms them into the frames F' dnd $(F'')'$, respectively, then the transformation that leads from F' to $(F'')'$ will tiffer from a'' by C. In this sense the commutator may be said to be the transformation law of a transformation. (This interpretation may also be applied to the finite commutator, which is the *ratio* between the original and the transformed transformation.) Such a transformation of a transformation maps every Lorentz transformation one-to-one on another (or the same) Lorentz transformation, in a manner that in group theory is called a similarity mapping. That means that the identity transformation is mapped on itself, the product of two transformations is mapped on the product of the maps of the two factor transformations, and the inverse of a transformation is mapped on the inverse of its map. In other words, under a similarity mapping a group is mapped on itself in such a manner that its structure remains undisturbed.

We may now ask under what conditions two Lorentz transformations may be each other's maps under a similarity transformation. A similar query concerning three-dimensional orthogonal transformations tells us that among rotations any two that involve a rotation through equal angles (regardless of the direction of the axis of rotation) are equivalent modulo a similarity transformation. As for Lorentz transformations, it can be shown that the two expressions

$$\left.\begin{aligned} J_1 &= \tfrac{1}{2}(\boldsymbol{\alpha}^2 + \boldsymbol{\beta}^2), \\ J_2 &= \boldsymbol{\alpha} \cdot \boldsymbol{\beta} \end{aligned}\right\} \tag{11.7}$$

are the only two invariants under a similarity transformation produced by a Lorentz transformation. It can be shown (though the proof will not be given here) that any two proper Lorentz transformations for which these two invariants J_1 and J_2 have the same numerical values can be mapped on each other by another Lorentz transformation. If for two transformations the lengths of the two vectors $\boldsymbol{\alpha}$ and $\boldsymbol{\beta}$ are the same and the angle between them, too, then they can be carried over into each other by a spatial rotation. A true Lorentz transformation will change the lengths of $\boldsymbol{\alpha}$ and $\boldsymbol{\beta}$ and the angle between them, but so that J_1 and J_2 remain unchanged. These assertions can be verified easily with the help of the infinitesimal transformation law (11.6).

One by-product of this discussion is that a Lorentz transformation with small values of the lengths of $\boldsymbol{\alpha}$ and $\boldsymbol{\beta}$ may be mapped on one with arbitrarily large values. Thus, not only does the proper Lorentz group not form a compact topological space (the n-dimensional orthogonal proper group does), the set of all equivalent Lorentz transformations with finite matrix coefficients is not compact, either.

12. Vectors and tensors in the Minkowski universe. The laws of nature must be covariant under Lorentz transformations, that is to say, they must take the same form in every inertial frame of reference This principle has been of considerable heuristic value in the conjecture of such laws. There has been considerable controversy about its ultimate significance, both in the special and in the general theory of relativity. It has been asserted that every conceivable relationship can be cast into Lorentz-covariant form, and that the principle of covariance, at least logically, is empty[1].

Without attempting to make a final adjudication, we should like to call attention to the distinction between the Lorentz covariance of a relationship and its tensor form. The tensor form is a special notation, which recommends itself by its simplicity for a broad class of covariant relations, but it is no more than a notation. Covariance of any relationship can be achieved, in principle, by the inclusion of a sufficient number of elements in the description, but such a relationship is not a law of nature unless the new elements are themselves subject to a dynamical law. For instance, we may describe the isotropic propagation of light in some optical medium first in the frame of reference in which the medium of transmission is at rest. Then, if we wish to go over to some other Lorentz frame, we must introduce into our description as a new set of quantities the components of the velocity of the medium of transmission with respect to our chosen frame. Our new description will be Lorentz-covariant in any deep sense only if we also state how this velocity may be determined (and what dynamical laws it obeys) independently of optical propagation experiments.

Conversely, a Lorentz-covariant law may assume a special form in a particular frame, in which the number of elements required for the description may be reduced. For instance, the description of any mechanical system consisting of several interacting particles is simplified in the center-of-mass frame of reference, which is defined as that frame in which the total linear momentum of the system

[1] For a critical discussion see, for instance, the article by Ilse Rosenthal-Schneider in P. A. Schilpp, Ed., Albert Einstein, Philosopher-Scientist Library of Living Philosophers, Evanston, 1949, p. 129—146 [*11*]. — E. Kretschmann: Ann. Phys. (Leipzig) **53**, 575 (1917).

vanishes. It would thus appear that the requirement of Lorentz covariance is meaningful only to the extent that the number, and the nature, of the variables to be related to each other is stated specifically.

Nevertheless, the requirement of Lorentz covariance has considerable practical importance. At the very least, the requirement tells us that the laws of nature will involve fewer descriptive elements if cast into a Lorentz-covariant form than they will in a Newtonian formulation. We shall now concern ourselves, in the next seven sections, with the structural elements that lend themselves for the Lorentz-covariant formulation of relationships. We shall start with the discussion of vectors and tensors at a fixed world point, or without reference to any location. In the following section we shall concern ourselves with vector and tensor fields.

The class of quantities designated as tensors includes a few subclasses normally referred to by special names, the *scalars* and the *vectors*. We shall begin with the scalars. A scalar is a single quantity which does not change its value under Lorentz transformations. A scalar may be a pure numeric, such as 3, or it may be a physical quantity, such as the total rest mass of a mechanical system, or finally the number of its constituent particles.

We shall define *vectors* in terms of their transformation properties with respect to Lorentz transformations. In component notation a Lorentz transformation will lead from a set of coordinates x^σ to new coordinates $x^{\varrho\prime}$ according to the equations

$$x^{\varrho\prime} = A^\varrho{}_\sigma x^\sigma + B^\varrho, \qquad \eta_{\varrho\sigma} A^\varrho{}_\mu A^\sigma{}_\nu = \eta_{\mu\nu}. \tag{12.1}$$

(Cf. Sect. 9.) If four quantities V^σ transform thus:

$$V^{\varrho\prime} = A^\varrho{}_\sigma V^\sigma, \tag{12.2}$$

remaining unaffected by the displacement parameters B^ϱ, we say that they are the components of a *contravariant vector*. A *covariant vector*, also, has four components, but their transformation law is

$$W_\varrho = A^\sigma{}_\varrho W_\sigma'. \tag{12.3}$$

We may add together two or several contravariant vectors by adding their respective components, and as the result of this *vector addition* obtain another contravariant vector. Likewise the sum of several covariant vectors is another covariant vector. The product of a contravariant vector and a covariant vector is defined as the sum of the algebraic products of corresponding components, $V^\varrho W_\varrho$; it is a scalar. Contravariant vectors may be converted into covariant vectors by means of the form $\eta_{\varrho\sigma}$,

$$V_\varrho = \eta_{\varrho\sigma} V^\sigma. \tag{12.4}$$

Likewise, by the use of the reciprocal form $\eta^{\varrho\sigma}$, whose components are equal to those of $\eta_{\varrho\sigma}$ for our choice of coordinates, a covariant vector may be changed into a contravariant vector,

$$V^\varrho = \eta^{\varrho\sigma} V_\sigma. \tag{12.5}$$

These two operations are usually referred to as the *lowering* and the *raising*, respectively, of indices. Two vectors V^ϱ and V_ϱ that are connected with each other by equations of the form (12.4), (12.5) are considered as the contravariant and covariant versions of the same intrinsic quantity, which may be referred to as a *vector*, without further qualifications.

The transformation law of a covariant vector may also be written as follows:

$$\left.\begin{aligned} V'_\varrho &= A_\varrho{}^{\sigma'} V_\sigma, \\ A_\varrho{}^{\sigma'} A^\tau{}_\sigma &= \delta^\tau_\varrho = \begin{cases} 1, & \tau = \varrho \\ 0, & \tau \neq \varrho. \end{cases} \end{aligned}\right\} \tag{12.6}$$

A' is the reciprocal matrix to A, but transposed. δ^ϱ_τ is called the Kronecker symbol. As a matter of convention, contravariant vectors (and tensors) are always written with superscripts, covariant vectors (and tensors) with subscripts. The reading and interpretation of equations is thereby greatly facilitated.

A *tensor* is a set of quantities, the *tensor components*, with several indices, superscripts, subscripts, or both, which transform as would a product of vectors, thus:

$$W_{\mu\nu\ldots}{}^{\varrho\sigma\ldots} = A_\mu{}^{\alpha'} A_\nu{}^{\beta'} \ldots A^\varrho{}_\iota A^\sigma{}_\varkappa \ldots W_{\alpha\beta\ldots}{}^{\iota\varkappa\ldots}. \tag{12.7}$$

The total number of indices is called the *rank* of a tensor. Depending on their character, the tensor is called contravariant, covariant, or mixed. Individual indices may be raised or lowered. Tensors of equal rank may be added, just as vectors, if corresponding indices have the same character. Tensors and vectors may be multiplied without any summations over indices, and the product will be a tensor of a rank equal to the sum of the ranks of the factors,

$$U_\varrho{}^\sigma V^\tau = W_\varrho{}^{\sigma\tau} \tag{12.8}$$

etc. Besides addition and multiplication, there is another simple operation, called *contraction*. It consists of the summation over one contravariant and one covariant index. The result of a contraction is a new tensor whose rank is equal to the rank of the original tensor minus two. Of course, contractions may be carried out several times in succession; but each individual contraction must involve one pair of indices only.

The Kronecker symbol is a mixed tensor of rank 2, whose components are the same in all coordinate systems. $\eta_{\varrho\sigma}$ and $\eta^{\varrho\sigma}$ are also tensors whose components are *invariant* in this sense. Henceforth we shall call δ^τ_ϱ the Kronecker tensor, $\eta_{\varrho\sigma}$ the covariant metric tensor, and $\eta^{\varrho\sigma}$ the contravariant metric tensor. These three tensors can be converted into each other by raising and lowering of appropriate indices.

Symmetry properties between like indices are an invariant property of a tensor. For instance, if

$$S^{\mu\nu}{}_\varrho = S^{\nu\mu}{}_\varrho, \qquad T_{\alpha\beta} = -T_{\beta\alpha} \tag{12.9}$$

in one coordinate system, the same equalities will hold in every other coordinate system.

Closely related to scalars, vectors, and tensors are the so-called densities. Indistinguishable from the previously mentioned quantities under proper Lorentz transformations, they will behave differently under improper Lorentz transformations. A scalar density, for instance, will obey the transformation law

$$\sigma' = \operatorname{Det}|A|\,\sigma \tag{12.10}$$

This density will change sign under a space reflection, and also under time reversal. If we perform space reflection and time reversal simultaneously, there will be no change in the sign of σ.

In addition to these densities, there are also other types of pseudotensors (a general designation, which includes the densities as one class out of three). A pseudoscalar may, for instance, be defined so that it changes sign under space reflections but not under time reversal; and the last class of pseudoscalars may do the reverse, i.e. change sign only under time reversal. What has been said about pseudoscalars applies, mutatis mutandis, also to pseudovectors and to pseudotensors. In every case we have three distinct classes of pseudoquantities. We shall not elaborate the rather obvious details.

13. Fields. The scalars, vectors, and other quantities in the preceding section were defined either without reference to a particular world point at all, or at one world point. A field is a function, or a set of functions, of the four coordinates. Hence a field must be defined on a four-dimensional domain in the Minkowski universe, and we shall in general assume (without repeating this assumption time and again) that the function(s) involved is (are) differentiable as many times as may be required for a particular discussion.

A single function will be called a *scalar field* if the value of the function at a particular world point (not for a fixed value of the coordinates) is the same in all inertial frames of reference. Likewise we shall call a set of four functions of the coordinates a contravariant vector field if at every fixed world point the values of these four functions in two different inertial frames are connected by the transformation law (12.2); similar definitions apply to tensor fields and to pseudotensor fields.

In a field all the algebraic operations of the preceding section are applicable. In addition we have available the partial differentiation with respect to the coordinates. We shall generally use the "comma notation",

$$\frac{\partial \sigma}{\partial x^{\varrho}} \equiv \sigma_{,\varrho}, \qquad \frac{\partial V^{\varrho}}{\partial x^{\tau}} \equiv V^{\varrho}{}_{,\tau}, \tag{13.1}$$

to denote differentiation with respect to a coordinate. In the event that a field is a function of another field as well as of the coordinates, the comma will denote the derivative which compares the values of the field at two neighboring points, whereas the usual symbol of partial differentiation will be employed to denote the *explicit* dependence of the field on the coordinates,

$$\left.\begin{aligned} z(x^{\varrho}) &= f[y(x^{\varrho}), x^{\sigma}], \\ z_{,\varrho} \equiv f_{,\varrho} &\equiv \frac{\partial f}{\partial x^{\varrho}} + \frac{\partial f}{\partial y} y_{,\varrho}. \end{aligned}\right\} \tag{13.2}$$

The comma operation applied to a tensor (or pseudotensor) field leads to a new tensor (or pseudotensor) field whose rank is greater than the original rank by one. The index of differentiation is covariant. For instance,

$$g_{\mu\nu} = h_{\mu,\nu} \tag{13.3}$$

is a covariant tensor of rank 2. Such a tensor is often called a *gradient*; but if the index of differentiation is contracted with some contravariant index of the tensor, the result is called a *divergence*.

As for the *infinitesimal transformation law* of a field, we shall uniformly employ the following notation. If the coordinates of a world point change by the infinitesimal amount

$$\delta x^{\varrho} = P^{\varrho}{}_{\sigma} x^{\sigma} + b^{\varrho}, \tag{13.4}$$

the "local" transformation law of a contravariant vector field, for instance, is

$$\delta V^{\varrho} = P^{\varrho}{}_{\sigma} V^{\sigma}. \tag{13.5}$$

This is the change of the components of the field at a fixed world point. We may also inquire how the field changes *as a function of its arguments, the coordinates.*

That is to say, we may compare the values of the field components not at a fixed world point but for fixed values of the coordinates. In this case we find

$$\bar{\delta} V^\varrho = P^\varrho{}_\sigma V^\sigma - V^\varrho{}_{,\sigma} (P^\sigma{}_\tau x^\tau + b^\sigma). \tag{13.6}$$

The symbol $\bar{\delta}$ shall always denote this "field transformation" law. Generally, the two types of transformation law are related to each other for all kinds of fields by the relationship

$$\bar{\delta} W = \delta W - W_{,\sigma} \delta x^\sigma. \tag{13.7}$$

The field transformation law has the property that generally

$$(\bar{\delta} W)_{,\sigma} \equiv \bar{\delta} (W_{,\sigma}), \tag{13.8}$$

for all types of fields.

14. Functions of several world points. Ordinarily the quantities discussed in Sects. 12 and 13 are all that is required for the "classical" treatment of relativistic physics. There are, however, other logical possibilities, of which we shall mention the following:

Fields may depend not only on the coordinates of one world point but on the coordinates of two, three, or more points. In that case they may still be scalars, vectors, tensors, or pseudotensors. In differentiating such a field, one must specify with respect to the coordinates of which point the derivatives are to be taken. Some of these fields are *invariant* fields in that they depend only on the invariant distance between two points or on the scalar product of coordinate differences between pairs of world points. In many physical applications such functions appear as Green's functions and as propagation kernels. Some invariant functions of two world points vanish outside the light cone, that is for all space-like distances between the two argument points. In that case the signature of the difference between their time coordinates is a further invariant (at least with respect to proper Lorentz transformations) available as an argument. An otherwise invariant function multiplied by such a sign (or, in the case of more than two argument points, by an odd product of such signs) will be an invariant pseudoscalar of the type that changes its sign under time reversal.

The gradients of invariant scalars and pseudoscalars are invariant vectors and pseudovectors, respectively, and the higher derivatives tensors or pseudotensors. The term *invariant* is used here uniformly to denote the fact that with the specified transformation laws these quantities reproduce themselves in every Lorentzian coordinate system as the same functions (or sets of functions) of their arguments. In other words, their field transformation laws ($\bar{\delta}$), Eqs. (13.6) ff. are zero.

15. Integrals. Integrals are another important non-local object useful in relativistic physics. An integral over a fixed four-dimensional domain will be a tensor density if its integrand is a tensor, and a tensor if its integrand is a tensor density. The two remaining types of pseudotensors are related to each other similarly. A fixed domain of integration requires a different description of its boundary in every coordinate system. Hence a four-dimensional integral over a fixed range of the coordinates of a scalar density will not be an invariant unless either the domain is the whole Minkowski universe or the integrand is non-zero only in certain isolated regions that remain inside the domain of integration for all the Lorentz frames considered. An example of the latter kind is the four-dimensional Dirac delta function, centered about some world point in the interior of the domain of integration. The integral will be 1 in every coordinate system that leaves the center point of the delta function in the interior of the domain of integration.

Relativistic physics also has occasion to deal with lower-dimensional integrals. In such cases it is advantageous to consider domains of integration of arbitrary shape, not just coordinate surfaces. Hence we shall describe the domain of integration parametrically, with the help of one, two, or three parameters u^s. In other words, the domain of integration consists all points whose coordinates have the values $x^\sigma(u^s)$, where the functions $x(u)$ are specified and the coordinates u take all values within their permitted range.

A line integral may be written in the form

$$J = \int V^{\cdots}_{\cdots\varrho} \frac{dx^\varrho}{du} \, du. \tag{15.1}$$

The dots indicate possible tensor indices of the quantity A in addition to the subscript ϱ, which is needed to form a scalar product with the line element of integration $(dx^\varrho/du)\, du$. Such a line integral is invariant with respect to substitutions of one parameter for another; with respect to Lorentz transformations it has the transformation properties described by the free indices (those indicated by dots) of the integrand field, provided the domain of integration is fixed and independent of the coordinate system. We can obtain *invariance* at least with respect to a limited range of Lorentz transformations with a domain of integration fixed with respect to the frame of reference if we restrict ourselves to an integrand which obeys the requirement

$$V_{\varrho,\sigma} - V_{\sigma,\varrho} = 0 \tag{15.2}$$

throughout the domain in which possible paths of integration are located and if the path of integration is either a closed curve or goes from infinity to infinity.

Higher-dimensional integrals can be constructed by a very slight refinement. A two-dimensional integral will have the form

$$\left.\begin{aligned} J_2 &= \iint V^{\cdots}_{\cdots\varrho\sigma} \frac{\partial x^\varrho}{\partial u'} \frac{\partial x^\sigma}{\partial u^2} \, du^1 \, du^2 \\ &= \frac{1}{2} \iint V^{\cdots}_{\cdots\varrho\sigma} \left(\frac{\partial x^\varrho}{\partial u^1} \frac{\partial x^\sigma}{\partial u^2} - \frac{\partial x^\varrho}{\partial u^2} \frac{\partial x^\sigma}{\partial u^1} \right) du^1 \, du^2, \\ & V^{\cdots}_{\cdots\varrho\sigma} + V^{\cdots}_{\cdots\sigma\varrho} = 0. \end{aligned}\right\} \tag{15.3}$$

Let $V^{\cdots}_{\cdots\varrho\sigma}$ be an antisymmetric tensor with respect to the two subscripts ϱ and σ. Then the integral as a whole will be invariant with respect to parameter transformations (i.e. the introduction of two new parameters $u^{1\prime}$, $u^{2\prime}$, as functions of u^1, u^2); with a fixed surface of integration it will have transformation properties with respect to Lorentz transformations that are determined by the remaining indices of V. Again, if we wish to construct an integral that is invariant (or covariant) with respect to a domain of integration fixed with respect to the coordinates (rather than in space), the condition will be that the surface of integration have no boundary (no edge) and that V satisfy the condition on and near the surface of integration:

$$V^{\cdots}_{\cdots\varrho\sigma,\tau} + V^{\cdots}_{\cdots\sigma\tau,\varrho} + V^{\cdots}_{\cdots\tau\varrho,\sigma} = 0. \tag{15.4}$$

The situation for a three-dimensional integral is quite analogous. Here the integrand must possess three covariant indices and be completely antisymmetric in them if the integral is to have any simple transformation properties. The condition for invariance with respect to a domain of integration fixed in terms of the range of coordinates is in this case:

$$V^{\cdots}_{\cdots\alpha\beta\gamma,\delta} - V^{\cdots}_{\cdots\beta\gamma\delta,\alpha} + V^{\cdots}_{\cdots\gamma\delta\alpha,\beta} - V^{\cdots}_{\cdots\delta\alpha\beta,\gamma} = 0. \tag{15.5}$$

In this last case, in particular, it is occasionally useful to carry out a process that is often referred to as "transition to the dual". We note, first, that the completely antisymmetric expressions

$$\delta_{\alpha\beta\gamma\delta} = 0, \pm 1, \qquad \delta^{\alpha\beta\gamma\delta} = 0, \pm 1, \tag{15.6}$$

[only those components whose indices are all different are non-zero; those whose indices are an even permutation of (1230) are 1, those whose indices are an odd permutation of (1230) are -1] are invariant tensor densities. They are called the tensor densities of LEVI-CIVITA. With their help one can construct the following "duals":

$$\left.\begin{aligned} &\tfrac{1}{2}\,\delta^{\alpha\beta\gamma\delta}\,V_{\gamma\delta} = V^{\alpha\beta}, \quad V^{\alpha\beta} + V^{\beta\alpha} = 0, \quad V_{\gamma\delta} = \tfrac{1}{2}\,\delta_{\alpha\beta\gamma\delta}\,V^{\alpha\beta} \\ &\tfrac{1}{6}\,\delta^{\alpha\beta\gamma\delta}\,V_{\beta\gamma\delta} = V^{\alpha}, \qquad V_{\beta\gamma\delta} = \delta_{\alpha\beta\gamma\delta}\,V^{\alpha}. \end{aligned}\right\} \tag{15.7}$$

Thus it is possible to construct the three-dimensional surface element, for instance, as a covariant vector density:

$$d\Sigma_{\varrho} = \delta_{\varrho\alpha\beta\gamma}\,\frac{\partial x^{\alpha}}{\partial u^1}\,\frac{\partial x^{\beta}}{\partial u^2}\,\frac{\partial x^{\gamma}}{\partial u^3}\,du^1\,du^2\,du^3. \tag{15.8}$$

Such products of differentials are known as Cartan forms. The three-dimensional integral may therefore be written in the form

$$J_3 = \int V^{\varrho}\,d\Sigma_{\varrho} \tag{15.9}$$

where the integrand is defined by Eq. (15.7). The condition for Lorentz invariance with fixed range of the coordinates (15.5) then goes over into the simple divergence condition

$$V^{\varrho}{}_{,\varrho} = 0. \tag{15.10}$$

The most important situation in which the domain of integration is fixed (or partly fixed) with respect to the coordinate system is integration over spatial domains at a fixed time. For such integrals to possess any Lorentz-covariant significance their integrands must obey the "integrability conditions" (15.2), (15.4), or (15.5) [equivalent to (15.10)], respectively, and the domain of integration must either be closed or extend to (spatial) infinity. These integrability conditions are, however, all equivalent to *equations of continuity* or *conservation laws*. Generally we shall find that integral relationships apply in relativistic physics to quantities that satisfy some conservation laws.

16. Two-component spinors[1]. In order to complete the discussion of geometric objects with linear transformation laws under Lorentz transformations, we must also treat *spinors*. Among the notations used to describe spinors there are two major classifications, four-component spinors and two-component spinors. The latter cannot be assigned linear transformation laws under all Lorentz transformations, including all improper transformations; that is why until 1957 the four-component spinors, which were introduced into physics by DIRAC, were considered

[1] W. PAULI: Die allgemeinen Prinzipien der Wellenmechanik, B. Relativistisches Einkörperproblem, S. 137ff.; In this Encyclopedia, vol. V, part 1. 1958. — B.L. VAN DER WAERDEN: Göttinger Nachr. **1929**, p. 100. — H. WEYE: Z. Physik **56**, 330 (1929). — T.D. LEE and C.N. YANG: Phys. Rev. **105**, 1671 (1957).

almost exclusively. When it was discovered that there are some physical particles (the neutrinos) that cannot reverse the sign of their intrinsic angular momentum (their spin), two-component spinors regained their interest. As two-component spinors are mathematically less complex than four-component spinors, we shall start with two-component spinors.

Let us begin with the three Pauli spin matrices σ_k ($k=1, 2, 3$), which have the following properties:

1. they are Hermitian two-rowed matrices;
1. each one of them has the same two eigenvalues, ± 1; and
3. they satisfy the anticommutation relations

$$\sigma_k \sigma_l + \sigma_l \sigma_k = 2\delta_{kl}. \tag{16.1}$$

One can show that three matrices with these properties are linearly independent of each other and from the unit matrix. Hence any two-rowed matrix (which has but four elements) may be represented uniquely as a linear combination of the unit matrix and the three Pauli matrices. Furthermore, the product of any two of the three Pauli matrices equals $\pm i$ times the third. Conventionally, one chooses the signs so that

$$\sigma_1 \sigma_2 = -\sigma_2 \sigma_1 = i\sigma_3, \quad \sigma_2 \sigma_3 = -\sigma_3 \sigma_2 = i\sigma_1, \quad \sigma_3 \sigma_1 = -\sigma_1 \sigma_3 = i\sigma_2. \tag{16.2}$$

With the help of the three-dimensional Levi-Civita symbol δ_{klm}, which is antisymmetric in all its indices and whose components have the values ± 1, 0 [cf. Eq. (15.6)], we may summarize Eqs. (16.1), (16.2) as follows

$$\sigma_k \sigma_l = \delta_{kl} + i\,\delta_{klm}\,\sigma_m. \tag{16.3}$$

In close analogy to the Pauli matrices we shall now establish a set of four 2×2 matrices in a complex two-dimensional space, the "spin space", by means of the following definitions: We introduce four Hermitian matrices σ^μ, defined in terms of the Pauli matrices and four mutually perpendicular unit vectors u^μ, v^μ_k ($k=1, 2, 3$),

$$\left.\begin{aligned} \sigma^\mu &= u^\mu + v^\mu_k \sigma_k, \\ u_\mu u^\mu &= 1, \qquad u_\mu v^\mu_k = 0, \\ \eta_{\mu\nu} v^\mu_k v^\nu_l &= -\delta_{kl}. \end{aligned}\right\} \tag{16.4}$$

Given these four matrices, and assuming that under Lorentz transformations they are to combine with each other linearly like the components of a contravariant vector, we may form the scalar product of the vector σ^μ by any ordinary covariant vector s_μ (whose components are ordinary numbers),

$$S = s_\mu \sigma^\mu = s_\mu (u^\mu + v^\mu_k \sigma_k). \tag{16.5}$$

The determinant of the scalar matrix S turns out to be the norm of s_μ,

$$|S| = \eta^{\mu\nu} s_\mu s_\nu, \tag{16.6}$$

as a brief calculation will show (not performed here).

This relationship between the norm of a vector and the determinant of the associated matrix may now be utilized to prove the correspondence between unimodular transformations in the spin space and Lorentz transformations. We shall consider a unimodular matrix a,

$$|a| = 1, \tag{16.7}$$

whose Hermitian conjugate and whose inverse are, of course, also unimodular. With its help we introduce a set of new matrices $\sigma^{\mu\prime}$,

$$\sigma^{\mu\prime} = a^{*-1} \sigma^{\mu} a^{-1} \tag{16.8}$$

and a new matrix S',

$$S' = s_\mu \sigma^{\mu\prime} = a^{*-1} S a^{-1}. \tag{16.9}$$

Its determinant, $|S',|$ must equal that of S; hence we have

$$\eta^{\mu\nu} s_\mu s_\nu = |S'| = |s_\mu \sigma^{\mu\prime}| \equiv (u^{\mu\prime} u^{\nu\prime} - v_k^{\mu\prime} v_k^{\nu\prime}) s_\mu s_\nu \tag{16.10}$$

where the new vectors $u^{\mu\prime}$, $v_k^{\mu\prime}$, are the coefficients of the representation of the new matrices

$$\sigma^{\mu\prime} = u^{\mu\prime} + v_k^{\mu\prime} \sigma_k. \tag{16.11}$$

Because the choice of the vector s_μ is arbitrary, it follows from Eq. (16.10) that

$$\eta^{\mu\nu} = u^{\mu\prime} u^{\nu\prime} - v_k^{\mu\prime} v_k^{\nu\prime}, \tag{16.12}$$

and hence that the new vectors $u^{\mu\prime}$, $v_k^{\mu\prime}$, are mutually perpendicular unit vectors, just as the original vectors u^μ, v_k^μ. We find, accordingly, that the unimodular spin space transformation a, applied according to Eq. (16.8) to the four matrices σ^μ, is equivalent to a linear transformation of four mutually perpendicular directions into a similar set, and hence equivalent to a Lorentz transformation.

An explicit calculation shows that the relationship between the primed and the unprimed unit vectors may be expressed fairly conveniently in terms of the complex numbers a_0, a_j, defined by the equation

$$a = a_0 + a_k \sigma_k, \qquad a_0^2 - a_k a_k = 1 \tag{16.13}$$

(a_μ must be a vector of norm 1 if the determinant of the matrix a is to be 1). We have

$$\left.\begin{aligned} u^{\mu\prime} &= (\bar{a}_0 a_0 + \bar{a}_k a_k) u^\mu - (\bar{a}_0 a_k + \bar{a}_k a_0 + i\,\delta_{ijk} \bar{a}_i a_j) v_k^\mu, \\ v_k^{\mu\prime} &= -(\bar{a}_0 a_k + a_0 \bar{a}_k + i\,\delta_{ijk} \bar{a}_i a_j) u^\mu - \\ &\quad + [(\bar{a}_0 a_0 - \bar{a}_i a_i)\,\delta_{kl} + (\bar{a}_k a_l + \bar{a}_l a_k) + i\,\delta_{klm} (\bar{a}_m a_0 - \bar{a}_0 a_m)]\, v_l^\mu. \end{aligned}\right\} \tag{16.14}$$

Without great effort it can be shown that the coefficients of these transformations represent *proper* Lorentz transformations, in that they permit neither time reversal nor inversion of space. On the other hand, *all* proper Lorentz coefficients can be constructed by appropriate choices of the values of the constants a_k, a_0 [consistent with the requirement of normality (16.13)]. Purely spatial rotations are obtained if we select a_0 to be real and all the a_k imaginary. The magnitude of the three-vector a_k equals the sine of half the angle of rotation. If we choose all four a_μ real, then we obtain a Lorentz transformation corresponding to two frames in relative motion without any spatial twist of the axes.

The representation of the Lorentz transformations by means of the unimodular matrices a is not unique; the Lorentz identity transformation may be represented either by the matrix 1 or by the matrix -1. In fact this ambiguity is unavoidable. Suppose we consider spatial rotations and wish to represent a rotation about the axis parallel to the unit vector $\boldsymbol{e}$ through 180°. The corresponding unimodular matrix is

$$a = i\,\boldsymbol{e} \cdot \boldsymbol{\sigma} \tag{16.15}$$

where the symbol $\boldsymbol{\sigma}$ stands for the three Pauli spin matrices. Clearly, the square of such a Lorentz transformation is the identity transformation, representing, as it does, a rotation through 360°. But the square of the matrix (16.15) is not the unit matrix, but -1. In other words, if we consider the one-parametric family of rotations about a fixed axis and correlate it with a continuous set of unimodular matrices, and if we identify originally the rotation through an angle zero with the identity matrix, then we must vary the angle of rotation ϑ from 0 to 4π before the unimodular matrix resumes the identity form. This ambiguity of the spin representation is not peculiar to the Lorentz group but is already encountered in the spin representation of the group of rotations in three-space.

If we link the unimodular transformations of spin space with the corresponding Lorentz transformations of the coordinates, we can arrange matters so that the form of the four matrices σ^μ is the same in all Lorentz-plus-spin frames considered. This is commonly done; the matrix σ^0 is chosen to be the unit matrix, whereas the remaining three matrices are set equal to the Pauli spin matrices. Accordingly, the matrices σ^μ in Lorentz-covariant theories are pure numerics (not dynamical variables), just as the metric tensor $\eta_{\mu\nu}$.

17. Spinor calculus. Improper Lorentz transformations. Given the spin-space and the group of complex unimodular transformations, the two-dimensional analog of the Levi-Civita object,

$$\varepsilon = \begin{pmatrix} 0 & 1 \\ -1 & 0 \end{pmatrix}, \tag{17.1}$$

is an invariant tensor, which may be used to change the transformation properties of a spin vector or tensor. Let us consider a two-component spinor (spin vector) u which satisfies the transformation law[1]

$$u' = a\,u, \quad \bar{u}' = \bar{a}\,\bar{u}, \quad \text{or} \quad u^{*\prime} = u^*\,a^*. \tag{17.2}$$

Because of the transformation character of ε,

$$\left.\begin{array}{ll} \varepsilon = a\,\varepsilon\,a^T, & \varepsilon = a^*\,\varepsilon\,\bar{a}, \\ \varepsilon = a^{T-1}\,\varepsilon\,a^{-1}, & \varepsilon = \bar{a}^{-1}\,\varepsilon\,a^{*-1}, \end{array}\right\} \tag{17.3}$$

we may construct a new type of spin vector from u, $v = \varepsilon u$, which transforms according to the rule

$$v' = a^{T-1}\,v, \quad v^{*\prime} = v^*\,\bar{a}^{-1}, \quad \bar{v}' = a^{*-1}\,\bar{v}. \tag{17.4}$$

In VAN DER WAERDEN's spinor notation, in which spin indices are written out explicitly, the four types of spin vectors u, $\bar{u}$, v, and $\bar{v}$ would be identified by the position of the index (subscript or superscript) and by the presence or absence of a dot on the index:

$$u \to u^A, \quad \bar{u}, u^* \to u^{\dot{A}} \quad v^T \to u_A, \quad \bar{v}, v^* \to u_{\dot{A}}. \tag{17.5}$$

A similar procedure may be applied to the spin matrices σ^μ: We shall define

$$\tau^\mu = \varepsilon\,\bar{\sigma}^\mu\,\varepsilon. \tag{17.6}$$

These new matrices turn out to have the value

$$\tau^\mu = -u^\mu + v_k^\mu\,\sigma_k, \quad \sigma^\mu\,\tau^\nu + \sigma^\nu\,\tau^\mu = \tau^\mu\,\sigma^\nu + \tau^\nu\,\sigma^\mu = -2\eta^{\mu\nu}. \tag{17.7}$$

The transformation equation for the τ^μ that is the analog of Eq. (16.8) for σ^μ turns out to be:

$$\tau^{\mu\prime} = a\,\tau^\mu\,a^*. \tag{17.8}$$

[1] In the following expressions the bar denotes transition to the conjugate complex without transposition, the superscript T transposition, and the asterisk the combination of both, i.e. transition to the Hermitian adjoint.

The τ^μ will be invariant under the same combination of proper Lorentz transformations and unimodular spin transformations that leaves the σ^μ unchanged.

There exists no unimodular matrix a that would induce an improper Lorentz transformation. There is however a possibility of connecting improper Lorentz transformations with spinor transformations in such a fashion that the form of the spin matrices remains unchanged; that is to permit the replacement of σ-matrices by τ-matrices and vice versa. If we wish to carry out a "time reversal" (i.e. multiply the time coordinate by -1), we may combine this transformation with a direct interchange $\sigma^\mu \to \tau^\mu$, $\tau^\mu \to \sigma^\mu$, and the form of the various spin matrices will be completely preserved, according to Eq. (17.7). Likewise, a complete spatial reflection, "parity transformation" ($x^k \to -x^k$) must be accompanied by the transition $\sigma^\mu \to -\tau^\mu$, $\tau^\mu \to -\sigma^\mu$. Thus the original form of the σ- and τ-matrices will be reproduced under the full Lorentz group.

In any covariant equations that are of interest in physics, the spin matrices σ^μ and τ^μ occur in combination with other quantities bearing one or several spin indices of the van der Waerden type; under proper Lorentz transformations these quantities obey linear transformation laws of the type (17.2), (17.4). Under improper Lorentz transformations spinors will have to be subjected to non-linear transformations. Consider two spinors u and U associated with each other by the relationship

$$\sigma^\varrho u_{,\varrho} = U. \tag{17.9}$$

This equation is covariant under proper Lorentz transformations if u and U obey the transformation equations

$$u' = a\,u, \qquad U' = a^{*-1}\,U. \tag{17.10}$$

Under time reversal Eq. (17.9) will be reproduced if we subject u and U to these transformations:

$$\left.\begin{aligned} u' &= -\varepsilon\bar{u}, & U' &= \varepsilon\,\overline{U},\\ x^{0\prime} &= -x^0, & x^{k\prime} &= x^k. \end{aligned}\right\} \tag{17.11}$$

Similarly, under a space reflection the corresponding transformation laws are:

$$\left.\begin{aligned} u' &= \varepsilon\bar{u}, & U' &= \varepsilon\,\overline{U},\\ x^{0\prime} &= x^0, & x^{k\prime} &= -x^k. \end{aligned}\right\} \tag{17.12}$$

It is not difficult to show that the proposed transformation laws (17.11), (17.12) are compatible with the structure of the full Lorentz group, in the following sense: It is possible to combine a proper Lorentz transformation (L) with a time reversal (T), (17.11) in such a manner as to obtain another proper Lorentz transformation L',

$$L' = T\,L\,T, \qquad T^2 = 1, \tag{17.13}$$

and to perform similar operations with a space reflection (P), (17.12), as well as with the complete coordinate reversal ($TP = PT$). In all these cases the resulting combined unimodular transformation a' represents correctly the corresponding Lorentz transformation L', provided that the ambiguity of the spin representation is allowed for in the improper Lorentz transformations as well as in the proper transformations; that is to say, the transformation $u \to -u$ for a spinor corresponds to the identical Lorentz transformation.

18. Four-component spinors[1]. Though we have found it possible to assign to spinor quantities transformation laws under improper Lorentz transformations that are consistent with the structure of the full Lorentz group, these transformation laws have certain pecularities that make them not only inconvenient for many purposes but which in particular restrict their usefulness for the physicist. The transformation laws are, for once, nonlinear. As a result it is impossible to extend the concept of contragredience (the relationship between contravariant and covariant vectors) to the improper Lorentz transformations, so that the scalar product in spin space should be invariant.

Under unimodular transformations the two spinors u and $\overline{U}$ of Eqs. (17.9), (17.10) are contragredient to each other; the expression $U^* u$ is invariant under unimodular transformations. Under improper Lorentz transformations this product changes its value: With spatial reflections the product goes over into its conjugate complex, under time reversal into its negative conjugate complex, and under the combined transformation it changes sign. Under proper Lorentz transformations U and $\varepsilon\bar{u}$ obey the same transformation law; they do not under time reversal, Eq. (17.11). Accordingly, depending on how a given set of spinors are related to each other by equations, one might assign them a number of slightly different transformation laws under improper transformations, whereas they obey identical transformation laws under proper Lorentz transformations (i.e. under unimodular spin transformations). This situation possesses an analog in tensor calculus (cf. the remarks at the end of Sect. 12), an analog that is imperfect insofar as all pseudotensor transformation laws are linear.

The nonlinearity of the transformation laws (17.11), (17.12) may be remedied formally by the introduction of four-component spinors. We may consider the spin space (with unimodular transformation a) and the conjugate complex spin space (with transformations $\bar{a}$) as two distinct spaces, with their transformation properties linked by the requirement that for all proper Lorentz transformations $\bar{a}$ is the conjugate complex of a. Each of these two spin spaces has its own "dual space", in which the unimodular transformations are accomplished with the help of the matrices a^{-1} and $\bar{a}^{-1}$, respectively. Under time reversal and under spatial reflections we have seen that we must go over from quantities that transform with a to those that transform with $\bar{a}^{-1}$. This nonlinear transition will become linear if we form the direct sum of each spin space with the dual of its conjugate complex space. We shall write the vectors of this new spin space with four complex dimensions in the form

$$\psi = \begin{pmatrix} u \\ v \end{pmatrix}. \tag{18.1}$$

Each of the two symbols u and v stands for a two-component spinor, with the following transformation laws under proper Lorentz transformations:

$$\left.\begin{aligned} u' &= a\,u, & v' &= a^{*-1}\,v, \\ \psi' &= A\,\psi, & A &= \begin{pmatrix} a & 0 \\ 0 & a^{*-1} \end{pmatrix}. \end{aligned}\right\} \tag{18.2}$$

The introduction of the 4×4 matrix A is purely formal as long as we restrict ourselves to proper Lorentz transformations. Representation of the proper Lorentz group by the matrices A is obviously reducible into the two irreducible representations afforded by a and a^{*-1}.

Under the improper Lorentz transformations T and P (time reversal and spatial reflections) we have found in the previous section that the two spin spaces

[1] P. A. M. DIRAC: Proc. Roy. Soc. Lond. **117**, 610; **118**, 341 (1928).

in which u and v are defined go over into each other, possibly with a change in sign. Accordingly we represent these transformations by means of linear transformations characterized by the following matrices:

$$\left.\begin{aligned} P&=\begin{pmatrix}0 & 1\\ 1 & 0\end{pmatrix}, \quad T=\begin{pmatrix}0 & -1\\ 1 & 0\end{pmatrix}\\ PT&=\begin{pmatrix}1 & 0\\ 0 & -1\end{pmatrix}.\end{aligned}\right\} \tag{18.3}$$

As before, the identity transformation is represented by either the identity 4×4 matrix I or by $-I$; likewise all other representations (18.2), (18.3) are defined only up to a factor ± 1.

If we are to construct the space that is "dual" to the space of the ψ-vectors, and to vectors that transform contragradiently to ψ under *all* Lorentz transformations, we are led to quantities of the type

$$\psi^\dagger = (v^*, u^*) = \psi^* \beta, \quad \beta = \begin{pmatrix}0 & 1\\ 1 & 0\end{pmatrix}, \quad \beta^2 = I. \tag{18.4}$$

$\psi^\dagger$ will in fact transform contragradiently to ψ under proper Lorentz transformations and at the same time retain its relationship to ψ, (18.4). That is to say, the transformation matrices A, Eq. (18.2), satisfy the requirement

$$\psi^\dagger A^{-1} = (A\,\psi)^* \beta, \quad \beta = A\,\beta\,A^*. \tag{18.5}$$

Among the improper Lorentz transformation matrices (18.3), the matrix P satisfies the condition (18.5) as well, but not the matrices T and PT. In a transformation involving time reversal the sign of a product such as $\psi^\dagger\,\psi$ is thus reversed[1].

We can construct 4×4 matrices that replace in effect the spin matrices σ^μ and τ^μ of the two-component formalism. In fact, in the four-component formalism, in which the transformation matrix A possesses a reciprocal matrix which is also its adjoint, $A^\dagger$,

$$A^\dagger \equiv \beta\,A^*\,\beta = A^{-1}, \tag{18.6}$$

we shall construct a set of four spin matrices that are self-adjoint and which transform under A-transformations in such a manner as to preserve this property. The matrices

$$\gamma^\mu = \begin{pmatrix}0 & -\tau^\mu\\ \sigma^\mu & 0\end{pmatrix}, \quad \gamma^{\mu\dagger} \equiv \beta\,\gamma^{\mu *}\,\beta = \gamma^\mu, \quad \gamma^{\mu\prime} \equiv A^\mu{}_\varrho\,A\,\gamma^\varrho\,A^\dagger = \gamma^\mu \tag{18.7}$$

have these properties. Because of the second set of Eqs. (17.7), the γ-matrices (which are a slightly modified version of DIRAC's original spin matrices) satisfy these anticommutation relations:

$$\gamma^\mu\gamma^\nu + \gamma^\nu\gamma^\mu = 2\,\eta^{\mu\nu}\,I. \tag{18.8}$$

---

[1] In quantum field theory such a statement requires modification on two grounds. First, bilinear products between annihilation and creation operators (ψ and $\psi^\dagger$, respectively) are related to the electric charge, and hence subject to charge conjugation. Second, for fermions these operators obey anticommutation relations; accordingly interchanging their sequence also reverses the sign. That is why the arguments above are not in contradiction to the TCP theorem.

If we wish to treat the γ^μ as absolute invariants under all Lorentz transformations, then for time reversal we must transform as follows:

$$\left.\begin{aligned} \gamma^{\mu\prime} &= A^\mu{}_\varrho\, T\gamma^\varrho T^{-1} = -A^\mu{}_\varrho\, T\gamma^\varrho T^\dagger = \gamma^\mu, \\ A^\mu{}_\varrho &= \begin{pmatrix} -1 & 0 & 0 & 0 \\ 0 & 1 & 0 & 0 \\ 0 & 0 & 1 & 0 \\ 0 & 0 & 0 & 1 \end{pmatrix}. \end{aligned}\right\} \tag{18.9}$$

With this choice we can also assure the scalar or vector character, respectively, of such formations as $\psi^\dagger\psi$, $\psi^\dagger\gamma^\mu\psi$, by transforming $\psi^\dagger$ with A^{-1} rather than with $A^\dagger$. Then, in order to maintain the validity of the defining equation (18.4), we must permit the matrix β to change its sign under time reversal.

B. Relativistic mechanics and electrodynamics.

I. The mechanics of mass points.

19. Scope of relativistic mechanics. Nonrelativistic conservation laws. Newtonian mechanics is based on the notion of action at a distance, the instantaneous exertion of a force by one body on another at a distance. According to the theory of relativity the concept of simultaneity has no absolute significance; a force that is exerted instantaneously in one frame of reference would be exerted either with a delay or even ahead of time in different frames. Thus action at a distance in its original meaning has no place in relativistic mechanics; accordingly relativistic mechanics is concerned primarily with physical systems in which interaction takes place only between adjacent regions in space. There is, first of all, the theory of impulsive interactions, or *collisions*, between mass points; then there is the mechanics of distributed matter, in which interaction takes place in terms of the stress of continuous matter. More recently, theories have been developed in which action at a distance takes place not simultaneously but at a finite rate of spreading bounded by the speed of light. Though these latter theories are relativistically invariant, they suffer from other conceptual difficulties, which have not been entirely resolved.

We shall treat the mechanics of mass points first, reserving the treatment of continuum mechanics for Part II (p. 159ff.). We begin with a resumé of the theory of collisions. In any system that consists of interacting mass points the details of the motion of each constituent particle depend both on the laws of interaction (the force laws) and on the trajectories of all other member particles. In addition, there are certain general laws applying uniformly to every assembly of mass points regardless of the particulars of the force laws; they are the conservation laws. Though the various conservation laws are interrelated, both in Newtonian and in relativistic mechanics, we shall enumerate them distinctly, as follows:

(a) The conservation of the total mass of the system, which holds even if in close encounters constituent particles should transfer part of their masses to each other;

(b) the conservation of the total linear momentum of the system;

(c) the conservation of the total energy of the system;

(d) the conservation of the total angular momentum of the system; and

(e) the conservation of the difference between the first moment of the mass distribution of the system and the product of the total linear momentum by the elapsed time (center-of-mass law).

Among these conservation laws, the energy law depends on the details of the force laws through the potential energy. However, we can assert that there always exists a conserved quantity which is the sum of the kinetic energy of the system (whose form is universal) and the potential energy, which latter is a function of the distances between the mass points only. Furthermore, for interaction by short-range forces, in which the potential energy between the mass points is negligible except during the brief periods of strong interaction (the "collisions"), we can say that the initial (kinetic) energy always equals the final (again purely kinetic) energy of the system[1].

In terms of equations the conservation laws are the following:

$$\frac{d}{dt}\Big(\sum_n m_n\Big) = 0, \tag{19.1}$$

$$\frac{d}{dt}\Big(\sum_n m_n \dot{\boldsymbol{x}}_n\Big) = 0, \tag{19.2}$$

$$\frac{d}{dt}\Big(\sum_n \frac{1}{2} m_n \dot{\boldsymbol{x}}_n^2 + V\Big) = 0, \tag{19.3}$$

$$\frac{d}{dt}\Big(\sum_n m_n \boldsymbol{x}_n \times \dot{\boldsymbol{x}}_n\Big) = 0, \tag{19.4}$$

$$\frac{d}{dt}\Big(\sum_n m_n \boldsymbol{x}_n - t\sum_n m_n \dot{\boldsymbol{x}}_n\Big) = 0. \tag{19.5}$$

The total number of these laws, if we count the different components of vector relations separately, is 11. They are interconnected, but none of them are redundant (i.e. satisfied identically modulo the remainder) if we admit the possibility of mass exchanges among the constituents of our system. To examine this interrelationship let us introduce the total mass of the system, M, and the coordinates of its center of mass, $\boldsymbol{X}$. Further, let us call the total linear momentum $\boldsymbol{P}$, so that, *by definition*,

$$\sum_n m_n = M, \qquad \sum_n m_n \boldsymbol{x}_n = M\boldsymbol{X}, \qquad \sum_n m_n \dot{\boldsymbol{x}}_n = \boldsymbol{P}. \tag{19.6}$$

With these definitions the conservation laws of mass and of linear momentum become simply

$$\dot{M} = 0, \qquad \dot{\boldsymbol{P}} = 0. \tag{19.7}$$

If we insert these two relationships into the last conservation law, Eq. (19.5), $\frac{d}{dt}(M\boldsymbol{X} - t\boldsymbol{P}) = 0$, we find that we obtain the additional relationship:

$$M\dot{\boldsymbol{X}} - \boldsymbol{P} = 0. \tag{19.8}$$

Combined with the defining equation for $\boldsymbol{P}$, this law implies:

$$\sum_n \dot{m}_n \boldsymbol{x}_n = 0. \tag{19.9}$$

Since this law must hold in particular for systems consisting of two mass points, we find that two mass points can transfer mass only when they are at the same location.

Whereas the conservation laws (19.1) through (19.5) are thus seen to be algebraically independent of each other, there is a strong interdependence if we require that these laws are to hold in all frames of references related to each other through Galilean transformations (1.1). Spatial rotations and reflections do no

[1] Inelastic collisions are excluded.

more than combine with each other the various components of the vector laws, i.e. Eqs. (19.2) as well as Eqs. (19.4) and (19.5). But if we shift the origin of the spatial coordinate system, and the origin of the time scale, as well as admit motion of two frames of reference with respect to each other,

$$\boldsymbol{x}' = \boldsymbol{x} + \boldsymbol{u}\,t + \boldsymbol{d}, \qquad t' = t + d^0, \tag{19.10}$$

then the quantities that are conserved according to our conservation laws are related to each other as follows (if we consider the masses to be invariant):

$$\left.\begin{aligned} &M' = M, \quad \boldsymbol{P}' = \boldsymbol{P} + \boldsymbol{u}\,M,\\ &E' = E + \boldsymbol{u}\cdot\boldsymbol{P} + \tfrac{1}{2} M\,\boldsymbol{u}^2,\\ &\boldsymbol{I}' = \boldsymbol{I} + M\boldsymbol{Y}\times\boldsymbol{u} + \boldsymbol{d}\times(\boldsymbol{P} + \boldsymbol{u}M),\\ &\boldsymbol{Y}' = \boldsymbol{Y} + \boldsymbol{d} - d^0\left(\frac{\boldsymbol{P}}{M} + \boldsymbol{u}\right), \end{aligned}\right\} \tag{19.11}$$

where the following additional symbols have been introduced:

$$\left.\begin{aligned} E &= \sum_n \tfrac{1}{2} m_n \dot{\boldsymbol{x}}_n^2 + V,\\ \boldsymbol{I} &= \sum_n m_n \boldsymbol{x}_n \times \dot{\boldsymbol{x}}_n,\\ \boldsymbol{Y} &= \boldsymbol{X} - \frac{t}{M}\boldsymbol{P}. \end{aligned}\right\} \tag{19.12}$$

From these transformation laws it is apparent that the law of conservation of linear momentum can hold in *all* inertial frames only if the total mass is conserved; that likewise energy can be conserved in all frames of reference only if both mass and linear momentum are; and that the conservation of angular momentum and of the vector denoted here by $M\boldsymbol{Y}$ (the moment of the mass distribution at the time $t=0$) in all inertial frames depend both on each other and on the remaining conservation laws.

If we are to discover the corresponding relativistic conservation laws, these interrelationships will prove of considerable value.

20. Mass, linear momentum, and energy. A proposed set of relativistic conservation laws will have to satisfy two principal requirements: For velocities of the constituent mass points that are small compared to the speed of light the classical conservation laws must be approximately valid; and the relativistic laws must be preserved by Lorentz transformations. The first requirement stems from the fact that for moderate velocities Newtonian mechanics and its conservation laws had been amply tested by experiment and observation. As for the second requirement, we can strengthen it by reference to the non-relativistic transformation laws listed in Sect. 19: As the relativistic laws are to go over into the classical laws for small velocities, and as the Galilean transformations (1.1) are good approximations of the Lorentz transformations for small values of u/c, the transformation laws of the relativistic laws under Lorentz transformations must be similar to the transformation laws of the classical conservation laws under Galilean transformations.

The mutual dependence of the conservation laws on each other under Galilean transformations suggests grouping together of the laws of conservation of linear momentum, mass, and energy on the one hand, and of the conservation laws of angular momentum and $M\boldsymbol{Y}$ on the other. Leaving aside the second group for the moment, we can certainly assert that in the first group the transformation law is *linear*, and that the relativistic analogs will very likely form the components of a vector or tensor (as spinors are excluded for quantities that possess a classical meaning). There is a world vector in relativistic kinematics that has a close similarity to the three-dimensional velocity vector. If we represent a four-dimensional trajectory of a mass point parametrically, through four functions $x^\varrho(\vartheta)$ (where ϑ is an arbitrarily chosen parameter along the curve in Minkowski space), then the ratios

$$U^\varrho = \frac{dx^\varrho/d\vartheta}{d\tau/d\vartheta} = \frac{dx^\varrho}{d\tau} \equiv \dot{x}^\varrho \tag{20.1}$$

form the components of a world vector. Their values are obviously independent of the choice of the parameter ϑ. For small values of the ratios $\left(\frac{dx^i}{d\vartheta}\Big/\frac{dx^0}{d\vartheta}\right)$ $(i \neq 0)$, the first three components U^i deviate but slightly from the values of the ordinary velocity components, whereas U^0 approximately equals unity. The exact expressions are:

$$U^i = \frac{u^i}{\sqrt{1-\frac{u^2}{c^2}}}, \qquad U^0 = \frac{1}{\sqrt{1-\frac{u^2}{c^2}}}. \tag{20.2}$$

If we now conjecture that the following Lorentz-covariant law holds:

$$\sum_n m_n \underset{n}{U^\varrho} \equiv \sum_n \underset{n}{p^\varrho} \equiv P^\varrho = \text{const}, \tag{20.3}$$

then we have the satisfaction that for moderate velocities three of these laws go over into the conservation law for linear momentum, whereas the fourth one represents the conservation law for mass.

However, a rigorous interpretation of the proposed law (20.3) is impossible unless we can remove the uncertainty as to when the contributions of the individual mass points are to be determined which, added together, make up the total value of the world vector P^ϱ. Classically, we should require that the individual $\underset{n}{p^\varrho}$ must be determined "at the same time", but relativistically this requirement would completely destroy the invariance of the law (20.3). There appears to be no possible relativistically invariant determination for the appropriate individual instances at which the $\underset{n}{p^\varrho}$ are to be determined, and the only alternative is to assert that the time of determination is to be immaterial. But that is possible only if each individual $\underset{n}{p^\varrho}$ remains constant except at any instance at which two (or more) of the particle trajectories intersect,—when a collision takes place. At such an instance it is possible that the two particles transfer to each other (four-dimensional) linear momentum. A similar restriction applies in classical mechanics, for entirely different reasons, to the transfer of mass. Our present conclusion must be that the conjectured conservation law (20.3) can hold only *at best* in the absence of long-range interactions. We shall find later that in action-at-a-distance theories there are conservation laws having the same general structure, but with different expressions for the $\underset{n}{p^\varrho}$.

The scalar quantities m_n occurring in Eq. (20.3) are usually referred to as *rest masses*. The quantities $\underset{n}{p^0}$, whose sum is to remain unchanged, and which individually represent the ratios between the linear momenta and the (three-dimensional) velocities of the particles,

$$\underset{n}{p^0} = \frac{m_n}{\sqrt{1-\frac{u_n^2}{c^2}}}, \tag{20.4}$$

are called *relativistic masses.* They alone can be added meaningfully to represent a property of the whole system, *its* relativistic mass P^0. The relativistic mass of a particle can never be smaller than its rest mass; for large velocities it increases until at the speed of light it becomes infinite.

If we expand the expression (20.4) into a power series with respect to c^{-1}, we obtain for the conservation law $P^0 = \text{const}$ the following:

$$P^0 = \sum_n m_n + \frac{1}{c^2} \sum_n \frac{1}{2} m_n u_n^2 + \cdots. \tag{20.5}$$

For processes in which the particles do not change their individual rest masses, the conservation of kinetic energy is a second-order result of the conservation of relativistic mass. Inasmuch as we must exclude long-range forces anyway, the potential energy vanishes, and the kinetic energy is the whole energy. The kinetic energy will be preserved rigorously if all rest masses remain unchanged and if we set the *relativistic kinetic energy* of a particle equal to the expression

$$K = m c^2 \left[\left(1 - \frac{u^2}{c^2}\right)^{-\frac{1}{2}} - 1\right]. \tag{20.6}$$

According to Eq. (20.3) the kinetic energy need not be conserved unless the rest masses remain unchanged. This conservation law leaves open the possibility for *inelastic collisions*, but not for a violation of the law of conservation of relativistic mass. For, if in one frame of reference the total linear momentum should remain conserved but not the total relativistic mass, then through a Lorentz transformation we could go over to another frame of reference in which the law of conservation for the linear momentum is violated. Hence if through an inelastic collision the total kinetic energy of a system [as measured by Eq. (20.6)] should change, it is unavoidable that there will be a compensating change in the rest masses, which, according to the power expansion (20.5), will be approximately c^{-2} times the (negative) change in kinetic energy.

In the whole experimental experience of the past fifty years the conservation law (20.3) has been confirmed, with the proviso that for electromagnetic interaction we must make separate allowance for the energy and linear momentum residing in the electromagnetic field. Energy and mass are not conserved separately (as they are in Newtonian physics) but only together, and in a ratio equal to c^2. The quantity mc^2 is, therefore, often referred to as the rest energy of a particle and expressed in units of energy, such as electron volts. Originally, this proportionality implied no more than a requirement that any change in intrinsic energy of a composite system must be compensated by a corresponding change of its mass. There was no immediate claim that *all* the mass of material particles is convertible into free energy. The possibility of such complete convertibility aroused both the curiosity of the physicist and the appetites of a society concerned with the availability of plentiful sources of energy.

By now all elementary particles have been involved in transmutation reactions in the laboratory. The majority of particles are inherently unstable and decay spontaneously into other particles; the only stable particles are the proton, the electron, and the neutrino. Protons combine with antiprotons and decay into unstable mesons, whereas the most common decay interaction between electrons and positrons is the decay into electromagnetic energy. The manner of creation of neutrinos suggests that a neutrino and an antineutrino may well combine into photons either directly or indirectly, though this experiment cannot yet be performed. Thus, from a fundamental point of view apparently all "material" particles are convertible (at least in pairs) into electromagnetic energy. So far, this complete convertibility has not led to technological applications.

21. Relativistic angular momentum. Just as we did with the linear momentum, we shall now propose a conservation law for the angular momentum, which will also include the constant we have designated by MY. In classical mechanics the angular momentum is a vector only with respect to rotations of the coordinate system but not with respect to displacements of the origin. As we have seen in Eq. (19.11), conservation of angular momentum in different inertial frames (in Newtonian physics) is possible only because the additional terms caused by a displacement of the origin are covered by the law of conservation of linear momentum.

Accordingly we should not expect the relativistic angular momentum to transform as a tensor with respect to displacements of the origin, either, but only with respect to *homogeneous* Lorentz transformations, i.e. with respect to transformations preserving the origin of the coordinate system in space and time If we permit ourselves to be guided again by the requirement that for small velocities the relativistic expression is to go over into the classical one, at least to zeroth order in c^{-1}, we are led to consider the following expression:

$$J^{\mu\nu} \equiv \sum_n \underset{n}{j}^{\mu\nu} \equiv \sum_n \left(\underset{n}{x}^{\mu} \underset{n}{p}^{\nu} - \underset{n}{x}^{\nu} \underset{n}{p}^{\mu}\right) = \text{const}. \tag{21.1}$$

This expression has altogether six algebraically independent components, because of its skewsymmetry with respect to the two superscripts. The independent index combinations are 12, 23, 31; 10, 20, and 30. The first three form, for moderate velocities, the components of the usual angular momentum. The latter three represent under the same conditions the three components of the vector MY. We shall call the whole expression $j^{\mu\nu}$ the relativistic angular momentum of a particle.

If we look at the transformation law for the angular momentum under (inhomogeneous) Lorentz transformations,

$$x^{\mu\prime} = A^{\mu}{}_{\varrho}\, x^{\varrho} + B^{\varrho}, \quad p^{\nu\prime} = A^{\nu}{}_{\sigma}\, p^{\sigma}, \tag{21.2}$$

we find:

$$j^{\mu\nu\prime} = A^{\mu}{}_{\varrho}\, A^{\nu}{}_{\sigma}\, j^{\varrho\sigma} + (B^{\mu} A^{\nu}{}_{\varrho} - B^{\nu} A^{\mu}{}_{\varrho})\, p^{\varrho}. \tag{21.3}$$

The relativistic angular momentum is not a tensor in the strict sense of the term. That it can be conserved in all Lorentz frames (if it is conserved in one) depends on the conservation of relativistic linear momentum.

Moreover, our comments concerning the conservation of *individual* linear momenta except in collisions also apply to angular momenta. Though there is a set of constants of the motion possessing the transformation law (21.3) in any known relativistic dynamics, the detailed dependence on particle rest mass and motion proposed in Eq. (21.1) must be supplemented by further terms in the presence of fields or other long-range interactions.

22. HAMILTON's principle. It is shown in analytical mechanics that the constants of the motion of a system are identical with the generators of invariant infinitesimal canonical transformations, i.e. those transformations that leave the form of the Hamiltonian (and hence the form of the Lagrangian) unchanged. The universal constants of the motion of classical mechanics which we have enumerated in Sect. 19 are in particular the generators of infinitesimal Galilean transformations, with respect to which any Newtonian theory is supposed to be invariant. It is tempting to connect the universal relativistic constants of the motion, which we have developed in Sects. 20 and 21, with the infinitesimal Lorentz transformations.

It might appear that such a program will be attractive only to the formalist, because we already have obtained satisfactory expressions for all the universal constants of the motion to be expected. However, the connection with Lorentz transformations reveals the appropriate form for the universal constants of the motion for relativistic theories that involve long-range interactions, provided only that such theories can be fitted into the very general framework of analytical mechanics.

In view of the fact that the Hamiltonian formulation of analytical mechanics is inappropriate for manifestly Lorentz-covariant developments, we shall begin by transferring the concept of infinitesimal generator to the Lagrangian version[1]. Given an action integral S,

$$S = \int L\left(\xi^n, \frac{d\xi^n}{d\vartheta}\right) d\vartheta, \tag{22.1}$$

with an arbitrarily chosen parameter ϑ, and the ξ^n forming the coordinates in some configuration space of indeterminate dimensionality (we do not assume the existence of a time axis outside this configuration space), the change in the value of S for an infinitesimal displacement of the path of integration is

$$\delta S = \int L_n \,\delta\xi^n \,d\vartheta + [p_n \,\delta\xi^n], \quad L_n \equiv \frac{\partial L}{\partial \xi^n} - \frac{d p_n}{d\vartheta}, \quad p_n \equiv \frac{\partial L}{\partial\left(\frac{d\xi^n}{d\vartheta}\right)}. \tag{22.2}$$

The expressions L_n will vanish if the original path of integration happened to be an extremal path, i.e. the solution of the variational problem of finding those curves in ξ^n-space along which the action S is stationary. In physical applications, we call the equations $L_n = 0$ the (Lagrangian) equations of motion of the problem.

Clearly, if the equations of motion are to be covariant under certain transformations, these transformations must be capable of reproducing the form of the integral (22.1) in detail. That is to say, after the coordinate transformation in configuration space, and substitution of the new coordinates for the original ones in the Lagrangian L, the new Lagrangian L' must be the same function of *its* arguments $\xi^{n\prime}$ as the original Lagrangian L was of the arguments ξ^n. That is not to say that L must be a scalar (in fact usually it is *not*). Ordinarily, invariance can be achieved only by taking advantage of the fact that for the form of the equations of motion the addition of an exact derivative $dQ/d\vartheta$ to L is immaterial. If the Lagrangians in two coordinate systems at the same point of ξ^n-space differ by a term of this type, then L' is to be the same function of $\xi^{n\prime}$, $d\xi^{n\prime}/d\vartheta$, as L is of ξ^n, $d\xi^n/d\vartheta$. For an infinitesimal transformation this means that not δL but $\bar{\delta} L$ must vanish, according to the notation explained in Sect. 13. A transformation with the property $\bar{\delta} L = 0$ will be called an *invariant transformation*.

Given a path of integration, its individual points will change their coordinate values under an infinitesimal transformation by amounts $\delta\xi^n(\vartheta)$. Depending on the character of the transformation, these infinitesimal changes in the coordinate values of a point along the trajectory may depend only on the values of the coordinates ξ^n themselves, or on their first (or even higher) derivatives as well. In the latter case, the transformation considered is not point-to-point but more

[1] R. E. PEIERLS: Proc. Roy. Soc. Lond., Ser. A **214**, 143 (1952).

general. It will merely transform one given curve uniquely into another one. If the change of L at a fixed world point is

$$\delta L = \frac{dQ}{d\vartheta}, \tag{22.3}$$

where the function Q is, for the time being, arbitrary, then, according to Eq. (13.7) we have for $\bar{\delta} L$

$$\bar{\delta} L = \frac{dQ}{d\vartheta} - \frac{\partial L}{\partial \xi^n} \delta \xi^n - p_n \frac{d}{d\vartheta} (\delta \xi^n), \tag{22.4}$$

and hence for $\bar{\delta} S$

$$\bar{\delta} S = - \int L_n \delta \xi^n d\vartheta - [C], \quad C = p_n \delta \xi^n - Q. \tag{22.5}$$

A transformation will be invariant if and only if the expression for $\bar{\delta} S$ vanishes[1]. We may, therefore, describe an invariant transformation as one in which the expression $L_n \delta \xi^n$ is an exact derivative. The function C, which normally depends on the arguments ξ^n, $d\xi^n/d\vartheta$, is characteristic for the infinitesimal invariant transformation. It is called the *generator* of that transformation. Because, according to Eq. (22.5),

$$L_n \delta \xi^n + \frac{dC}{d\vartheta} \equiv 0, \tag{22.6}$$

the generator of an invariant infinitesimal transformation will be constant along any path that satisfies the Lagrangian equations of motion. Hence we may obtain constants of the motion if we know infinitesimal invariant transformations.

One trivial transformation that always exists is an infinitesimal parameter transformation. Under it we have, for fixed parameter values,

$$\delta \xi^n (\vartheta) = - \frac{d\xi^n}{d\vartheta} d\vartheta. \tag{22.7}$$

The function $\delta\vartheta(\vartheta)$ is arbitrary. The condition

$$\frac{dC}{d\vartheta} \equiv L_n \frac{d\xi^n}{d\vartheta} \delta\vartheta \tag{22.8}$$

can be given the form

$$\frac{dC}{d\vartheta} = \delta\vartheta \frac{d}{d\vartheta} \left(L - p_n \frac{d\xi^n}{d\vartheta} \right). \tag{22.9}$$

It is not difficult to show (though the proof will be omitted here) that any Lagrangian satisfying this requirement differs only by an exact derivative from one that obeys the condition

$$L \equiv p_n \frac{d\xi^n}{d\vartheta} \tag{22.10}$$

and which is, therefore, homogeneous of the first degree in the (first) derivatives of the coordinates. This result is also intuitively reasonable. It is universally assumed to hold whenever line integrals are put into parametric form. Hence the generator of the parameter transformation (22.7) vanishes.

23. The generators of infinitesimal Lorentz transformations. Under an infinitesimal coordinate transformation the coordinates of any world point change according to Eq. (13.4),

$$\delta \underset{n}{x}^\varrho = P^\varrho{}_\sigma \underset{n}{x}^\sigma + b^\varrho, \quad P_{\varrho\sigma} + P_{\sigma\varrho} = 0. \tag{23.1}$$

[1] Strictly speaking, it is sufficient that $\bar{\delta} S$ vanish "strongly" [in the sense in which this term is being used by DIRAC, Can. J. Math. **2** 129 (1950); **3**, 1 (1951)] modulo the equations of motion. That is to say that all *variations* of $\bar{\delta} S$ must vanish modulo the equations of motion. This refinement is, however, not pertinent for the developments described here.

Suppose we introduce a Lagrangian that depends on all *four* coordinates of every mass point in a system, as well as on their derivatives with respect to an arbitrarily chosen parameter ϑ; we shall from now on characterize such derivatives simply by a prime, thus $\underset{n}{x}^{\varrho\prime}$. For the Lorentz covariance (with respect to proper transformations) of the resulting equations of motion it is necessary and sufficient that the transformation (23.1), applied to all particle coordinates at once, be an invariant transformation. That is to say, the expression

$$\sum_n \underset{n}{L}_\varrho \left(P^\varrho{}_\sigma \underset{n}{x}^\sigma + b^\varrho\right) \equiv - C' \tag{23.2}$$

must be an exact time derivative. We can write the expression for C' in the form

$$C' \equiv \tfrac{1}{2} P_{\sigma\varrho} \sum_n \left(\underset{n}{L}^\varrho \underset{n}{x}^\sigma - \underset{n}{L}^\sigma \underset{n}{x}^\varrho\right) - b_\varrho \sum_n \underset{n}{L}^\sigma = 0 \tag{23.3}$$

and obtain the following constants of the motion:

$$J^{\varrho\sigma\prime} = \sum_n \left(\underset{n}{L}^\sigma \underset{n}{x}^\varrho - \underset{n}{L}^\varrho \underset{n}{x}^\sigma\right) \tag{23.4}$$

and

$$P^{\varrho\prime} = \sum_n \underset{n}{L}^\varrho. \tag{23.5}$$

These expressions will coincide with those we have derived earlier for relativistic angular and linear momentum, respectively, if we adopt as our action the expression

$$S = - \sum_n m_n \int \sqrt{\eta_{\varrho\sigma} \underset{n}{x}^{\varrho\prime} \underset{n}{x}^{\sigma\prime}}\, d\vartheta, \tag{23.6}$$

in other words the sum of the particles' individual proper times, each multiplied by its respective rest mass. The corresponding Lagrangian is

$$L = \sum_n \underset{n}{L}, \qquad \underset{n}{L} = - m_n \sqrt{\eta_{\varrho\sigma} \underset{n}{x}^{\varrho\prime} \underset{n}{x}^{\sigma\prime}} = - m_n \tau_n', \tag{23.7}$$

and the equations of motion

$$\underset{u}{L}_\varrho = - \frac{d \underset{n}{p}_\varrho}{d\vartheta}, \qquad \underset{n}{p}_\varrho = - \eta_{\varrho\sigma} m_n \frac{\underset{n}{x}^{\sigma\prime}}{\tau'} = - \eta_{\varrho\sigma} m_n \frac{d\underset{n}{x}^\sigma}{d\tau}. \tag{23.8}$$

For different Lagrangians and action integrals different expressions would result.

Some further comments apply to the form of the action integral (23.6). The dynamical behavior of a mechanical system consisting of several mass points may be described in Minkowski space as a collection of curves, each of them representing a separate particle of the system. This same behavior may also be represented by a single curve in a $4N$-dimensional configuration space, N being the total number of constituent mass points. From the point of view of transformation theory this configuration space is artificial. Under the Lorentz group the transformation matrices which we can introduce formally in that configuration space decompose into N irreducible representations of rank 4. Accordingly the representation in the four-dimensional physical space-time continuum (Minkowski universe) is not only more intuitive but also preferable if we are to obtain some insight into the possible structure of Lorentz-invariant variational principles.

Because of the unavailability of a relativistically invariant concept of simultaneity, it is impossible to introduce a Lagrangian that would in an unambiguous manner sample "corresponding" points on the various particle trajectories. Hence a relativistically invariant integral that might be suitable as a Hamiltonian principle may be a sum over single line integrals, each involving integration along one particle path; a sum of double integrals, each involving integration over a pair of particles, etc. The action (23.6) is an example for the first type. The second type of action principle, presumably combined with the first, would provide a possibility for interaction between pairs of particles. In point mechanics no more involved action principles have as yet been studied.

The general formal theory of collision processes is exhausted once we have obtained all relativistic constants of the motion. As we have seen, two particles that collide with each other are capable of exchanging mass and linear momentum at that instant and, by the same token, relativistic angular momentum. A detailed law of collision dynamics would tell us for two particular particles colliding at a given relative speed the probability of their exchanging a particular amount of linear momentum, a total of four individual quantities; such a law must be Lorentz-covariant. There is not much physical interest in the exploration of this problem, but there is considerable interest in the corresponding quantum problem. The determination of the so-called collision matrix of interacting particles has become one of the most active aspects of high-energy physics. A discussion of this work would, however, lead beyond the confines of this article[1].

In the following three sections we shall be concerned with the theory of particles that are interacting with each other at a distance, but in a Lorentz-invariant manner. We shall develop this theory starting from an invariant principle of least action that involves cross terms between pairs of particles.

24. Relativistic action at a distance[2]. We shall consider a dynamical system with the action integral

$$\left.\begin{aligned} S &= \tfrac{1}{2}\sum_{m\neq n}\sum S_{mn}^{\text{int}} + \sum_{n} S_n^{\text{part}},\\ S_{mn}^{\text{int}} &= -q_m q_n \iint_{\vartheta\,\bar\vartheta} F(\tau_{mn}^2)\,\eta_{\varrho\sigma}\,\underset{m}{x}^{\varrho\prime}(\vartheta)\,\underset{n}{x}^{\sigma\prime}(\bar\vartheta)\,d\vartheta\,d\bar\vartheta,\\ S_n^{\text{part}} &= -m_n\int d\tau_n. \end{aligned}\right\}\tag{24.1}$$

The function F is to be a Lorentz-invariant function of the argument τ_{mn}^2, the squared proper time interval between two world points. For the time being we shall not specify whether this function is non-zero for time-like intervals, on the light cone (for null intervals), or for space-like intervals, or for a combination of these three domains. The constant parameters q_n are to be characteristic of the "strengths" of the individual particles.

For a variation of the paths of the particles we obtain the expressions:

$$\left.\begin{aligned} \delta S_{mn}^{\text{int}} &= 2q_m q_n \eta_{\alpha\beta}\eta_{\varrho\sigma}\int_\tau\int_{\bar\tau}\frac{dF}{d\tau_{mn}^2}\Big\{\Big[\underset{n}{\dot x}^\beta\big(\underset{m}{x}^\varrho-\underset{n}{x}^\varrho\big)-\underset{n}{\dot x}^\varrho\big(\underset{m}{x}^\beta-\underset{n}{x}^\beta\big)\Big]\underset{m}{\dot x}^\sigma\,\delta\underset{m}{x}^\alpha+\\ &\quad+\Big[\underset{m}{\dot x}^\alpha\big(\underset{n}{x}^\sigma-\underset{m}{x}^\sigma\big)-\underset{m}{x}^\sigma\big(\underset{n}{x}^\alpha-\underset{m}{x}^\alpha\big)\Big]\underset{n}{\dot x}^\varrho\,\delta\underset{n}{x}^\beta\Big\}\,d\tau\,d\bar\tau-\\ &\quad-q_m q_n\,\eta_{\alpha\beta}\Big\{\Big[\int F\underset{n}{\dot x}^\beta\,d\bar\tau\,\delta\underset{m}{x}^\alpha\Big]_{\tau\,\bar\tau}+\Big[\int F\underset{m}{\dot x}^\alpha\,d\tau\,\delta\underset{n}{x}^\beta\Big]_{\bar\tau\,\tau}\Big\},\\ \delta S_n^{\text{part}} &= m_n\int_\tau\frac{d}{d\tau}\big(\eta_{\alpha\beta}\,\underset{n}{\dot x}^\alpha\big)\,\delta\underset{n}{x}^\beta\,d\tau-m_n\Big[\eta_{\alpha\beta}\,\underset{n}{\dot x}^\alpha\,\delta\underset{n}{x}^\beta\Big]_\tau. \end{aligned}\right\}\tag{24.2}$$

---

[1] See vols. V/2 and XLIII of this Encyclopedia.

[2] J. Rzewuski: Field Theory, Part I, Classical Theory, Państwowe Wydawnictwo Naukowe, Warszawa, 1958. — A.D. Fokker: Z. Physik **58**, 386 (1929). — J.A. Wheeler and R.P. Feynman: Rev. Mod. Phys. **21**, 425 (1949).

In these expressions the dot denotes differentiation with respect to the proper time. Next we shall add the different contributions together and factor the variations of the individual particle coordinates. The result of this operation is:

$$\left.\begin{aligned} \delta S &= \sum_n \Big\{ \int_{\tau_n} \underset{n}{L}_\alpha \, \delta \underset{n}{x}^\alpha \, d\tau_n + \Big[\underset{n}{P}_\alpha \, \delta \underset{n}{x}^\alpha\Big]_{\tau_n} \Big\}, \\ \underset{n}{L}_\alpha &= \eta_{\alpha\beta} \Big(m_n \underset{n}{\ddot{x}}^\beta + q_n \underset{n}{\dot{x}}^\varrho \underset{n}{\Phi}_\varrho{}^\beta\Big), \qquad \underset{n}{P}_\alpha = -\, \eta_{\alpha\beta}\Big(m_n \underset{n}{\dot{x}}^\beta + q_n \underset{n}{\Phi}^\beta\Big). \end{aligned}\right\} \tag{24.3}$$

$\underset{n}{\Phi}_\varrho$ and $\underset{n}{\Phi}_{\varrho\sigma}$ have the following meaning:

$$\left.\begin{aligned} \underset{n}{\Phi}_\varrho &= \eta_{\varrho\tau} \sum_{m \neq n} q_m \int_{\tau_m} F(\tau_{mn}^2) \, \underset{m}{\dot{x}}^\tau(\tau_m) \, d\tau_m \equiv \eta_{\varrho\tau} \sum_{m \neq n} q_m \int_{\tau_m} F(\tau_{mn}^2) \, d\underset{m}{x}^\tau(\tau_m), \\ \underset{n}{\Phi}_{\varrho\sigma} &= \underset{n}{\Phi}_{\varrho,\sigma} - \underset{n}{\Phi}_{\sigma,\varrho}. \end{aligned}\right\} \tag{24.4}$$

If we were to use the language of electromagnetic theory, we should call these quantities the "incident" electromagnetic potentials and field strengths, respectively, incident on the n-th particle.

The action (24.1) is manifestly Lorentz-invariant. If we assume that the equations of motion, $\underset{n}{L}_\alpha \equiv 0$, are satisfied, then it follows that

$$\sum_n \Big[\underset{n}{P}_\alpha \, \delta \underset{n}{x}^\alpha\Big]_{\tau_n} = 0 \tag{24.5}$$

if the $\delta \underset{n}{x}^\alpha$ now represent the expression (23.1), corresponding to an infinitesimal Lorentz transformation. Separating the coefficients of the ten constants that describe such a transformation we obtain the relations:

$$\sum_n \Big[\underset{n}{P}^\alpha\Big]_{\tau_n} = 0, \qquad \sum_n \Big[\underset{n}{x}^\alpha \underset{n}{P}^\beta - \underset{n}{x}^\beta \underset{n}{P}^\alpha\Big]_{\tau_n} = 0. \tag{24.6}$$

These relations hold regardless of the $2N$ arbitrary limits we impose on the N individual trajectories. The limits τ_n are independent of each other insofar as the joint parameter ϑ, which we introduced originally, may, in the range of values (say from 0 to 1) that we assign to it in the expression (24.1), run through quite arbitrary pieces of the particles' individual trajectories. However, each of the $\underset{n}{P}^\alpha$, computed at the end points of the chosen range of the proper time τ_n, depends also on the ranges τ_m established for the other particles, according to the defining expressions (24.3), (24.4). We can speak of 10 N constants of the motion only in the sense that any of the ten expressions (24.6) taken at the upper limit equals the same expression taken at the lower limit (but each of the potentials $\underset{n}{\Phi}_\varrho$ depends on both limits) and that, accordingly, the derivative of any of these ten expressions with respect to either of the limits of any one τ_n vanishes.

But fundamentally the situation here is quite different from that prevailing in classical mechanics. Because the force exerted on one particle by all the others at one point of its trajectory depends on the whole of the trajectories of the latter, our equations of motion are profoundly "non-local". If the ranges over which we extend the action integral S of Eq. (24.1) are limited, then the incident potentials and field strengths, Eq. (24.4), depend on the ranges adopted. If the ranges of integration extend from the infinite past to the infinite future, then we must in general know the whole past and future history of each particle to calculate the instantaneous force on one.

Thus the equations of motion are integro-differential equations that do not lend themselves to the usual treatment in terms of a Cauchy initial-value problem. Very little is known about the actual variety of solutions, though KRZYWICKI,

RZEWUSKI, ZAMORSKI, and ZIEBA[1] have offered an argument according to which the solutions of this type of equations depend on $6N$ arbitrarily chosen parameters[2].

Typically, relativistic actions at a distance have been obtained from field theories when the differential equations that describe the propagation of the field were replaced by those GREEN's functions (or propagators) that describe the field in terms of an integral transform of the source distribution, assuming some reasonable boundary conditions. If the field propagates with the speed of light, then the function F is non-zero only on the light cone, and the integrals (24.4) actually sample the trajectory of every other particle only at two points, the one in the past from which a light signal could have reached the particle under consideration, and the point in the future which a light signal from the particle being considered could have reached. These two points correspond to the retarded and the advanced LIÉNARD-WIECHERT potentials.

If the field to be eliminated from the description is of the Klein-Gordon type, the corresponding propagator is a well-defined function $\overline{\Delta}(x^{\varrho})$ ([*10*], p. 286), which vanishes outside the light cone but is non-zero both on the light cone and for all positive values of the argument τ^2. The total incident potential between two particles permanently at rest relative to each other is of the Yukawa type, $e^{-\varkappa r}/r$.

There is no a priori grounds against considering functions F that are non-zero for space-like distances, i.e. for negative values of the argument τ^2. In such a theory, however, one particle would be sensitive to the details of the trajectory of another particle in a region where no light signal could travel in either direction; this case has not been considered. The lack of "causality" that attends this hypothetical case should, however, perhaps not be taken too seriously, as the solutions of non-local equations of motions do not lend themselves to the analysis that is normally associated with Cauchy problems, in which the notions of "cause" and "effect" have an immediate and intuitive significance.

With these remarks we shall leave the topic of relativistic action at a distance and turn to the further formal development of relativistic mechanics in analogy to its classical predecessor.

25. Relativistic Hamiltonian formalism. In developing relativistic analytical mechanics, we may follow two formally distinct paths, whether it be Lagrangian or Hamiltonian mechanics. One approach retains the distinct role of the time coordinate and assures relativistic invariance by starting out with a Lorentz-invariant action integral. This approach can be justified on the grounds that the requirement of Lorentz covariance does not increase the number of degrees of freedom of a system; thus by representing the $3N$ configuration coordinates of a mechanical system, or its $6N$ canonical coordinates, as functions of the independent variable t, the time, we have just the right number of separate variables which, for instance, may be chosen at will at an initial instant t_0, to assure a unique continuation for the whole range of values of t. In the Lagrangian formalism the Lorentz covariance is assured by the Lorentz invariance of the action integral; in the Hamiltonian formalism we must construct the generators of infinitesimal Lorentz transformations, confirm that they are constants of the motion, and also confirm that among themselves they obey a Lie algebra (have Poisson brackets) that corresponds to the commutator structure of the infinitesimal Lorentz group.

---

[1] Annales Polonici Mathematici II **1**, 77 (1955).

[2] JAN RZEWUSKI: Field Theory I, Warszawa, Państwowe Wydawnictwo Naukowe, 1958, discusses theories with relativistic actions at a distance fairly extensively [*10*].

However, if we wish to make Lorentz covariance of the theory manifest in a more direct manner, we shall choose a four-dimensional formalism. Whereas in the formalism that involves the configuration coordinates and the time differently, the only "natural" Cauchy problem is one in which all the dependent variables are given at the same coordinate instant t_0, the four-dimensional formalism is suited equally well to all Cauchy problems in which the initial data are given on some space-like hypersurface in a $(3N+1)$-dimensional, or even in a $4N$-dimensional configuration space. But to set up any Cauchy problem at all, we must introduce, in addition to the space-time coordinates, which appear symmetrically, an additional parameter ϑ, or perhaps a separate parameter ϑ_n for each particle, which will permit the parametric representation of the trajectories. As the choice of parameter is not restricted, it follows that the equations of motion must be parameter-invariant, in addition to being Lorentz-covariant.

Even in the Lagrangian formalism it must be impossible to solve the equations of motion for the highest occurring derivatives, as the transformation of the parameter will change the value of these highest derivatives. Thus, the Lagrangian equations of motion will satisfy one algebraic identity for each parameter introduced. Take the force-free particle as an example. Its equations of motion are:

$$x^{\varrho\prime\prime} - \dot{x}^{\varrho}\,\dot{x}_{\alpha}\,x^{\alpha\prime\prime} = 0, \tag{25.1}$$

where the prime denotes differentiation with respect to the parameter ϑ, and the dot denotes differentiation with respect to proper time. If we multiply this set of four equations by $\dot{x}_{\varrho}$, the product vanishes identically. In other words, one of the four equations may be expressed identically in the three remaining equations. Similar identities hold in the general case as well. For instance, the equations of motion of Sect. 24 contain one identity for each set of four equations belonging to one particle.

If we proceed to construct a Hamiltonian theory based on a four-dimensional Lagrangian theory, we find that the existence of identities between the Lagrangian equations of motion leads to pecularities of the Hamiltonian theory as well. Starting with a Lagrangian that satisfies the Euler relation (22.10) we find that the Hamiltonian, formed according to the usual prescription, vanishes. Nevertheless, we can introduce a function of the arguments x^{ϱ}, p_{ϱ} (for one particle, or a correspondingly greater range of the index for several mass points) that yields the equations of motion.

Given a function of N arguments $\xi^{n\prime}$ that is homogeneous of the first degree (such as our Lagrangian); then its derivatives with respect to these arguments will be functions homogeneous of the zeroth degree. It is a well known theorem that there are only $(N-1)$ algebraically independent homogeneous, zeroth-degree functions of N arguments. Given the N zeroth degree functions

$$p_n = \frac{\partial L}{\partial \xi^{n\prime}}, \tag{25.2}$$

the canonical momentum components, their N^2 partial derivatives with respect to the "velocities" $\xi^{n\prime}$ form N linearly dependent "vectors" (in the vector space contragradient to the one of which the "velocities" form the base vectors) as evidenced by the identities

$$\xi^{n\prime}\frac{\partial p_n}{\partial \xi^{m\prime}} \equiv \xi^{n\prime}\frac{\partial^2 L}{\partial \xi^{m\prime}\,\partial \xi^{n\prime}} \equiv \xi^{n\prime}\frac{\partial p^{m\prime}}{\partial \xi^{n\prime}} \equiv 0, \tag{25.3}$$

which are a direct consequence of the homogeneity of the Lagrangian function L.

It follows that the $2N$ variables ξ^n, p_n satisfy an algebraic relationship of the form

$$H(\xi^n, p_n) = 0, \tag{25.4}$$

which turns into an identity if we substitute the expressions (25.2) for the momentum components. In other words, we have the two sets of relationships

$$\left.\begin{aligned} \frac{\partial H}{\partial p_m} \frac{\partial^2 L}{\partial \xi^{m\prime} \partial \xi^{n\prime}} &= 0, \\ \frac{\partial H}{\partial \xi^n} + \frac{\partial H}{\partial p_m} \frac{\partial^2 L}{\partial \xi^n \partial \xi^{m\prime}} &= 0, \end{aligned}\right\} \tag{25.5}$$

identically satisfied modulo the Eqs. (25.2). With their help we can demonstrate the existence of Hamiltonian equations of motion.

We shall assume that the quadratic matrix of second derivatives

$$L_{mn} \equiv \frac{\partial^2 L}{\partial \xi^{m\prime} \partial \xi^{n\prime}} \tag{25.6}$$

possesses only one null vector, the one given by Eq. (25.3)[1]. In that case comparison of the first set of Eqs. (25.5) with (25.3) leads to the conclusion

$$\frac{\partial H}{\partial p_n} = \varkappa \xi^{n\prime}. \tag{25.7}$$

In view of the fact that Eq. (25.4) determines H only up to an arbitrary nonzero factor, we shall set the constant of proportionality equal to unity; thus we obtain the first set of Hamiltonian equations,

$$\xi^{n\prime} = \frac{\partial H}{\partial p_n}. \tag{25.8}$$

But because of the Euler relation (22.10), we can now interpret the second set of Eqs. (25.5) to read:

$$\frac{\partial H}{\partial \xi^n} + \frac{\partial L}{\partial \xi^n} = 0. \tag{25.9}$$

Out of the original Lagrangian equations of motion,

$$\frac{\partial L}{\partial \xi^n} - \frac{d p_n}{d \vartheta} = 0, \tag{25.10}$$

we obtain the second set of Hamiltonian equations of motion,

$$p_n' = -\frac{\partial H}{\partial \xi^n}. \tag{25.11}$$

Thus we find that in a theory with homogeneous Lagrangian the Hamiltonian is that vanishing function of the canonical variables (25.4), that exists by virtue of the homogeneity itself. In addition to the usual equations of motion we have the requirement that the Hamiltonian itself vanish, a requirement that will continue to hold by virtue of the equations of motion if it is satisfied in phase space at one point along the trajectory.

The phase space itself appears $2N$-dimensional. But, because of the algebraic constraint (25.4), physically meaningful trajectories are confined to a $(2N-1)$-

[1] A Lagrangian that does not satisfy this assumption will lead to Euler-Lagrange equations with identities in addition to the one required by its homogeneity, of the type (25.1). The consideration of such theories involves complications unrelated to the task of developing a relativistic analytical mechanics.

dimensional hypersurface. The Hamiltonian is determined only up to an arbitrary factor; the equations of motion generated by two possible Hamiltonians will differ from each other in so far as each time derivative in one set of equations will equal its corresponding time derivative in the other set except for a joint factor of proportionality (which may vary along the trajectory). Hence the choice of a particular Hamiltonian among all those permissible amounts to the adoption of a particular parameter ϑ.

Finally, a particular variable ξ^n may be given the role of independent variable by that choice of Hamiltonian which makes $(\partial H/\partial p_n)$ for the chosen n equal to unity. H will then take the form:

$$H = p_n + H^*(\xi^m, p_m; \xi^n), \qquad m \neq n. \tag{25.12}$$

It is evident that adoption of the new function H^* as the Hamiltonian of a theory in which ξ^n is considered the independent variable and the canonical coordinates number $2(N-1)$ leads to equations of motion equivalent to the original parametric equations and that in this version we have recovered a theory that resembles the conventional non-relativistic formalism.

26. Dirac commutators. In ordinary analytical mechanics the canonical transformations are defined as those coordinate transformations in phase space that preserve the Hamiltonian form of the equations of motion. The Poisson brackets represent the commutator algebra of the infinitesimal group of canonical transformations, in terms of their generators. In the parametrized theory the Poisson brackets lead to a certain degree of ambiguity, because of the appearance of the Hamiltonian constraint $H=0$. Suppose that two functions of the canonical variables, $A(\xi, p)$, $\bar{A}(\xi, p)$, differ only by some multiple of H, so that their values are equal in every permissible situation. Then the two Poisson brackets (A, B) and $(\bar{A}, B)$, B being arbitrary, will still differ from each other, and in general by a non-vanishing amount.

In the presence of one or several constraints it is possible to modify the concept of Poisson brackets so that this ambiguity, and incidentally other difficulties, are eliminated[1]. The principle of these Dirac commutators can be illustrated for the case of the Hamiltonian constraint, which appears as a result of the parametrization of the theory. Because of the constraint, we are not concerned with the whole $(2N)$-dimensional phase space spanned by the coordinates (ξ, p) but with a subspace, or hypersurface, with $2N-1$ dimensions, defined by the constraint $H=0$. All physical trajectories lie wholly within this subspace; in fact, any trajectory that has one point in common with the constraint subspace lies wholly within it, as the constraint itself is a constant of the motion. Accordingly, we may restrict ourselves to the consideration of canonical transformations that map the constraint subspace on itself.

Infinitesimal canonical transformations in general will change the value of H in accordance with the transformation law

$$\bar{\delta} H = (C, H). \tag{26.1}$$

To require, then, that the subspace $H=0$ be mapped on itself is equivalent to subjecting generators to the condition

$$(C, H) = 0. \tag{26.2}$$

[1] P. A. M. DIRAC: Canad. J. Math. **2**, 129 (1950); **3**, 1 (1951). — P. G. BERGMANN and I. GOLDBERG: Phys. Rev. **98**, 531 (1955).

In both these equations the symbol (,) represents the usual Poisson bracket. There are $2(N-1)$ non-trivial and algebraically independent functions C that satisfy the condition (26.2).

The constraint subspace is mapped on itself by the totality of the constants of the motion of the mechanical system. Because time no longer plays a special role, we need make no distinction between time-dependent and time-independent constants of the motion; the parameter ϑ has no physical significance, hence no dynamical variable depends on ϑ. The constants of the motion form a function group, in the sense of LIE, that is to say, commutators between constants of the motion are again constants of the motion. Their "prediction" from given initial-value data is trivial.

The remaining two dimensions of phase space are described by the function H (which is constrained to vanish) and by its canonical conjugate, the "time". Clearly, time as a function of ϑ cannot be predicted, as the choice of ϑ is arbitrary. Hence our set of $(2N-2)$ permissible variables represents the physically interesting range of quantities, and nothing is lost if we restrict our attention to them. These variables are now often referred to as the "observables" of the theory. They alone turn into Hilbert space operators in the quantum theory of the mechanical system.

In the section on quantum mechanics, and in the article on general relativity, we shall consider systems with more elaborate sets of constraints. In the more general case the construction of the transformation group that preserves the constraints, and their commutator algebra, requires more powerful methods; in particular the dimensionality of the transformation space depends not merely on the number of constraints imposed but also on their classification according to DIRAC (first-class and second-class constraints). This complete theory is unnecessary for our present purposes. Its development will therefore be deferred to Sect. 39.

II. Continuum mechanics.

27. Relativistic perfect gas. In classical mechanics, we treat a mechanical system that consists of a very large number of interacting particles (molecules) in terms of certain macroscopic parameters, such as the density, flow velocity, and stress, to which we frequently add such thermodynamic quantities as temperature, entropy density, and the like. Presumably, the macroscopic dynamical laws may be obtained from the detailed theory of the microscopic interactions by means of an averaging procedure that permits us to discard those microscopic dynamic variables that are not ordinarily observed. The relativistic mechanics of continuous matter resembles the non-relativistic continuum mechanics insofar as its laws are also assumed to be consistent with the microscopic dynamics of the constituent particles. The chief complication stems from the circumstance that readily available relativistic laws of interaction are somewhat harder to come by than the corresponding Newtonian laws. Essentially, the only cases that lend themselves to treatment are the perfect gas, which consists of non-interacting mass points, and the plasma, which involves electromagnetic interactions.

In Sect. 29 we shall present a micro-model that involves interaction. Before doing so we shall concern ourselves with the much simpler case of the perfect gas. It is, of course, tempting to forgo the construction of a micro-model and to construct continuum mechanics from conservation laws, augmented by some assumed equation of state. Such a procedure, however, fails to assure us that the conjectured equation of state is consistent with any micro-model, regardless of whether this model is known to us. It is conceivable that a conjectured equation of state may lead to elastic waves that permit the propagation of signals with a velocity greater than the speed of light. Even if this contingency be avoided by the assumption of a sufficiently small elastic modulus, other difficulties might arise, not readily recognizable but nevertheless fatal to the model. That is why EINSTEIN has warned against this macroscopic-axiomatic approach.

Let us start with a perfect gas that consists of a mixture of mass points of all non-negative masses. The Hamiltonian constraint of a single constituent particle may be given the form

$$H = \tfrac{1}{2}(\eta^{\varrho\sigma} p_\varrho p_\sigma - m^2), \qquad p_\varrho = \eta_{\varrho\sigma} m \dot{x}^\sigma. \tag{27.1}$$

In that form the parameter turns out to be the proper time along the trajectory. If we consider a collection of particles of different masses m, the covariant vectors p_ϱ are all time-like but otherwise not confined to a particular hypersurface. We shall consider such a collection, and their trajectories, in an eight-dimensional phase space whose coordinates are x^ϱ, p_ϱ. This phase space corresponds to the μ-space in the terminology of statistical mechanics. Because we consider particles of different masses, trajectories may pass through every point of that phase space that corresponds to time-like linear momentum. Just as in ordinary phase spaces, trajectories cannot intersect. Within the domain of time-like p-vectors, one and only one possible trajectory passes through every point. We have thus a congruence of trajectories that covers the accessible portion of our phase space.

The actual collection of particles will occupy this family of trajectories with a variable density. As we are not measuring the density of points but of curves (trajectories), the density is not a scalar but a contravariant vector field (strictly speaking, a vector density field; but as canonical transformations in phase space have the determinant 1, the distinction between vectors and vector densities is moot). Moreover, the vector must at each point of phase space have the direction of the curves themselves. It may thus be put into the form $f(x^\varrho, p_\varrho)\,\dot{\xi}^A$, where ξ^A is used as the symbol for the *eight* coordinates of our phase space and the index A goes from 1 to 8. The symbol $\dot{\xi}^A$ thus stands for the eight time derivatives $\dot{x}^\varrho, \dot{p}_\varrho$, of which the latter four, of course, vanish if the equations of motion are satisfied.

If we wish to count the number of particles in any domain, we must carry out an integration over a seven-dimensional hypersurface. We introduce as the parameters of our hypersurface the four components of linear momentum and any three parameters describing a three-dimensional hypersurface in ordinary Minkowski space. We shall denote the three-dimensional hypersurface element in Minkowski space by the usual symbol dS_ϱ. Our seven-dimensional surface element $d\Sigma_A$ then has the components

$$d\Sigma_A = (0, d^4p\, dS_\varrho). \tag{27.2}$$

The first four components vanish, as our choice of parameters determines the orientation of the seven-dimensional hypersurface to that extent. The number of trajectories intersected by our seven-dimensional surface then is

$$N = \iint f(x, p)\, d^4p\, \dot{x}^\varrho\, dS_\varrho \tag{27.3}$$

and will, of course, depend on the domain of integration. However, this number must not change if we deform our hypersurface without changing its boundary, and that implies that the seven-dimensional divergence of the integrand vanishes, because of GAUSS's theorem. Accordingly, we have, as a special case of LIOUVILLE's theorem,

$$p^\varrho \frac{\partial f}{\partial x^\varrho} = 0, \qquad \dot{x}^\varrho \frac{\partial f}{\partial x^\varrho} = 0, \tag{27.4}$$

as a condition on the function f, which otherwise may be chosen at will. Physically, this restriction means that the density of trajectories cannot change along the direction of the trajectories themselves.

If we endow the particles with some permanent property, such as some function of their rest mass (which will be different for different particles), to be designated by σ, then the total amount of that property will be

$$Q = \iint f \sigma \, d^4 p \, \dot{x}^\varrho \, dS_\varrho . \tag{27.5}$$

There will be a local flux j^ϱ, which we obtain by integrating $f\sigma\dot{x}^\varrho$ over the momentum components but not the Minkowski hypersurface element dS_ϱ, and we get:

$$j^\varrho = \int f \sigma \dot{x}^\varrho \, d^4 p = \int \frac{1}{m} f \sigma p^\varrho \, d^4 p . \tag{27.6}$$

According to Eq. (27.4), this flux satisfies the *equation of continuity*.

$$j^\varrho{}_{,\varrho} = 0 . \tag{27.7}$$

One of the permanent properties of a particle in the force-free case is its linear momentum. Moreover we know that the vector sum of the linear momentum over the individual particles remains conserved even in collisions. The total linear momentum of our collection of particles is

$$P^\alpha = \iint f p^\alpha \dot{x}^\varrho \, d^4 p \, dS_\varrho = \int t^{\alpha\varrho} \, dS_\varrho . \tag{27.8}$$

The integrand,

$$\left.\begin{aligned} t^{\alpha\varrho} &= \int f p^\alpha \dot{x}^\varrho \, d^4 p \\ &= \int f \frac{p^\alpha p^\varrho}{m} d^4 p , \qquad t^{\alpha\varrho} = t^{\varrho\alpha} , \end{aligned}\right\} \tag{27.9}$$

again satisfies the equation of continuity,

$$t^{\alpha\varrho}{}_{,\varrho} = 0 . \tag{27.10}$$

The symmetric tensor $t^{\mu\nu}$, which represents the flux of linear momentum, is called the energy-momentum tensor. If we place our seven-dimensional hypersurface of integration so that its normal is parallel to the p_0-direction, then its only nonvanishing component has the form $d^4 p \, d^3 x$, and the only non-vanishing component of dS_ϱ is $dS_0 = d^3 x$. Hence the expression $t^{\alpha 0}$ represents the ordinary spatial density of the quantity P^α; t^{00} is the density of (relativistic) mass or energy, and t^{n0} the three components of the density of relativistic three-dimensional momentum.

If we modify our force law in such a way as to permit the exchange of linear momentum in collisions but not otherwise, Eqs. (27.9), (27.10) remain valid. As for the latter of these two relationships, Eq. (27.10) merely expresses the conservation of (total) linear momentum in configuration space. As long as exchange of linear momentum is confined to particles in close proximity to each other, the rate of change of linear momentum density at any (three-dimensional) space point is determined by the divergence of the (three-dimensional) linear momentum flux.

We can also set up expressions for the angular momentum flux. The tensor

$$W^{\mu\nu\varrho} = x^\mu t^{\nu\varrho} - x^\nu t^{\mu\varrho} , \qquad W^{\mu\nu\varrho}{}_{,\varrho} = 0 , \tag{27.11}$$

represents this (four-dimensional) flux and satisfies the required equatoin of continuity, because of Eq. (27.10) and because of the symmetry of $t^{\mu\nu}$.

28. Local velocity and stress. In non-relativistic fluid mechanics we assign to a streaming fluid a local velocity, which is defined as the linear momentum density divided by the mass density. The local velocity not only appears in the equation of continuity for the mass, it is also required to separate in the total flux of momentum density that portion that is accounted for by the macroscopic motion of the fluid from the remainder, the stress. The macroscopic stress, in turn, contains both the contributions to the linear momentum flux by the random motion of the constituent particles and the transfer of linear momentum by the forces of interaction, which in the case of the perfect gas are of course zero.

In relativistic mechanics the introduction of a local velocity concept is somewhat ambiguous. Following the procedure of the non-relativistic theory, we may define the local velocity $\boldsymbol{u}$, with components u_k, as the ratio between the linear momentum density t^{0k} and the relativistic mass density t^{00},

$$u_k = \frac{t^{0k}}{t^{00}}. \tag{28.1}$$

To this three-dimensional vector there corresponds a four-dimensional formation

$$U^k = \frac{u_k}{\sqrt{1-\frac{\boldsymbol{u}^2}{c^2}}}, \qquad U^0 = \frac{1}{\sqrt{1-\frac{\boldsymbol{u}^2}{c^2}}} \tag{28.2}$$

whose norm equals 1. This definition for the local velocity has been adopted on occasion, but U^ϱ is not a world vector. Worse yet, its four components do not form a "geometric object"; that is to say, knowledge of the four components U^ϱ in one Lorentz frame does not enable us to determine their values in another Lorentz frame, either by the law of the transformation of vectors or by any other transformation law. The underlying geometric object is the energy-momentum tensor $t^{\mu\nu}$; we need to know all of its 10 components in one Lorentz frame in order to calculate the four U^ϱ in the same, or any other, Lorentz frame.

Under the circumstances, the usefulness of the definition (28.1) for the local velocity is open to question. As an alternative, we may define a time-like unit vector V^ϱ by requiring it to be an eigenvector of the symmetric tensor $t^{\mu\nu}$:

$$t^{\mu\nu} V_\nu = \mu V^\mu, \qquad \eta_{\varrho\sigma} V^\varrho V^\sigma = 1. \tag{28.3}$$

The eigenvalue μ might be called the local proper mass density. If we adopt a Lorentz frame in which V^ϱ has the components (0, 0, 0, 1), then in that Lorentz frame the component t^{00} equals μ and the components t^{0s} ($s=1, 2, 3$) vanish. The components t^{rs}, however, will normally be different from zero. The difference

$$\sigma^{\mu\nu} = t^{\mu\nu} - \mu V^\mu V^\nu \tag{28.4}$$

is a tensor and may be considered the stress. Obviously, its scalar product by the vector V^ϱ vanishes.

The definition (28.4) for the stress tensor may appear very formal. For its defense one of its properties may be cited: For the perfect gas the stress is a non-negative quantity, that is, the scalar $\sigma^{\mu\nu} w_\mu w_\nu$, with arbitrarily chosen vector w_μ, is non-negative. For the proof, which will not be presented here, the use of the explicit defining equation (27.9) is essential. In other words, the non-negativity of the stress tensor (28.4) depends on the stress being wholly kinetic. If there were attractive forces, it is perfectly possible for the stress to become negative (cohesive).

With the introduction of the stress tensor, the divergence relationship (27.10) may be given a special form. We have

$$0 = (\mu V^\mu V^\varrho)_{,\varrho} + \sigma^{\mu\varrho}{}_{,\varrho} = \mu V^\mu{}_{,\varrho} V^\varrho - (\mu V^\varrho)_{,\varrho} V^\mu + \sigma^{\mu\varrho}{}_{,\varrho}. \tag{28.5}$$

If we multiply this equation by V_μ, we obtain the equation

$$(\mu V^\varrho)_{,\varrho} + V_\mu \sigma^{\mu\varrho}{}_{,\varrho} = 0, \tag{28.6}$$

which in turn may again be substituted into Eq. (28.5). The end result of this manipulation is

$$\mu V^a{}_{,\varrho} V^\varrho + (\delta^\alpha_\beta - V^\alpha V_\beta)\, \sigma^{\beta\varrho}{}_{,\varrho} = 0. \tag{28.7}$$

Eqs. (28.6), (28.7) together constitute the equations of motion in a manner easy to interpret. In a Lorentz frame whose time axis is parallel to V_ϱ, only three components of Eqs. (28.7) are not empty. These state that the mass density times the acceleration is determined by the divergence of the (three-dimensional) stress tensor,

$$\mu \dot{u}_k + \sigma_{kl,l} = 0. \tag{28.8}$$

The single Eq. (28.6) in this special frame determines the rate of change of μ,

$$\dot{\mu} + (\mu u_l + u_s \sigma_{sl})_{,l} = 0. \tag{28.9}$$

The last term represents the divergence of the flux of mechanical work, which appears side by side with the divergence of the flux of proper mass.

29. Interactions. Though ordinarily the relativistic theory of continua with internal forces is developed either with the help of fields or with the assumption that non-relativistic mechanics is valid at each world point in a "co-moving" frame of reference, neither of these procedures is necessary. Treatment with the help of fields as the intermediaries of interaction is probably physically most realistic, as very likely all interactions between particles take place through intermediary fields[1]. The assumption of the validity of non-relativistic continuum mechanics in co-moving coordinate systems is, however, quite objectionable. Physically, such an assumption implies that in any reasonably small spatial domain the range of velocities of the constituent particles is small compared to the speed of light, in other words, that matter there is fairly cold. Further, it is assumed that intersecting jets of matter, as they might be anticipated in the collision of cosmic clouds, do not lead to velocity gradients comparable to c over distances that are not large compared to the mean free path. Clearly, such assumptions and restrictions remove from scrutiny just those situations in which a consistently relativistic treatment might be needed. In this section we shall give a treatment that is based on relativistic action at a distance.

We begin with a collection of particles of varying states of motion which are capable of interacting with each other by means of potentials of the type introduced in Sect. 24. If we consider the law of motion of one single particle under the influence of the cloud of surrounding particles, we may write it in the form

$$\left.\begin{aligned} \frac{dp_\varrho}{d\tau} &= -\frac{\partial H}{\partial x^\varrho}, \qquad \frac{dx^\varrho}{d\tau} = \frac{\partial H}{\partial p_\varrho} = \eta^{\varrho\mu}(p_\mu - F_\mu), \\ H &= \frac{1}{2m}\left[\eta^{\mu\nu}(p_\mu - F_\mu)(p_\nu - F_\nu) - m^2\right], \end{aligned}\right\} \tag{29.1}$$

[1] This question can be discussed fully only in the article on general relativity.

where the vector F_ϱ is given by the integral

$$F^\varrho(x) = \iint f(\bar{x}, \bar{p})\, F(x, \bar{x})\, \frac{\partial H(\bar{x}, \bar{p})}{\partial p_\varrho}\, d^4\bar{p}\, d^4\bar{x}. \tag{29.2}$$

The density function $f(x, p)$ has the same significance as in Sect. 28. F is to be an invariant function of the space-time interval between the two world points x and $\bar{x}$, of whom the first one is fixed, whereas $\bar{x}$ is one of the two sets of variables of integration. We have again LIOUVILLE's theorem, to the effect that $(f, H) = 0$.

In order to obtain a dynamical law, we shall consider a three-dimensional hypersurface S. This three-dimensional surface will intersect a certain amount of linear momentum, P_α,

$$P_\alpha = \iint\limits_{S, p} f(x, p)\, p_\alpha \frac{\partial H}{\partial p_\varrho}\, d^4 p\, d S_\varrho, \tag{29.3}$$

according to Eq. (27.8). If we displace a limited domain of this surface by a small amount described by the displacement vector field ξ^ν, which by assumption vanishes on the boundary of S, then the linear momentum intersected by the new surface will differ by an amount δP_α. If there were no change in the linear momentum of any member particle, this change would vanish, as the distorted surface intersects exactly the same trajectories as the original surface does. Hence the total of δP_α is given by the change of all the individual linear momenta through interaction, and we have the relationship

$$\delta P_\alpha = - \iint f \frac{\partial H}{\partial x^\alpha}\, d^4 p\, \xi^\sigma\, d S_\sigma. \tag{29.4}$$

On the other hand, because of GAUSS's theorem, we can express the change in P_α as the divergence of the integrand of Eq. (29.3),

$$\delta P_\alpha = \int\limits_S \frac{\partial}{\partial x^\varrho} \left[\int\limits_p p_\alpha f \frac{\partial H}{\partial p_\varrho}\, d^4 p \right] \xi^\sigma\, d S_\sigma. \tag{29.5}$$

We conclude that at a given world point we have the relationship:

$$\frac{\partial}{\partial x^\varrho} \left(\int\limits_p p_\alpha f \frac{\partial H}{\partial p_\varrho}\, d^4 p \right) + \int\limits_p f \frac{\partial H}{\partial x^\alpha}\, d^4 p = 0. \tag{29.6}$$

We shall now convert the second term into the divergence of a tensor as well. We have, because of the form (29.1) of H,

$$\frac{\partial H}{\partial x^\alpha} = -\frac{1}{m} (p^\varrho - F^\varrho)\, F_{\varrho,\alpha} = -\frac{\partial H}{\partial p_\varrho} F_{\varrho,\alpha} \tag{29.7}$$

and hence, because of the expression (29.2),

$$\left.\begin{aligned} &\int\limits_p f \frac{\partial H}{\partial x^\alpha}\, d^4 p = \\ &\quad - \int\limits_p \int\limits_{\bar{p}} \int\limits_{\bar{x}} f(x, p)\, f(\bar{x}, \bar{p})\, \frac{\partial H(x, p)}{\partial p_\varrho}\, \eta_{\varrho\sigma}\, \frac{\partial H(\bar{x}, \bar{p})}{\partial \bar{p}_\sigma}\, \frac{\partial}{\partial x^\alpha} F(x, \bar{x})\, d^4 p\, d^4\bar{p}\, d^4\bar{x}. \end{aligned}\right\} \tag{29.8}$$

In the twelve-fold integral on the right we shall for abbreviation introduce the quantity

$$\int\limits_p f(x, p)\, \frac{\partial H(x, p)}{\partial p_\varrho}\, d^4 p = u^\varrho(x). \tag{29.9}$$

With its help we may write

$$\left.\begin{aligned}\int_p f \frac{\partial H}{\partial x^\alpha}\, d^4 p &= + \int_{\bar{x}} \eta_{\varrho\sigma}\, u^\varrho(x)\, u^\sigma(\bar{x})\, \frac{dF(x_1, \bar{x})}{d\tau}\, \frac{\bar{x}_a - x_a}{\tau}\, d^4\bar{x},\\ \tau^2 &= \eta_{\mu\nu}\, (\bar{x}^\mu - x^\mu)\, (\bar{x}^\nu - x^\nu) = \eta_{\mu\nu}\, \xi^\mu \xi^\nu.\end{aligned}\right\} \tag{29.10}$$

We shall now introduce the coordinate difference $(\bar{x} - x)$ as the new variable of integration:

$$\int_p f \frac{\partial H}{\partial x^\alpha}\, d^4 p = \int_\xi u^\varrho(x)\, u_\varrho(x + \xi)\, \frac{dF(\tau)}{d\tau}\, \frac{\xi_a}{\tau}\, d^4\xi. \tag{29.11}$$

The last expression cannot be simplified further without some assumption concerning the structure of our population, that is concerning the dependence of the function $f(x, p)$ on its arguments. The nature of f reflects itself in the dependence of $u^\varrho(x)$ on *its* argument. We shall now assume that u^ϱ has a possibly strong local dependence on x, which reflects such physical situations as the periodicity of a crystalline arrangement of the atoms, or the structure induced in a viscous fluid by shear motion, but that this local dependence is modified by a further long-range dependence on x, which makes itself felt over distances large compared to the dimensions of the local structure. In other words, we shall consider that the product $u^\varrho(x)\, u_\varrho(x+\xi)$ if averaged over a patch which in each direction measures several distances between closest neighbors[1] depends strongly on ξ but only much more weakly, and over relatively large distances, on x. We shall use this local average and introduce it into the expression (29.11). The implication is that the net transfer of momentum into and out of a large (macroscopic) four-dimensional domain is caused by the dependence of the local average on macroscopic changes in the coordinates. We shall write:

$$\int_p f \frac{\partial H}{\partial x^\alpha}\, d^4 p = \int_\xi U(\xi, x)\, \frac{dF}{d\tau}\, \frac{\xi_\alpha}{\tau}\, d^4\xi, \qquad U(\xi, x) = \langle u^\varrho(x)\, u_\varrho(x + \xi)\rangle. \tag{29.12}$$

By renaming the variable ξ into $-\xi$ and taking half the sum of the original and the changed expressions for the integral, we can also write:

$$\int_p f \frac{\partial H}{\partial x^\alpha}\, d^4 p = \frac{1}{2} \int_\xi \frac{dF}{d\tau}\, \frac{\xi_\alpha}{\tau}\, [U(\xi, x) - U(-\xi, x)]\, d^4\xi. \tag{29.13}$$

U is, by its definition, an autocorrelation function. If it were independent of x, it would have to be an even function of ξ, and hence the integral (29.13) would vanish. In other words, that integral is non-zero only to the extent that there is some long-range dependence on x which cannot be eliminated by "local averaging". To bring out this deviation, we establish that, by virtue of its definition (29.12), the function U satisfies the relationship

$$U(-\xi, x) = U(\xi, x - \xi) \approx U(\xi, x) - \frac{\partial}{\partial x^\mu}\, U(\xi, x)\, \xi^\mu, \tag{29.14}$$

where we also have utilized the assumption of weak dependence of U on x and

[1] The distance between two particles may be defined invariantly as the proper distance between one fixed point on one trajectory and that point on the other trajectory for which the (space-like) interval is maximized.

replaced the function itself by the first two terms of its Taylor expansion. Our final result is:

$$\left.\begin{aligned} \int\limits_{p} f\frac{\partial H}{\partial x^{\alpha}}\,d^4p &= \frac{\partial}{\partial x^{\mu}}\left\{\frac{1}{2}\int\limits_{\xi}\frac{dF}{d\tau}\frac{\xi_{\alpha}}{\tau}\,U(\xi,x)\,\xi^{\mu}\,d^4\xi\right\} = \vartheta_{\alpha}{}^{\mu}{}_{,\mu}, \\ \vartheta^{\alpha\mu} &= \frac{1}{2}\int\frac{dF}{d\tau}\frac{\xi^{\alpha}\xi^{\mu}}{\tau}\,U(\xi,x)\,d^4\xi. \end{aligned}\right\} \tag{29.15}$$

We have now shown that it is possible to represent the dynamical law of a material continuum by a four-dimensional divergence relationship,

$$\frac{\partial}{\partial x^{\varrho}}\left\{\int\limits_{p} p_{\alpha}f\frac{\partial H}{\partial p_{\varrho}}\,d^4p + \frac{1}{2}\int\limits_{\xi}\frac{dF}{d\tau}\frac{\xi^{\alpha}\xi^{\varrho}}{\tau}\,U(\xi,x)\,d^4\xi\right\} = 0, \tag{29.16}$$

on the assumptions (1) of particles with relativistic action at a distance of the form (29.1), (29.2) with arbitrary local momentum distribution, and (2) of the existence of suitable "local averages", which make possible the formulation of a quasi-continuous flux of linear momentum along the lines we have indicated. This derivation takes into account only short-range forces but does not require that interacting particles have similar velocities. All integrations in momentum space have been carried out without restrictions.

Naturally, all objections raised in statistical mechanics against Boltzmann's „Stoßzahlansatz" hold also against the development given here. What we have achieved is to obtain an energy-momentum-stress tensor whose divergence vanishes and which, by virtue of its definition, represents the total transfer of momentum across a three-dimensional hypersurface. Our derivation is relativistically invariant at each step. Hence we have been relieved of the necessity of introducing locally non-relativistic conditions, e.g. small relative velocities of the constituent particles.

Herewith we shall conclude our presentation of relativistic particle mechanics. In the next part of this article we shall introduce the electromagnetic field, and with it another possible way of constructing an energy-momentum tensor.

III. Electrodynamics

30. Preliminaries. To carry out a complete formulation of Lorentz-covariant electrodynamics we shall start with Maxwell's field equations and the Lorentz-Maxwell expression for the ponderomotive force (2.3). We shall supplement these laws with the defining equations for the electromagnetic potentials,

$$\boldsymbol{B} = \operatorname{curl}\boldsymbol{A}, \qquad \boldsymbol{E} = -\operatorname{grad}\varphi - \frac{1}{c}\frac{\partial\boldsymbol{A}}{\partial t}, \tag{30.1}$$

and with the equation of continuity for electric charges,

$$\frac{\partial\sigma}{\partial t} + \operatorname{div}\boldsymbol{j} = 0, \tag{30.2}$$

which is identically satisfied if Maxwell's field equations hold. We shall find that the field equations and Eqs. (30.1), (30.2) are Lorentz-covariant as they stand provided we make some very plausible assumptions concerning the transformation properties of the quantities that occur in them. As for the ponderomotive law, on the left-hand side there is an intrinsically mechanical quantity, the force; provided we define the force appropriately, the ponderomotive equation is also Lorentz-covariant without modification. For many purposes, however, it is advantageous to give it a slightly different form.

It is most convenient to begin with the equation of continuity, partly because of the very intuitive and fundamental significance of the density of electric charge and electric current. The equation of continuity tells us that electric charge cannot be created or annihilated, that the rate of change of electric charge contained in a spatial volume is determined by the current passing through its surface. An incidental consequence of the conservation of charge is that the total charge of a body must be an invariant: If that body is brought from one state of motion to another by means of external forces, its charge cannot change in the absence of electric currents passing across its surface. In particular the electric charge of a particle (such as an electron or proton) cannot depend on its state of motion. Accordingly it is the same for all observers.

The total electric charge of a body is expressible as an integral having the form

$$Q = \int \sigma \, d^3 x . \tag{30.3}$$

Such an integral can have the same value in all Lorentz coordinate systems only if it is a special instance of the more general integral over a three-dimensional hypersurface in Minkowski space having the form

$$Q = \int j^\varrho \, d S_\varrho , \tag{30.4}$$

where j^ϱ is a contravariant vector subject to the divergence relationship

$$j^\varrho{}_{,\varrho} = 0 , \tag{30.5}$$

according to Sect. 15.

The expression (30.4) will go over into (30.3) if j^0 is identified with the electric charge density σ. Comparison of the divergence relationship (30.5) with the equation of continuity (30.2) then causes us to identify the remaining components of j^ϱ with the spatial electric current density. We shall make this identification; the contravariant vector j^ϱ is commonly referred to as the *charge-current density*. If this identification is physically correct, it follows that for a Lorentz transformation involving motion along the x-axis at the rate u the charge and current densities in the two frames of reference must be related to each other by the transformation equations:

$$\left.\begin{aligned} j'_x &= \left(1 - \frac{u^2}{c^2}\right)^{-\frac{1}{2}} (j_x + u\,\sigma) , \\ j'_y &= j_y , \; j'_z = j_z , \\ \sigma' &= \left(1 - \frac{u^2}{c^2}\right)^{-\frac{1}{2}} \left(\sigma + \frac{u}{c^2} j_x\right) . \end{aligned}\right\} \tag{30.6}$$

These equations are to be compared with the non-relativistic equations (2.5), last line. It is evident that the relativistic equations will go over into the non-relativistic equations if u/c is so small that all terms having any power of c in the denominator may be disregarded.

The transformation law (30.6) has one rather startling consequence. If a wire carries current while having zero charge density, a moving observer will ascertain a non-vanishing charge density as well. This seeming paradox (electric charge is invariant!) is to be explained by the relativity of simultaneity. The moving observer in determining the total charge in a piece of wire must add up all the charges inside that piece at a given instant in time. An instant in time at the two ends of that piece of wire for the moving observer will represent two different instants for the laboratory-connected observer. Hence the total number of conduction electrons found simultaneously within the specified piece of wire (which forms part of a closed electric circuit) will not be the same for the two observers.

31. Field intensities and potentials. If electric charge and current density form a contravariant world vector, then those of MAXWELL'S equations whose right-hand side consist of these quantities must also form the components of a four-vector relationship. This argument suggests that the magnetic field strength $\boldsymbol{H}$ and the electric displacement $\boldsymbol{D}$ together form a single world tensor, and, as a consequence, $\boldsymbol{B}$ and $\boldsymbol{E}$ another. Either of these two world tensors must have six algebraically independent components, which under spatial orthogonal transformations arrange themselves into one vector and one pseudovector. All these properties are possessed by skewsymmetric tensors of rank 2. A tensor $f_{\mu\nu}$ with the components

$$f^{\mu\nu} = \begin{pmatrix} 0 & f^{12} & -f^{31} & f^{10} \\ -f^{12} & 0 & f^{23} & f^{20} \\ f^{31} & -f^{23} & 0 & f^{30} \\ -f^{10} & -f^{20} & -f^{30} & 0 \end{pmatrix} = \begin{pmatrix} 0, & H_3, & -H_2, & -\frac{1}{c}D_1 \\ -H_3, & 0, & H_1, & -\frac{1}{c}D_2 \\ H_2, & -H_1, & 0, & -\frac{1}{c}D_3 \\ +\frac{1}{c}D_1, & +\frac{1}{c}D_2, & +\frac{1}{c}D_3, & 0 \end{pmatrix} \tag{31.1}$$

with

$$H_k = \tfrac{1}{2}\,\delta_{kmn} f^{mn}, \tag{31.2}$$

will enable us to write one set of MAXWELL'S equations in the form

$$f^{\mu\varrho}{}_{,\varrho} = 4\pi j^{\mu}, \tag{31.3}$$

where we have, as before, taken the liberty of introducing a time scale on which c becomes unity.

Again the hypothesis that $\boldsymbol{D}$ and $\boldsymbol{H}$ together form a tensor according to (31.1) leads to definite transformation laws. Just as in the preceding section for charge and current density, we shall write down the transformation equations for a relative motion along the x-axis. The expressions are:

$$\left.\begin{aligned} D'_x &= B_x, & H'_x &= H_x, \\ D'_y &= \left(1-\frac{u^2}{c^2}\right)^{-\frac{1}{2}}\left(D_y + \frac{u}{c}H_z\right), & H'_y &= \left(1-\frac{u^2}{c^2}\right)^{-\frac{1}{2}}\left(H_y - \frac{u}{c}D_z\right), \\ D'_z &= \left(1-\frac{u^2}{c^2}\right)^{-\frac{1}{2}}\left(D_z - \frac{u}{c}H_y\right), & H'_z &= \left(1-\frac{u^2}{c^2}\right)^{-\frac{1}{2}}\left(H_z + \frac{u}{c}D_y\right). \end{aligned}\right\} \tag{31.4}$$

If we compare this set of transformation equations with the corresponding expressions in (2.5), we find a considerable measure of correspondence in the transformation law for $\boldsymbol{H}$ (where the deviations are of the second order in u/c), but somewhat grosser deviations in the transformation equations for $\boldsymbol{D}$, where even the first-order terms in u/c are different. The electric displacement is characterized by the fact that its divergence is proportional to the charge density. The charge density in non-relativistic theory is invariant with respect to Galilean transformations, but we have seen in Sect. 30 that it transforms under Lorentz transformations, and in such a manner that a pure current density in one frame of reference can give rise to a charge density in another frame. Accordingly, even though $\boldsymbol{D}$ may vanish in one frame of reference, it cannot possibly vanish in the other.

In empty space $\boldsymbol{E}$ equals $\boldsymbol{D}$, and $\boldsymbol{B}$ equals $\boldsymbol{H}$, in the units we have employed (unrationalized Gaussian). Consequently, the tensor consisting of $\boldsymbol{E}$ and $\boldsymbol{B}$ must have the identical transformation properties as the tensor (31.1). We shall introduce, with a convenient factor of powers of c, the tensor

$$\varphi^{\mu\nu} = \begin{pmatrix} 0, & c^3 B_3, & -c^3 B_2, & -c^2 E_1 \\ -c^3 B_3, & 0, & c^3 E_1, & -c^2 E_2 \\ c^3 B_2, & -c^3 B_1, & 0, & -c^2 E_3 \\ c^2 E_1, & c^2 E_2, & c^2 E_3, & 0 \end{pmatrix}. \tag{31.5}$$

Its covariant equivalent is:

$$\varphi_{\mu\nu} = \begin{pmatrix} 0, & \frac{1}{c} B_3, & -\frac{1}{c} B_2, & E_1 \\ -\frac{1}{c} B_3, & 0, & \frac{1}{c} E_1, & E_2 \\ \frac{1}{c} B_2, & -\frac{1}{c} B_1, & 0, & E_3 \\ -E_1, & -E_2, & -E_3, & 0. \end{pmatrix}. \tag{31.6}$$

The transformation law of the quantities appearing in the 4×4 array (31.5) is, of course, exactly the same as that of the quantities appearing in (31.1). Comparison, once more, with the non-relativistic transformation law (2.5) shows that the discrepancy in the case of $\boldsymbol{E}$ is of the second order, but in $\boldsymbol{B}$ of the first order.

If we introduce the components of the (covariant) $\varphi_{\mu\nu}$ into Eqs. (30.1), we find that the electromagnetic potentials form a world vector

$$\varphi_\mu = \left(-\frac{1}{c}\boldsymbol{A}, \varphi\right), \tag{31.7}$$

with the resulting covariant tensor equations

$$\varphi_{\mu\nu} = \varphi_{\mu,\nu} - \varphi_{\nu,\mu}, \qquad \varphi_{\alpha\beta,\gamma} + \varphi_{\beta\gamma,\alpha} + \varphi_{\gamma\alpha,\beta} = 0. \tag{31.8}$$

Herewith we have shown that all of MAXWELL's field equations may be cast into relativistic form, provided, of course, that the resulting transformation laws are physically reasonable. This reasonableness may be explored up to a certain point. Physically, the field equations tell us that whereas the electric field strength and the magnetic induction are governed by equations that insure their being derivable from a set of potentials, the magnetic field strength and the electric displacement are generated by the distribution of charges and currents. In vacuum these two sets of fields are pair-wise equal. The postulated transformation laws (1) permit this equality to hold for every observer if it holds for one; (2) provide for a relationship between charge and current density that is consistent with the physical hypothesis that all currents are nothing but charges in motion; and (3) maintain the relationship (31.8) between field quantities and potentials uniformly for all observers. The non-relativistic laws (2.5) have the same properties but one: Even in the absence of dielectrics and magnetizable materials the pair-wise equality between the two types of fields (f and φ) holds only for an observer "at rest". Hence the crucial tests of the correctness of the relativistic transformation laws are precisely the type of experiments that historically have contributed to the birth and growth of the theory of relativity.

It remains to state the four-dimensional version of *gauge covariance*: We may add to the potential vector φ_μ an arbitrary gradient field $\psi_{,\mu}$ without changing the numerical values of the $\varphi_{\mu\nu}$. This fact is evident from Eq. (31.8).

32. The ponderomotive law. Given the covariant tensor $\varphi_{\mu\nu}$, Eq. (31.6), we may write the last Eq. (2.3), the expression for the force on a moving electric charge, as follows:

$$\left(1-\frac{u^2}{c^2}\right)^{-\frac{1}{2}} f_s = e\, U^\varrho \varphi_{s\varrho}. \tag{32.1}$$

The right-hand side consists of the first three components of a world vector, to which we may formally adjoin a fourth component,

$$\left(1-\frac{u^2}{c^2}\right)^{-\frac{1}{2}} f_0 = e\, U^\varrho \varphi_{0\varrho} = -\, e\, \boldsymbol{u}\cdot\boldsymbol{E} = -\,\boldsymbol{u}\cdot\boldsymbol{f}. \tag{32.2}$$

This last expression represents the negative rate at which the field performs work on the particle, and hence the rate of decrease of the particle's energy.

The left-hand sides will also form the components of a world vector if we define the force as the rate of change of relativistic linear momentum or, more exactly, the rate of change of its product of rest mass and world velocity. We should have, then:

$$c^2 \frac{d}{d\tau}(m\, U_\mu) = e\, U^\varrho \varphi_{\varrho\mu}. \tag{32.3}$$

This relationship is both Lorentz-covariant and gauge-invariant, the latter because no undifferentiated potentials are involved. It agrees with the non-relativistic expression for the change in linear momentum and energy up to terms quadratic in $\boldsymbol{u}/c$. The factor c^2 on the left may be omitted if we employ coordinates in which $c=1$.

Having accepted the relationship (32.3) we may immediately decide whether or not the rest mass of a particle is changed as a result of its motion in an electromagnetic field. We can decide this question on physical grounds as follows. For a sufficiently strong magnetic field at right angles to the magnetic induction there exists a frame of reference in which the electric field is zero. In that frame the particle does not change its speed, hence not its energy, and its rest mass must remain unchanged. But as the rest mass is a scalar, its conservation cannot depend on the frame of reference chosen. This argument is not completely satisfactory, because only electric fields that are weaker than the magnetic induction at the same world point and which are perpendicular to it can be transformed away. A simple formal proof of the conservation of rest mass requires the multiplication of Eq. (32.3) by U^μ. The ensuing product on the right vanishes identically, because of the skew-symmetry of $\varphi_{\varrho\mu}$. On the left, the scalar product of U^μ by its own proper-time derivative vanishes identically, because U^μ is a unit vector. What is left is the set of dynamic equations:

$$\frac{dm}{d\tau} = 0, \qquad m\, c^2 \frac{d U_\mu}{d\tau} = e\, U^\varrho \varphi_{\varrho\mu}. \tag{32.4}$$

The equations of motion can also be obtained by means of a variational principle. We define the action along a trajectory in Minkowski space as follows

$$\left.\begin{aligned} S &= -\int\left(m\frac{d\tau}{d\vartheta} + e\,\varphi_\varrho \frac{dx^\varrho}{d\vartheta}\right) d\vartheta \\ &= -\int (m\, d\tau + e\,\varphi_\varrho\, dx^\varrho). \end{aligned}\right\} \tag{32.5}$$

The notation is the same as in Sects. 23 and 24. m and e are to be considered constants and not to be varied. Likewise, the potentials are to be considered given functions of the four coordinates. What is to be varied are the four coordinates x^μ of the particle as functions of the (arbitrary) parameter ϑ. The principle is obviously Lorentz-covariant (in that the integrand is a scalar), and parameter-invariant, because the Lagrangian L,

$$L = -\left(m \frac{d\tau}{d\vartheta} + e\,\varphi_\varrho \frac{dx^\varrho}{d\vartheta}\right), \tag{32.6}$$

is homogeneous of the first degree in the first derivatives of the coordinates with respect to the parameter. Moreover, though the action integral (32.5) is not itself gauge-invariant, it leads to gauge-invariant differential equations; this is because a gauge transformation adds to the Lagrangian the expression

$$L' - L = -e\,\psi_{,\varrho} \frac{dx^\varrho}{d\vartheta} = -e \frac{d\psi}{d\vartheta} \tag{32.7}$$

and hence the change in the action integral S itself depends only on the end points,

$$S' - S = -e\,[\psi]. \tag{32.8}$$

Actual calculation of the Euler-Lagrange equations leads back to Eqs. (32.3).

33. Variational principle for the field equations. Given an electromagnetic field, we have just seen that the law of motion of an electric point charge may be obtained as the extremal condition of a variational principle. In this section we shall see that the equations of the electromagnetic field itself can also be derived from an action principle; the action integral for the field equations is, however, not a line integral but a four-dimensional volume integral. The existence of such an action integral is of considerable importance, as it permits one to construct, in a natural manner, pairs of canonically conjugate field variables; their commutator brackets lead ultimately to the commutator brackets of quantum electrodynamics and, by analogy, to the quantization of other fields. In this section we shall deal only with the c-number (non-quantum) theory. Moreover, we shall assume that the two tensors $f^{\mu\nu}$ and $\varphi^{\mu\nu}$ are identical, that is to say we adopt the point of view of LORENTZ's theory of electrons. For the time being, we shall make the assumption that the sources of the field are continuously distributed, that, in other words, the charge-current density world vector j^μ is continuous. Later on, we shall drop this assumption, but the consideration of point sources involves a classical mass renormalization that we shall defer until later. To simplify the notation we shall again work with units of space and time in which $c=1$.

Under these circumstances, we may introduce the action integral

$$S = -\frac{1}{16\pi} \int (\varphi^{\varrho\sigma} \varphi_{\varrho\sigma} + 16\pi\, j^\varrho \varphi_\varrho)\, d^4x, \tag{33.1}$$

in which the variables to be varied are the potentials φ_ϱ as functions of the four arguments x^ϱ, whereas the charge-current distribution is now to be considered as given. The field strengths $\varphi_{\varrho\sigma}$ are to denote the expressions (31.8). The variation of the action integral (33.1) may be written in the form

$$\delta S = \int \left(\frac{1}{4\pi} \varphi^{\mu\varrho}{}_{,\varrho} - j^\mu\right) \delta\varphi_\mu\, d^4x - \frac{1}{4\pi} \oint \varphi^{\mu\varrho}\, \delta\varphi_\mu\, dS_\varrho. \tag{33.2}$$

The action integral (33.1) is manifestly Lorentz-invariant, as both integrands are separately scalars. Though not invariant under gauge transformations, it changes under gauge transformations only by a surface integral,

$$\left.\begin{aligned} S' = S - \int j^{\varrho}\psi_{,\varrho}\, d^4x &= S - \int (j^{\varrho}\psi)_{,\varrho}\, d^4x + \int j^{\varrho}{}_{,\varrho}\,\psi\, d^4x \\ &= S - \oint \psi j^{\varrho}\, d S_{\varrho} + 0, \end{aligned}\right\} \tag{33.3}$$

the last equality being true because of GAUSS's theorem, and because of the equation of continuity for the charge-current density. This equation of continuity must hold if the Euler-Lagrange equations

$$L^{\varrho} \equiv \frac{1}{4\pi}\,\varphi^{\varrho\sigma}{}_{,\sigma} - j^{\varrho} = 0 \tag{33.4}$$

are to possess solutions. Because of Eq. (33.3), the Euler-Lagrange equations must be gauge-invariant, as indeed they are. This result is a perfectly general one. Given an action S which under gauge transformations changes only by a surface integral, even though in other respects it may be quite different from the Lagrangian of Eq. (33.1). Then we have

$$\delta S \equiv \int L^{\varrho}\,\delta\varphi_{\varrho}\, d^4x + \oint Q^{\varrho}\, d S_{\varrho}, \tag{33.5}$$

where the set of functions Q^{ϱ} need not be specified. If we assume in particular that the variation $\delta\varphi_{\varrho}$ is to be the result of an infinitesimal gauge transformation, then we have by assumption

$$\left.\begin{aligned} \delta S &\equiv \int L^{\varrho}\psi_{,\varrho}\, d^4x + \oint Q^{\varrho}\, d S_{\varrho} \equiv - \oint C^{\varrho}\, d S_{\varrho}, \\ (Q^{\varrho} + C^{\varrho})_{,\varrho} &\equiv - L^{\varrho}\psi_{,\varrho} = - (L^{\varrho}\psi)_{,\varrho} + \psi L^{\varrho}{}_{,\varrho}. \end{aligned}\right\} \tag{33.6}$$

The function ψ is arbitrary. However, if we integrate the last Eq. (33.6) over some four-dimensional domain, then we find that the integral over the term $\psi L^{\varrho}{}_{,\varrho}$ is to equal a surface integral; it must be independent of the value that might be assigned to the function ψ in the interior. Clearly this requirement can be satisfied only if

$$L^{\varrho}{}_{,\varrho} \equiv 0, \tag{33.7}$$

that is to say, if the left-hand sides of the field equations satisfy a differential identity. The time derivative of the expression L^0 vanishes *identically* provided the spatial divergence of the remaining three field equations, $L^s{}_{,s}$ is zero. Accordingly the field equations of a gauge-invariant theory cannot determine completely time derivatives of all orders of the electromagnetic potentials out of the potentials and their derivatives on some initial three-dimensional hypersurface. *Gauge-invariant Euler-Lagrange equations never form a Cauchy system for the potentials*, but only for the field intensities.

34. Conservation laws. Both Lorentz covariance and gauge invariance imply characteristic conservation laws that are comparable to those met with in mechanics. The Lagrangian being a scalar, the action integral (33.1) will not change its value under a Lorentz transformation except insofar as the domain of integration is affected. Accordingly we have:

$$\bar{\delta} S = \int L^{\varrho}\,\bar{\delta}\varphi_{\varrho}\, d^4x + \oint Q^{\varrho}\, d S_{\varrho} - \int \varphi_{\varrho}\,\bar{\delta} j^{\varrho}\, d^4x = - \oint L\,\delta x^{\varrho}\, d S_{\varrho}, \tag{34.1}$$

where, according to Eq. (33.2), the symbol L^{ϱ} denotes the expression (33.4), Q^{ϱ} stands for

$$Q^{\varrho} = - \frac{1}{4\pi}\,\varphi^{\mu\varrho}\,\bar{\delta}\varphi_{\mu}, \tag{34.2}$$

and the variation $\bar{\delta}\varphi_\mu$ equals

$$\bar{\delta}\varphi_\mu = - P^\nu{}_\mu \varphi_\nu - \varphi_{\mu,\varrho}\,\delta x^\varrho. \tag{34.3}$$

δx^ϱ is the expression (13.4), and

$$\bar{\delta} j^\varrho = P^\varrho{}_\sigma j^\sigma - j^\varrho{}_{,\sigma}\,\delta x^\sigma.$$

The relationship (34.1) is satisfied identically, whether or not we consider a situation in which MAXWELL's field equations hold. If they do, then we have the following equality, which is no longer an identity:

$$\oint (Q^\varrho + L\,\delta x^\varrho)\, d S_\varrho = \int \varphi_\varrho\, \bar{\delta} j^\varrho\, d^4x. \tag{34.4}$$

When we substitute into this equality the various expressions for Q^ϱ, δx^ϱ, and for $\bar{\delta} j^\varrho$, then we obtain some terms in which $P^\sigma{}_\tau$ is a factor and other terms that contain b^σ. The coefficients of these arbitrary constants on the left must equal those on the right; but because of the skewsymmetry of the P's we must also skewsymmetrize their coefficients. Accordingly we obtain two sets of equalities,

$$\left.\begin{aligned} \oint t^\varrho_\sigma\, d S_\varrho &= \int \varphi_\alpha j^\alpha{}_{,\sigma}\, d^4x, \\ t^\varrho_\sigma &= \frac{1}{4\pi}\varphi^{\varrho\alpha}\varphi_{\alpha,\sigma} - \delta^\varrho_\sigma L, \end{aligned}\right\} \tag{34.5}$$

and

$$\left.\begin{aligned} &\oint \left[(x_\tau t^\varrho_\sigma - x_\sigma t^\varrho_\tau) + \frac{1}{4\pi}(\varphi_\sigma \varphi^\varrho{}_\tau - \varphi_\tau \varphi^\varrho{}_\sigma)\right] d S_\varrho \\ &\quad = \int [(\varphi_\tau j_\sigma - \varphi_\sigma j_\tau) + \varphi_\alpha (x_\tau j^\alpha{}_{,\sigma} - x_\sigma j^\alpha{}_{,\tau})]\, d^4x. \end{aligned}\right\} \tag{34.6}$$

The expression t^ϱ_σ is often called the canonical energy-momentum tensor. Because it is neither gauge-invariant nor symmetric, another formulation of the same two sets of relationships is frequently preferred. If we subject the volume integrals on the right-hand sides to a number of integrations by parts and transfer the resulting surface integrals to the left, we may bring the integral equalities (34.5), (34.6) into the form:

$$\left.\begin{aligned} \oint \vartheta^{\varrho\sigma}\, d S_\varrho &= \int j^\alpha \varphi_{\alpha\sigma}\, d^4x, \\ \vartheta^{\varrho\sigma} &= \frac{1}{4\pi}\left(\varphi^{\varrho\alpha}\varphi^{\sigma\alpha} - \frac{1}{4}\eta^{\varrho\sigma}\varphi^{\alpha\beta}\varphi_{\alpha\beta}\right), \end{aligned}\right\} \tag{34.7}$$

and

$$\oint (x^\tau \vartheta^{\sigma\varrho} - x^\sigma \vartheta^{\tau\varrho})\, d S_\varrho = \int (x^\tau \varphi^{\alpha\sigma} - x^\sigma \varphi^{\alpha\tau})\, j_\alpha\, d^4x. \tag{34.8}$$

The tensor $\vartheta^{\varrho\sigma}$ is both symmetric and gauge-invariant. It represents, so to speak, the contribution of the electromagnetic field to the total linear momentum density, energy density, and stress of a system. It has the further attractive property that in the expression for the angular momentum density (34.6) the same symmetric energy-momentum tensor appears in a completely straight-forward manner.

An intuitive interpretation of our integral equalities may be obtained by an appropriate choice of the domain of integration. We shall choose as this domain a four-dimensional slice that extends to infinity in all three directions of space and is bounded by two three-dimensional hypersurfaces. If these two hypersurfaces are chosen to be hyperplanes normal to the time axis, then each of them represents all space at one instant in time. In any event, the integral equalities tell us that the difference between the three-dimensional integrals taken over these two bounding hypersurfaces is given by the total amount of work done on the electric charges between these two instants in time, or by the total amount of linear or angular momentum imparted to the charges, respectively. In the absence of free electric charges and currents, the three-dimensional integrals over the left-hand sides, extended over all space, are conserved in the course of time.

35. Singular charge distributions. In the last two sections we have treated the charge-current density as a continuous distribution, though in nature electric charge occurs only as an attribute of particles, such as electrons, protons, and mesons. Even though we shall forego the quantum-theoretical treatment of elementary particles in this article, we cannot close our eyes entirely to their existence. Not only must be we able to construct electromagnetic fields that arise out of the presence of charged particles, but in turn the ponderomotive equations of these particles themselves depend on their being mass points. In any theory which endows the elementary particles with finite extension (a form factor), we are immediately confronted with a host of new problems, foremost among them the question how the internal degrees of freedom of a relativistic particle react to an externally applied stress. These questions have been treated extensively in the literature. We do know that particles have some internal structure, as manifested by their spin and by their intrinsic magnetic dipole moment. To some extent this internal structure can be described in terms of classical concepts; but probably it becomes fully accessible only in a consistently quantum-theoretical approach. To discuss these questions here would be out of place. We shall confine our attention to point-like charged particles.

We shall now consider a dynamical system in which the charge-current distribution consists of point charges subject to the ponderomotive laws. We must formulate dynamical laws for this system that are analogous to the known field equations and to the ponderomotive equations but which retain their meaning in the face of the unavoidable singularities associated with point charges.

The location of each point charge is associated with an infinity of the electromagnetic field, which is sufficiently strong that the integral of the energy density, taken over a three-dimensional domain that intersects a particle trajectory, diverges. That is the significance of the so-called infinite self-energy of a point charge. Taken by itself this infinity would not be fatal, as the energy density is not required for the formulation of the dynamic laws of our system, and because the definition of energy density is somewhat ambiguous (cf. the two possible expressions t_σ^ϱ and $\vartheta^{\varrho\sigma}$ that we obtained in Sect. 34). However, the ponderomotive equations that determine the motion of each particle involve the electromagnetic field at the location of the particle itself. If the field there is infinite, these equations are not meaningful without further specifications. To give the equations of motion of particles some meaning, one associates with each particle a "self-field", subtracts this self-field from the total field, and introduces the (finite) difference, the "external field", into the ponderomotive law. A discussion of this procedure belongs into our article on relativity insofar as we need to assure ourselves that the "subtraction procedure" to be employed is relativistically invariant, or can be made relativistically invariant. Otherwise the results of the separation of the field into self-field and external field would lead to a different law of motion in every Lorentz frame.

The self-field should be a solution of the Maxwell equations by itself and should be Lorentz-covariant; the external field will then also be Lorentz-covariant. Further, the self-field should have a singularity at the location of the particle, so that its associated charge-current density corresponds precisely to the moving point charge actually present. The external field will then satisfy the Maxwell equations without sources and can therefore be assumed to remain finite and differentiable at the location of the point charge. Finally, we should be able to construct expressions for the energy density and for the linear and angular momentum density that remain finite and are subject to reasonable conservation laws.

We shall accomplish these aims in two stages. First we shall construct a suitable self-field to be associated with each point charge in our system. Then, by subtracting all our self-fields from the total field, we shall obtain an external field that is completely free of infinities and that interacts with the particles in such a manner that the total dynamic system may be described again with the help of a single variational principle. If this variational principle is mathematically sound, then the construction of an energy-momentum tensor and of an angular momentum density are fairly straight-forward.

If we rewrite MAXWELL's equations in terms of the potentials, we get

$$\eta^{\varrho\sigma}(\varphi_{\mu},_{\varrho\sigma}-\varphi_{\varrho},_{\mu\sigma})=4\pi j_{\mu}. \tag{35.1}$$

We may *separate* these equations by adopting a *gauge condition*, the so-called *Lorentz gauge*

$$\varphi^{\varrho},_{\varrho}=0. \tag{35.2}$$

This gauge condition can always be satisfied by means of a gauge transformation, which does not change the values of the electromagnetic field strengths. The remaining set of equations,

$$\eta^{\varrho\sigma}\varphi_{\mu},_{\varrho\sigma}=4\pi j_{\mu}, \tag{35.3}$$

together with the Lorentz condition (35.2), may now be solved by means of an integral representation *(quellenmäßige Darstellung)* if we introduce a kernel function $\bar{D}(x, x')$ satisfying the differential equation

$$\eta^{\varrho\sigma}\bar{D}(x, x'),_{\varrho\sigma}=4\pi\,\delta_4(x, x'), \tag{35.4}$$

which is invariant under Lorentz transformations (δ_4 denotes the four-dimensional Dirac function). With the help of such a kernel function $\bar{D}$ we obtain a solution in the form

$$\varphi_{\mu}(x)=\int_{x'}\bar{D}(x, x')\,j_{\mu}(x')\,d^4x' \tag{35.5}$$

for any charge-current distribution that satisfies the equation of continuity. The properties of the solution depend on the choice we make for $\bar{D}$.

A well-known solution for $\bar{D}$ is

$$D^{\mathrm{ret}}=\frac{1}{r}\,\delta_1(\xi^0-r),\qquad \xi^0=x^0-x^{0\prime},\qquad r=\sqrt{(x^k-x^{k\prime})^2}, \tag{35.6}$$

which we shall call the *retarded solution*. This function is invariant with respect to proper Lorentz transformations and with respect to spatial reflections, but it goes over into the *advanced solution*

$$D^{\mathrm{adv}} = \frac{1}{r}\,\delta_1(\xi^0 + r) \tag{35.7}$$

under time reversal. A completely invariant solution is the *symmetric kernel function*

$$\overline{D} = \frac{1}{2}\,(D^{\mathrm{adv}} + D^{\mathrm{ret}}) = \frac{1}{2r}\left[\delta_1(\xi^0 - r) + \delta_1(\xi^0 + r)\right] = \delta_1(\xi^{0\,2} - r^2)\,. \tag{35.8}$$

This kernel is, moreover, an even function of its arguments $(x^\varrho - x^{\varrho\prime})$. From now on we shall reserve the symbol $\overline{D}$ for the particular solution (35.8).

With its help we may now proceed to construct solutions of MAXWELL's equations belonging to a point charge in arbitrary motion. Suppose the parametric representation of the trajectory is $X^\varrho(\vartheta)$ (we must use distinctive notation for the coordinates of a particle and for the coordinates of arbitrary world points, because we now have to deal with a system consisting both of particles and of fields; hence the capitals), then the charge-current distribution may be given the form:

$$j^\varrho(x) = e\int_\vartheta \delta_4[x, X(\vartheta)]\,U^\varrho(\vartheta)\,d\vartheta = e\int_\vartheta \delta_4(x, X)\,dX^\varrho\,. \tag{35.9}$$

If we substitute that expression in Eq. (35.5), we obtain for the potentials

$$\varphi^\varrho(x) = e\int_\vartheta \overline{D}\,(x, X)\,dX^\varrho\,. \tag{35.10}$$

Had we used D^{ret}, (35.6), instead of $\overline{D}$, (35.8), we should have obtained the conventional Liénard-Wiechert potentials, in a four-dimensional notation.

We shall adopt the expression (35.10) as the self-potential.

36. Separation of the field. When the charge-current distribution consists of n separate point charges, we may separate the total field $\varphi^\mu(x)$ into the "external field" $\chi^\mu(x)$ and the "self-field" $\mathring{\varphi}^\mu(x)$,

$$\left.\begin{aligned} \varphi^\mu(x) &= \chi^\mu(x) + \mathring{\varphi}^\mu(x)\,,\\ \mathring{\varphi}^\mu(x) &= \int \overline{D}\,(x, \bar{x})\,j^\mu(\bar{x})\,d^4\bar{x}\,,\\ j^\mu(x) &= \sum_{k=1}^{n} e_k \int \delta_4(x, X_k)\,dX_k^\mu\,, \end{aligned}\right\} \tag{36.1}$$

where $\overline{D}\,(x, \bar{x})$ is the kernel function (35.8). The external field will be free of singularities.

We shall now combine the action principles (32.5) and (33.1) into one. As a preliminary, we shall conjecture that this combined action principle has the form:

$$\left.\begin{aligned} S &\equiv S_f + S_i + S_p\,,\\ S_f &= -\frac{1}{16\pi}\int \varphi^{\mu\nu}\,\varphi_{\mu\nu}\,d^4x\,,\\ S_i &= -\int \varphi_\mu\,j^\mu\,d^4x = -\sum_k e_k \int \varphi_\mu(X_k)\,dX_k^\mu,\\ S_p &= -\sum_k m_k \int d\tau_k\,. \end{aligned}\right\} \tag{36.2}$$

We shall find that this integral has one infinite term, which must be eliminated. When we substitute the separated field (36.1) into the action integral (36.2), we obtain the expression

$$\left.\begin{aligned} S &= K + \Gamma + J + S_p, \\ K &= -\frac{1}{16\pi}\int \chi^{\mu\nu}\chi_{\mu\nu}\,d^4x, \\ \Gamma &= -\frac{1}{2}\sum_{k,l=1}^{n}\sum e_k e_l \iint \overline{D}(X_k, X_l)\,\eta_{\mu\nu}\,dX_k^\mu\,dX_l^\nu \\ J &= -\sum_{k=1}^{n} e_k \int \chi_\mu(X_k)\,dX_k^\mu + \frac{1}{4\pi}\int \chi^{\mu\varrho}{}_{,\varrho}\,\mathring{\varphi}_\mu\,d^4x - \\ &\quad - \frac{1}{8\pi}\oint \mathring{\varphi}_\mu(\mathring{\varphi}^{\mu\varrho} + 2\chi^{\mu\varrho})\,dS_\varrho, \\ S_p &= -\sum_{k=1}^{n} m_k \int d\tau_k. \end{aligned}\right\} \tag{36.3}$$

The individual terms may be interpreted as follows: K represents the action of the external field by itself; it is the only term in the total action free of references to the point charges. S_p is, as before, the action describing free and non-interacting particles. The term Γ contains the interaction of the point charges with each other; aside from the relativistically modified Coulomb interaction it also contains an infinite self-energy contribution, which we shall later remove. Finally, the terms J contain the interaction between the external field and the particles; presumably only the first term in J makes any significant contribution. We shall now take up the term Γ in some more detail.

Γ may be given the form

$$\left.\begin{aligned} \Gamma = &-\frac{1}{2}\sum_{k=1}^{n} e_k^2 \iint \overline{D}(\overline{X}_k, \overline{\overline{X}}_k)\,\eta_{\mu\nu}\,d\overline{X}_k^\mu\,d\overline{\overline{X}}_k^\nu - \\ &- \sum_{k<l}\sum e_k e_l \iint \overline{D}(X_k, X_l)\,\eta_{\mu\nu}\,dX_k^\mu\,dX_l^\nu. \end{aligned}\right\} \tag{36.4}$$

The last double sum on the right is a special case of the interaction term in the expression (24.1). The first term might be interpreted as the action of the self-field on the particle responsible for that self-field; it is infinite. We shall show that in structure this self-energy term resembles the action of a free particle, S_p. To this end we shall evaluate its contribution to a small segment of trajectory. We define:

$$\left.\begin{aligned} e^2 \iint \overline{D}(X, \overline{X})\,\eta_{\mu\nu}\,dX^\mu\,d\overline{X}^\nu &= \int \mu\,d\tau, \\ \mu(X) &= e^2\,\eta_{\mu\nu}\frac{dX^\mu}{d\tau}\int \overline{D}(X, \overline{X})\,d\overline{X}^\nu. \end{aligned}\right\} \tag{36.5}$$

We shall now evaluate μ, the "electromagnetic mass" of a point charge. The purpose of this evaluation will be to demonstrate that this quantity, which is infinite, is independent of the state of motion of the point charge, its velocity as well as its acceleration and higher derivatives.

We shall introduce a parametric representation of the particle trajectory, $\overline{X}^\nu(\vartheta)$, with the understanding that $\vartheta = 0$ at $\overline{X} = X$. We shall also assume that $\overline{X}(\vartheta)$ may be expanded into a power series in ϑ, at least in the immediate vicinity of the point X. We thus put:

$$\overline{X}^\nu - X^\nu = u^\nu\vartheta + \sum_{2}^{\infty} a_k^\nu\,\vartheta^k \tag{36.6}$$

and

$$d\overline{X}^\nu = \left(u^\nu + \sum_2^\infty k\, a_k^\nu\, \vartheta^{k-1}\right) d\vartheta, \qquad \frac{dX^\mu}{d\tau} = \frac{u^\mu}{u}, \qquad u = \sqrt{\eta_{\alpha\beta}\, u^\alpha u^\beta}. \tag{36.7}$$

When we substitute these power series expressions into the integral (36.5), we find for μ:

$$\left.\begin{aligned}\mu = e^2 \int \delta_1 \Big\{u^2 \vartheta^2 + 2 \sum_{r=2}^\infty (\boldsymbol{u}\cdot\boldsymbol{a}_r)\, \vartheta^{r+1} + \sum_{r,r'=2}^\infty \sum (\boldsymbol{a}_r \cdot \boldsymbol{a}_{r'})\, \vartheta^{r+r'}\Big\} \times \\ \times \Big[u + \sum_2^\infty k\, u^{-1} (\boldsymbol{u}\cdot\boldsymbol{a}_r)\, \vartheta^{r-1}\Big] d\vartheta, \qquad (\boldsymbol{u}\cdot\boldsymbol{a}_r) \equiv \eta_{\alpha\beta}\, u^\alpha a_r^\beta, \quad \text{etc.}\end{aligned}\right\} \tag{36.8}$$

We shall now introduce the argument of the delta function,

$$(\overline{X} - X)^2 \equiv \xi, \tag{36.9}$$

as a new variable of integration and endeavor to substitute it throughout. Clearly, ξ has a double zero at the point $\vartheta = 0$, and hence ϑ as a function of ξ has a branch point. Accordingly, we set

$$\vartheta = \sqrt{\xi} \sum_0^\infty \alpha_r\, \xi^r + \xi \sum_0^\infty \beta_r\, \xi^r \tag{36.10}$$

and determine the two sets of expansion coefficients α_r and β_r directly from the defining equation (36.9). The only two numerical values required for our calculation are those of the coefficients α_0 and β_0; they are

$$\vartheta = u^{-1} \sqrt{\xi} - u^{-4} (\boldsymbol{u}\cdot\boldsymbol{a}_2)\, \xi + \cdots. \tag{36.11}$$

When this series is inserted into the integral, we find, as our final result,

$$\mu = e^2 \int_\xi \delta_1(\xi)\, [1 + O(\xi)]\, \frac{d\xi}{2\sqrt{\xi}}. \tag{36.12}$$

The infinite term is proportional to e^2 and independent of all kinematic parameters; the coefficient of a finite term vanishes; all remaining terms in the expansion make zero contributions.

Having established that the self-energy term makes a contribution to the action integral that except for being infinite resembles that of the mass term S_p, we shall omit this term on the grounds that the *observed* mass is actually the sum of the "electromagnetic" mass μ and the remaining mass m and that this combined mass, which alone is observable, for reasons we admittedly do not understand, is a finite positive quantity for all known particles. In what follows, the symbol m_k will then denote the combined (observable) mass. This procedure of identifying the sum of the divergent electromagnetic self-energy and of the non-electromagnetic mass with the phenomenological mass is known as (non-quantum) *mass renormalization.* With renormalization, our action integral now takes the form:

$$\left.\begin{aligned}S &= K + \Gamma^* + J + S_p, \\ \Gamma^* &= -\tfrac{1}{2} \sum_{k \neq l} \sum e_k\, e_l \iint \overline{D}(X_k, X_l)\, \eta_{\mu\nu}\, dX_k^\mu\, dX_l^\nu.\end{aligned}\right\} \tag{36.13}$$

The integrals K, J, and S_p have the same form as in Eq. (36.3). In fact, the only difference in the form of the expressions (36.3) and (36.13) is that in the latter integral in the relativistic Coulomb-like term the self-interaction no longer appears and that the quantity m_k in the term S_p is to be interpreted as the renormalized mass.

37. Variation of the new action principle. The action principle (36.13) involves both a field, $\chi_\mu(x)$, and particle coordinates, $X_k^\mu(\vartheta)$, as dynamical variables. Their variations are independent of each other, whereas those of $\mathring{\varphi}_\mu$ are not, because of the defining equation (36.1). Some of the integrals appearing in the expression (36.13) are multiple, involving as factors both dynamical variables in the interior and on the boundary of the domain of integration. Hence, the boundary conditions to be imposed on the variations of the variables are not as obvious as they are in conventional local field theories. Usually, one simply sets the variations on the boundary of the domain of integration zero and leaves the variations in the interior unrestricted. If this prescription were to be applied to the action (36.13), then we should not obtain the correct dynamical laws. In fact, because the relationship between the field φ_μ [occurring in the action (36.2)] and the field χ_μ, the relationship (36.1), is non-local, a requirement that the variation of one, or the other, should vanish on the boundary leads to different results. This inequivalence, by the way, has nothing to do with the renormalization we carried out in order to arrive at the finite action (36.13), but is caused solely by the non-local character of the relationship between the original variables and the variables that we have called separated.

It turns out that reasonable results are obtained if in the expression (36.13) we suppress the surface integral forming part of the "interaction" J. Hence we shall begin with the revised action principle

$$\left.\begin{aligned} S = &-\frac{1}{16\pi}\int \chi^{\mu\nu}\chi_{\mu\nu}\,d^4x - \sum_{k<l}\sum e_k e_l \iint \overline{D}(X_k, X_l)\,\eta_{\mu\nu}\,dX_k^\mu\,dX_l^\nu + \\ &+\frac{1}{4\pi}\int \chi^{\mu\varrho}{}_{,\varrho}\,\mathring{\varphi}_\mu\,d^4x - \sum_k \left[e_k\int \chi_\mu(X_k)\,dX_k^\mu + m_k\int d\tau_k\right]. \end{aligned}\right\} \tag{37.1}$$

In the variation of this action we shall separate the terms containing variations in the interior from those having variations on the boundary. The result is:

$$\left.\begin{aligned} \delta S = &\frac{1}{4\pi}\int \chi^{\mu\varrho}{}_{,\varrho}\,(\delta\chi_\mu + \delta\mathring{\varphi}_\mu)\,d^4x + \sum_k \Big\{e_k\Big[\int \chi_{\mu\varrho}(X_k)\,U_k^\varrho\,\delta X_k^\mu\,d\tau_k + \\ &+\sum_{l\neq k} e_l \iint [\overline{D}(X_k - X_l)]_{,\varrho}\,\eta_{\mu\varrho}\,(\delta X_k^\mu U_k^\sigma - \delta X_k^\sigma U_k^\mu)\,U_l^\varrho\,d\tau_k\,d\tau_l\Big] + \\ &+\eta_{\mu\nu}\,m_k\int \frac{dU_k^\mu}{d\tau_k}\,\delta X_k^\nu\,d\tau_k\Big\} + \\ &+\frac{1}{4\pi}\oint (\mathring{\varphi}_\mu\,\delta\chi^{\mu\varrho} - \varphi^{\mu\varrho}\,\delta\chi_\mu)\,dS_\varrho - \\ &-\sum_k \Big\{e_k\Big[\chi_\mu(X_k)\,\delta X_k^\mu + \delta X_k^\mu \sum_{l\neq k} e_l\int \overline{D}(X_k - X_l)\,\eta_{\mu\nu}\,U_l^\nu\,d\tau_l\Big]_{\tau_k} + \\ &+m_k\Big[\eta_{\mu\nu}\,U_k^\nu\,\delta X_k^\mu\Big]_{\tau_k}\Big\}, \qquad U_k^\mu \equiv \frac{dX_k^\mu}{d\tau_k}. \end{aligned}\right\} \tag{37.2}$$

Thus the Euler-Lagrange equations take the following form:

$$\left.\begin{aligned} L^\mu(x) &\equiv \frac{1}{4\pi}\chi^{\mu\varrho}{}_{,\varrho} = 0, \\ L_\mu^k(\tau_k) &\equiv m_k\,\eta_{\mu\nu}\frac{dU_k^\nu}{d\tau_k} + e_k\Big\{\chi_{\mu\varrho}(X_k) + \\ &\quad+\sum_{l\neq k} e_l\int \big([\overline{D}(X_k - X_l)]_{,\varrho}\,\eta_{\mu\sigma} - [\overline{D}(X_k - X_l)]_{,\mu}\,\eta_{\sigma\varrho}\big)\,U_l^\sigma\,d\tau_l\Big\}\,U_k^\varrho = 0. \end{aligned}\right\} \tag{37.3}$$

Whereas the field equations govern the behavior of the χ-field alone, which is unaffected by the motions of the point charges, the point charges themselves experience forces that are caused in part by the χ-field. It might appear difficult to reconcile the conservation of total energy density and linear momentum density with such a one-way interaction. To clarify that question we construct these quantities in the usual fashion: We consider the change in the action integral S that is brought about by an infinitesimal shift of the origin of our four-dimensional coordinate system. The dynamical variables are changed as follows:

$$\left.\begin{aligned}\bar{\delta}\chi^\mu &= -\chi^\mu{}_{,\alpha}\, a^\alpha,\\ \bar{\delta}X^\mu_k &= +a^\mu,\end{aligned}\right\} \tag{37.4}$$

where the change in $\chi^\mu(x)$ is determined at a world point with fixed coordinate values and the change in X^μ_k for a fixed value of the parameter ϑ. The total change in the action S, determined for a finite four-dimensional domain, may now be expressed alternatively either as a surface integral, taking account of the change in the domain of integration as a result of the shift of coordinate system, or as the expression (37.2), with the variations substituted from Eq. (37.4). These two expressions must equal each other. Hence, if the Euler equations (37.3) are satisfied, it follows that a certain surface integral must vanish. This surface integral contains the components of the constant vector a^α as factors. Accordingly, we obtain four vanishing surface integrals. A straightforward calculation leads to the following identity:

$$\left.\begin{aligned}0 &= \sum_k \int L^k_\alpha\, d\tau_k - \frac{1}{4\pi}\int L^\mu(\chi_{\mu,\alpha} + \mathring{\varphi}_{\mu,\alpha})\, d^4x\\ &\equiv \frac{1}{4\pi}\oint\left[\frac{1}{4}\delta^\varrho_\alpha\,\chi^{\mu\nu}\chi_{\mu\nu} + \mathring{\varphi}^{\mu\varrho}{}_{,\alpha}\,\chi_\mu - \chi^{\mu\varrho}(\chi_{\mu,\alpha} + \mathring{\varphi}_{\mu,\alpha}) - \delta^\varrho_\alpha\chi_\mu\,\mathring{\varphi}^{\mu\sigma}{}_{,\sigma}\right] dS_\varrho +\\ &\quad + \sum_k \int_{\tau_k}\Big[\eta_{\alpha\sigma}\, m_k\, U^\sigma_k + e_k\,\chi_\alpha(X_k) + \eta_{\alpha\sigma}\sum_{l\neq k} e_k\, e_l \int \bar{D}\,(X_k, X_l)\, dX^\sigma_l\Big].\end{aligned}\right\} \tag{37.5}$$

The integrand in this form is not manifestly gauge-invariant. By adding a term that vanishes modulo the field equations and by making a few transformations, one can cast the vanishing surface integral into the form:

$$\left.\begin{aligned}&\frac{1}{4\pi}\oint\left[\frac{1}{2}\delta^\varrho_\alpha\,\chi^{\mu\nu}\chi_{\mu\nu} - \chi^{\mu\varrho}\chi_{\mu\alpha} + \frac{1}{2}\delta^\varrho_\alpha\,\chi^{\mu\nu}\mathring{\varphi}_{\mu\nu} - \chi^{\mu\varrho}\mathring{\varphi}_{\mu\alpha} - \mathring{\varphi}^{\mu\varrho}\chi_{\mu\alpha}\right] dS_\varrho +\\ &\quad + \eta_{\alpha\sigma}\sum_k \int_{\tau_k}\Big[m_k\, U^\sigma_k + \sum_{l\neq k} e_k\, e_l \int \bar{D}(X_k, X_l)\, dX^\sigma_l\Big] = 0.\end{aligned}\right\} \tag{37.6}$$

Accordingly, the total energy and the total linear momentum of our system, defined as the expressions

$$\left.\begin{aligned}P_\alpha &= \frac{1}{4\pi}\int\left[\frac{1}{4}\delta^0_\alpha\,\chi^{\mu\nu}(\chi_{\mu\nu} + 2\,\mathring{\varphi}_{\mu\nu}) + (\chi^{s0}\chi_{s\alpha} + \chi^{s0}\mathring{\varphi}_{s\alpha} + \mathring{\varphi}^{s0}\chi_{s\alpha})\right] d^3x +\\ &\quad + \eta_{\alpha\sigma}\Big[\sum_k m_k\, U^\sigma_k + \sum\sum_{l\neq k} e_k\, e_l \int \bar{D}(X_k, X_l)\, dX^\sigma_l\Big]\end{aligned}\right\} \tag{37.7}$$

are constants of the motion.

Because the field equations are independent of the particles, the χ-field satisfies conservation laws of its own, which may be separated from the remainder. If we make this separation, we obtain the following two sets of conservation laws:

$$\left.\begin{aligned}&\frac{1}{4\pi}\oint\left(\frac{1}{4}\delta^\varrho_\alpha\,\chi^{\mu\nu}\chi_{\mu\nu} - \chi^{\mu\varrho}\chi_{\mu\alpha}\right) dS_\varrho = 0,\\ &\eta_{\alpha\sigma}\sum_k \int_{\tau_k}\Big[m_k\, U^\sigma_k + \sum_{l\neq k} e_k\, e_l \int \bar{D}(X_k, X_l)\, dX^\sigma_l\Big] +\\ &\quad + \frac{1}{4\pi}\oint\left[\frac{1}{2}\delta^\varrho_\alpha\,\chi^{\mu\nu}\mathring{\varphi}_{\mu\nu} - (\chi^{\mu\varrho}\mathring{\varphi}_{\mu\alpha} + \mathring{\varphi}^{\mu\varrho}\chi_{\mu\alpha})\right] dS_\varrho = 0.\end{aligned}\right\} \tag{37.8}$$

Thus one set of conservation laws is associated with the external field, the other primarily with the motion of the particles. The partial separability of the conservation laws reflects the corresponding properties of the dynamical laws themselves.

The equations of motion $L_\mu^k = 0$, the second set (37.3), contain no reference to radiation damping. A radiation damping term may be obtained if we replace the kernel function $\overline{D}$ by the retarded function D^{ret}, which equals $2\overline{D}$ on one-half of the light cone (the one on which the sources of the field lie in the past) and vanishes on the other half. By considering the retarded field as the "self-field", we change, of course, the definition of the free field, which now is χ^{ret}. As for the effects of the other charges, $l \neq k$, the transfer of part of the field from the self-field to the free field does not change the situation significantly. However, the self-field of the k-th particle also undergoes a change. This change, which turns out to be non-zero but finite at the location of the k-th particle, requires a separate term, different from all those appearing in our notation; this separate term is known as the radiation damping term.

C. Relativistic quantum theory.

I. Quantum mechanics.

38. Foundations. In the remainder of this article we shall be concerned with specific features of quantum mechanics that arise in connection with the Lorentz covariance of the theory. We are particularly concerned with formulations that make the Lorentz covariance *manifest*. This is to say that we like to use formulations that exhibit Lorentz covariance at each step of the development of the theory. The requirement of manifest Lorentz covariance is far more than a purely aesthetic desideratum; in the renormalization of quantum field theories, such as quantum electrodynamics, the original theory is modified by the subtraction of divergent (infinite) terms. Such cut-offs were also performed in the thirties, but then the procedures were not manifestly Lorentz-covariant; as a result, they were not Lorentz-covariant at all, and the remaining finite terms led to different physical results in different Lorentz frames.

The obstacle to manifest Lorentz covariance of quantum theory rests in the fact that in the usual formulations of the dynamical law, be it in Schrödinger or in Heisenberg representation, the time coordinate appears in an entirely different role from the space coordinates. In quantum mechanics, these differences consist of the following:

a) In contrast to the configuration coordinates of a mechanical system, the time is not an operator in Hilbert space but a c-number, and it is not an observable. It is physically meaningless to assign to the time an expectation value.

b) In the normalization of a state vector, integration is to be performed over the range of a complete set of commuting observables, of which the time is not one. Physically, the complete set of commuting operators corresponds to a set of observations that need to be made "simultaneously".

c) Because time is not an observable, the uncertainty relationship between time and energy differs basically from those connecting other pairs of canonically conjugate variables.

Our task will be to place time and space coordinates on an even footing. We shall first take up relativistic quantum mechanics, reserving the latter part of our discussion for relativistic quantum field theory.

Conventionally, quantum theories are formally available in a Hamiltonian rather than a Lagrangian formalism, and that for good reason. In classical mechanics, the Hamiltonian (canonical) equations of motion are formally simpler than the Lagrangian equations of motion; in fact in the Hamiltonian formalism we have available a general law of motion for *any* dynamical variable,

$$\frac{dQ}{dt} = (Q, H) + \frac{\partial Q}{\partial t}, \tag{38.1}$$

which utilizes the *Poisson bracket*, a commutator that is readily correlated with quantum-theoretical commutators. From the point of view of relativity, however, the Lagrangian formulation of mechanics is more natural than the Hamiltonian formulation, because in the former it is always possible to include the time coordinate among the coordinates of configuration space and to express the action integral over a trajectory with the help of an arbitrary parametric representation of the trajectory. As a matter of fact, a separate parameter may be introduced for the trajectory of each constituent particle ("many-time formalism"), and that is what we did in Sect. 22ff.

However, in Sect. 25 we showed that even in relativistic mechanics one can construct a Hamiltonian formalism without violating the basic symmetry between the four coordinates of each particle, by retaining the arbitrary parameter of the Lagrangian formulation and expressing the dynamical law in terms of this parameter.

In Sect. 39 we shall first develop the Hamiltonian non-quantum formalism a little farther than we did in Sect. 26. We shall in particular study the group of transformations that belongs naturally to the hypersurface in phase space on which the Hamiltonian constraint (25.4) is satisfied. In this manner we shall obtain a special set of dynamical variables, the generators of permissible transformations, which alone will be assumed to possess quantum-theoretical analogues. With their help we shall then go over to a discussion of the properties of a Lorentz-covariant quantum mechanics.

39. Canonical transformation groups with constraints. We shall begin by developing a slightly more general formal theory than is required for our immediate purpose. Let us consider a mechanical system whose dynamical law is described by a Hamiltonian $H(q, p)$, with the usual equations of motion (25.8), (25.11), and whose canonical coordinates $q_1, \ldots, p_n$ also satisfy a set of N constraints $C_1, \ldots, C_N$,

$$C_s(q, p) = 0, \qquad s = 1, \ldots, N. \tag{39.1}$$

For the requirements to be consistent, the time derivatives of the constraints, i.e. their Poisson brackets with the Hamiltonian H, must vanish, at least modulo the constraints themselves. Later on, we shall consider specifically the case in which there is only one constraint, $H=0$. But there are other applications of the theory, and so we shall develop first the general case, indicating later what simplifications result from the specialization to the Hamiltonian constraint.

For a more efficient notation, we shall combine the canonical coordinates (q, p) into a single set, ξ^ϱ, $\varrho = 1, \ldots, 2n$. The equations of motion then read

$$\varepsilon_{\varrho\sigma} \frac{d\xi^\sigma}{d\vartheta} = \frac{\partial H}{\partial \xi^\varrho}, \qquad \varepsilon = \begin{pmatrix} 0 & -I \\ +I & 0 \end{pmatrix}, \qquad \varepsilon_{\varrho\sigma} + \varepsilon_{\sigma\varrho} = 0. \tag{39.2}$$

The system is not capable of any state in which the constraints (39.1) are violated. Of the whole $2n$-dimensional phase space only the $(2n-N)$-dimensional subspace (or hypersurface) (39.1) possesses physical interest. We can introduce on this subspace a new coordinate system x^k, $k=1, \ldots, (2n-N)$, and rewrite the canonical equations of motion in terms of these new coordinates. We may use the new coordinates to describe the constraint-subspace parametrically, by expressing the values of the original canonical coordinates on the subspace as functions of the x, i.e. $\xi^\varrho(x^k)$. Accordingly, we have

$$\frac{d\xi^\varrho}{d\vartheta} = \frac{\partial\xi^\varrho}{\partial x^k}\frac{dx^k}{d\vartheta}, \qquad \varepsilon_{\varrho\sigma}\frac{\partial\xi^\sigma}{\partial x^k}\frac{dx^k}{d\vartheta} = \frac{\partial H}{\partial\xi^\varrho}, \tag{39.3}$$

and

$$\left.\begin{aligned} &\varepsilon_{mn}\frac{dx^n}{d\vartheta} = \frac{\partial H}{\partial x^m}, \qquad \varepsilon_{mn} \equiv \frac{\partial\xi^\varrho}{\partial x^m}\frac{\partial\xi^\sigma}{\partial x^n}\varepsilon_{\varrho\sigma}, \qquad \frac{\partial H}{\partial x^m} \equiv \frac{\partial H}{\partial\xi^\varrho}\frac{\partial\xi^\varrho}{\partial x^m}, \\ &\varepsilon_{mn} + \varepsilon_{nm} = 0. \end{aligned}\right\} \tag{39.4}$$

In general ε_{mn} is some function of the x^k. It is an antisymmetric matrix, and it may be regular or singular. If the number of constraints is odd, then ε_{mn} must be a singular matrix, as any skewsymmetric square matrix with an odd number of rows (and columns) has a vanishing determinant.

We shall now construct the (infinitesimal) group of all coordinate transformations in phase space that map the constraint-hypersurface on itself and which preserve the canonical form of the equations of motion on the hypersurface. (We require nothing concerning the equations of motion in the remainder of phase space.) If a transformation maps the constraint hypersurface on itself, then that means that with fixed functions $\xi^\varrho(x^k)$ a given point on the constraint-hypersurface x^k will now be identified by different values of the x^k-coordinates. Hence our infinitesimal transformation can be described in terms of the δx^k only,

$$\delta\xi^\varrho = \frac{\partial\xi^\varrho}{\partial x^k}\delta x^k, \tag{39.5}$$

and the ε_{mn} remain unchanged functions of their arguments x^k. It turns out that the requirement that the equations of motion should be reproduced on the constraint-hypersurface (possibly with a changed Hamiltonian) leads to the requirement that the δx^k are related to a *generator* Γ by the equations

$$\varepsilon_{mn}\delta x^n = \frac{\partial\Gamma}{\partial x^m}\left(= \frac{\partial\xi^\varrho}{\partial x^m}\frac{\partial\Gamma}{\partial\xi^\varrho}\right). \tag{39.6}$$

The same equation is also the necessary and sufficient condition for the skew-symmetric tensor ε_{mn} to remain an invariant set of functions of the arguments x^k.

Eqs. (39.6) do not require that the generator be defined off the hypersurface. If the matrix ε_{mn} is regular, then any function of the x^k may serve as generator. However, if ε_{mn} is singular, so that there exist one or several vectors $U^n_{(s)}$

$$\varepsilon_{mn}U^n_{(s)} = 0, \tag{39.7}$$

then multiplication of Eq. (39.6) by any one of these *null vectors* $U^m_{(s)}$ yields the set of conditions on the generators:

$$\frac{\partial\Gamma}{\partial x^k}U^k_{(s)} = 0. \tag{39.8}$$

If we define, in the usual manner, the commutator of two infinitesimal transformations generated by A and B, respectively, each of which satisfies Eq. (39.8), and if we construct the generator of that commutator, then we find the expression

$$\left.\begin{aligned} \Gamma = \{A, B\} = \frac{\partial A}{\partial x^k}\delta_B x^k &= -\frac{\partial B}{\partial x^k}\delta_A x^k \\ &= \varepsilon_{mn}\delta_A x^m \delta_B x^n. \end{aligned}\right\} \tag{39.9}$$

This expression contains explicit reference to the transformations themselves, rather than merely to their generators; this distinction is not entirely trivial, because for singular ε_{mn} each permissible generator determines an infinity of infinitesimal transformation δx^k, which differ from each other by arbitrary linear combinations of the null vectors $U^k_{(s)}$. Nevertheless, the commutator bracket (39.9) is actually independent of the particular choice of the transformations, as a brief calculation will show.

In particular there are transformations belonging to the generator zero, which have the form of a set of linear combinations of the null vectors $\boldsymbol{U}_{(s)}$, Eq. (39.7),

$$\delta_B x^k = \beta^{(s)} U^k_{(s)}. \tag{39.10}$$

Inserted into any one of the forms (39.9) they yield the commutator zero, so that we may conclude that the commutator bracket of any permissible generator with any generator that vanishes (modulo the constraints) vanishes itself (also modulo the constraints). Hence the transformations of the type (39.10), i.e. the transformations belonging to the generator zero, form an invariant (normal) subgroup of the group of all transformations (39.6). If we form the *factor group* (or *quotient group*) of the large group with respect to its invariant subgroup, then this factor group will be in a one-to-one relationship to the set of all permissible generators: To each member of the factor group belongs exactly one generator, provided we consider generators differing only by constraints as "the same generator"; to each generator belongs exactly one member of the factor group. Herewith a commutator algebra has been constructed among the generators that satisfy the condition (39.8). This commutator algebra is the natural point of departure for any relativistic quantum mechanics.

In conclusion we shall return to the situation that arises when we represent the motion of a particle or of a mechanical system parametrically in order to treat the time coordinate on the same footing as the configuration coordinates. We found in Sect. 26 that parametrization leads to canonical momentum components that are homogeneous of the zeroth degree in the velocities and hence not algebraically independent of each other. Their algebraic dependence may be expressed as a constraint, Eq. (26.4), which also serves as the Hamiltonian of the theory. If the system originally had n degrees of freedom, then the phase space of the parametrized system will be $(2n+2)$-dimensional, hence the constraint hypersurface $(2n+1)$-dimensional. To describe this constraint hypersurface we shall introduce parameters $x^r (r=1, \ldots, 2n+1)$ and construct the form ε_{rs}

$$\varepsilon_{rs} = \sum_m \left(\frac{\partial q_m}{\partial x^r} \frac{\partial p_m}{\partial x^r} - \frac{\partial p_m}{\partial x^r} \frac{\partial q_m}{\partial x^s} \right). \tag{39.11}$$

This form is singular; its null vector can be found easily:

$$\varepsilon_{rs} U^s = 0, \qquad U^s = (x^s, H). \tag{39.12}$$

As before, the parenthesis (,) denotes the ordinary Poisson bracket, with q_m, p_m as the canonical coordinates. The condition for prospective generators of infinitesimal canonical transformations that leave the constraint(s) intact, (39.8), reduces here to the single requirement that the generators have vanishing Poisson brackets with the Hamiltonian, i.e. that they be constants of the motion. This result is not unreasonable. As our constraint and the Hamiltonian are identical, only invariant transformations will preserve the constraint; and the generators of invariant transformations are the constants of the motion.

At first sight it might appear that the restriction to constants of the motion prevents us from considering Poisson (or rather Dirac) commutators between most variables of physical interest, but this is not the case. The condition (39.8) does not rule out constants of the motion that depend explicitly on the time coordinate. As a matter of fact, in a formalism in which we have reduced the role of the time coordinate to that of a configuration coordinate, the separation of the time-dependent from the time-independent constants of the motion would be unnatural. But every physical variable A at some particular time t_0 equals a constant of the motion, namely that function of the canonical coordinates (including the time) which permits us to determine the value of the physical variable $A(t_0)$ from t, $q_m(t)$, and $p_m(t)$. If we restrict ourselves to constants of the motion, we may even give meaning to brackets between variables at different times, something we must do if we wish to examine the transformation properties of our brackets under Lorentz transformations. The bracket between two dynamical variables $A(t_1)$ and $B(t_2)$ is to be interpreted as the bracket between the two constants of the motion which at times t_1 and t_2 equal A and B, respectively. If A and B themselves are scalar, then the bracket between them will be scalar as well.

A major application of Dirac commutators, to the gravitational field, will be presented in the following article on general relativity, Sect. 29.

40. Spin zero particles. We shall consider first the non-quantum properties of a mass point of rest mass m. The Lagrangian of such a particle may be given the form

$$L = -m\sqrt{\eta_{\varrho\sigma}\frac{dx^\varrho}{d\vartheta}\frac{dx^\sigma}{d\vartheta}} \tag{40.1}$$

[identical with Eqs. (23.6), (23.7)]. The resulting canonical momentum components,

$$p_\varrho = -m\,\eta_{\varrho\sigma}\frac{dx^\sigma}{d\tau} \tag{40.2}$$

[cf. Eq. (23.8)], satisfy the Hamiltonian constraint (25.4), which in our case turns out to be

$$H \equiv \frac{1}{2m}(\eta^{\varrho\sigma}p_\varrho p_\sigma - m^2) = 0. \tag{40.3}$$

Of the four components of the relativistic linear momentum, all have vanishing Poisson brackets with the Hamiltonian constraints, but only three (any three) are algebraically independent. Thus any three functions of the linear momentum components form a *complete set of commuting observables*. Their joint eigenstates span the Hilbert space.

We shall denote these three arbitrary functions by the symbols $f_k(p_\varrho)$, $k = 1, \ldots, 3$. The joint eigen state of these three functions that belongs to the eigen values f_1', f_2', f_3' shall be denoted by the symbol $|f_k'\rangle$, which is consistent with DIRAC's symbolism. In general a state $|y\rangle$ will be normalized if it satisfies the integral relationship

$$\int |\langle \boldsymbol{f}' | y\rangle|^2\, d^3f' = 1. \tag{40.4}$$

Suppose we wish to go over to another description of the quantum states of a free particle, say in terms of the three functions $g_l(p_\varrho)$. Then the condition of normalization takes the form

$$\int |\langle \boldsymbol{g}' | y\rangle|^2\, d^3g' = 1. \tag{40.5}$$

Furthermore, the three functions f_k are reversibly unique functions of the three functions g_l. It follows that

$$|\langle \boldsymbol{g}' | y \rangle|^2 = J \, |\langle \boldsymbol{f}' | y \rangle|^2, \tag{40.6}$$

where

$$J = \det \left| \frac{\partial f'}{\partial g'} \right|, \tag{40.7}$$

the Jacobian of the transition from f to g.

Let us consider in particular a transition to a new description that corresponds to an infinitesimal Lorentz transformation. In that case the new parameters g are in the same relationship to the Lorentz-transformed p'_ϱ as the f are to the original p_ϱ. For an infinitesimal transformation we must have

$$P_\mu{}^\varrho p_\varrho = \delta p_\mu = \frac{\partial p_\mu}{\partial f_k} \delta f_k, \tag{40.8}$$

and these four equations we can solve for the three quantities δf_k. A brief cal-calculation which utilizes the linear dependence of the Eqs. (40.8), yields[1]:

$$\left.\begin{aligned} \delta f_k &= F_k^\nu \, \delta p_\nu, \\ F_k^\nu &= \frac{1}{\Delta} \delta^{\nu\sigma\alpha\beta} \delta_{kmn} \, p_\sigma \frac{\partial p_\alpha}{\partial f_m} \frac{\partial p_\beta}{\partial f_n}, \\ \Delta &= \frac{1}{3} \delta^{\varrho\sigma\alpha\beta} \delta_{kmn} \, p_\sigma \frac{\partial p_\alpha}{\partial f_m} \frac{\partial p_\beta}{\partial f_n} \frac{\partial p_\varrho}{\partial f_k}. \end{aligned}\right\} \tag{40.9}$$

The Jacobian of this infinitesimal transformation of parameters is:

$$\delta J = P_\mu{}^\nu p_\nu F_k^\mu F_l^\varrho \frac{\partial^2 p_\varrho}{\partial f_k \, \partial f_l}. \tag{40.10}$$

We may substitute this expression into Eq. (40.6), modifying that equation for an infinitesimal parameter transformation. We find that the infinitesimal change of the probability density $|\langle \boldsymbol{f}' | y \rangle|^2$ as a function of the parameters is

$$\bar{\delta} |\langle \boldsymbol{f}' | y \rangle|^2 = \frac{1}{2} P_{\varrho\sigma} \frac{\partial}{\partial f_k} (p^\sigma F_k^\sigma - p^\varrho F_k^\varrho) \, |\langle \boldsymbol{f}' | y \rangle|^2. \tag{40.11}$$

We must now discover a linear Hermitian transformation of the matrix element $\langle \boldsymbol{f}' | y \rangle$ alone that is consistent with the transformation law (40.11) for its absolute square. The most general transformation of this kind is

$$\left.\begin{aligned} \bar{\delta} \langle \boldsymbol{f}' | y \rangle &= -\frac{i}{\hbar} C \langle \boldsymbol{f}' | y \rangle = -\frac{i}{\hbar} P_{\varrho\sigma} C^{\varrho\sigma} \langle \boldsymbol{f}' | y \rangle, \\ C^{\varrho\sigma} &= \tfrac{1}{4} [X^k (p^\varrho F_k^\sigma - p^\sigma F_k^\varrho) + (p^\varrho F_k^\sigma - p^\sigma F_k^\varrho) X^k] + A^{\varrho\sigma}(\boldsymbol{f}). \end{aligned}\right\} \tag{40.12}$$

X^k is the Hermitian operator

$$X^k = -\frac{\hbar}{i} \frac{\partial}{\partial f_k}. \tag{40.13}$$

$A^{\varrho\sigma}(\boldsymbol{f})$ is a set of six arbitrary functions of the parameters f_k. They are determined by the further requirement that the commutator of two operators $C^{\varrho\sigma}$ correspond to the group-theoretical commutator of the two infinitesimal Lorentz transformations; they are found to vanish in the course of a calculation which is straightforward but somewhat lengthy and which we shall not reproduce here.

[1] The symbols δ_{kmn} and $\delta^{\varrho\sigma\alpha\beta}$ are alternators as in Sect. 15, p. 137.

In terms of three arbitrarily chosen, algebraically independent functions of the four components of the linear momentum of a free particle obeying the Klein-Gordon equation, the six expressions (40.12) generate infinitesimal Lorentz transformations. Hence they are the quantum-mechanical operators corresponding to the components of the relativistic angular momentum. For illustration we shall determine the explicit expressions for $C^{\varrho\sigma}$ that result if we adopt as our three parameters the three spatial components of the angular momentum. In that case a brief calculation leads to the expressions

$$\left.\begin{aligned} C^{kl} &= \tfrac{1}{2}(p^k X^l - p^l X^k), \\ C^{k0} &= -\tfrac{1}{4}(p^0 X^k + X^k p^0). \end{aligned}\right\} \tag{40.14}$$

p^0 is to be understood as the function $(m^2 + p_k p_k)^{\frac{1}{2}}$ of the three chosen parameters. In conclusion, it might be noted that the three operators X^k are to be interpreted as operators that represent the spatial coordinates of a particle on some (arbitrary) space-like hypersurface.

We have now constructed a Hilbert space that is appropriate to the actual number of degrees of freedom of our system, the free mass point of zero spin. Our construction is relativistic in that it does not depend on the accidental choice either of a particular Lorentz frame or of a set of f-parameters. All observables of our system, i.e. physical quantities that can in principle be determined, are representable by Hermitian operators in the Hilbert space. In the next section we shall carry out a similar construction for a mass point of spin $\frac{1}{2}$.

41. Neutrinos[1]. The wave equation of the (free) neutrino, in terms of two-component spinors, has the form

$$\sigma^\varrho p_\varrho u = \frac{\hbar}{i}\sigma^\varrho u_{,\varrho} = 0. \tag{41.1}$$

The notation is the same as in Sects. 16 and 17. u is the wave function for a single particle. Under space reflections and under time reversal this equation goes over into the form

$$\tau^\varrho p_\varrho v = 0, \qquad v = \pm\varepsilon\bar{u}. \tag{41.2}$$

In one and the same frame of reference, the two wave equations (41.1) and (41.2) represent two different types of particle, the free neutrino and the antineutrino, respectively. Both have the rest mass zero and travel at the speed of light. Both particles have a spin which for every observer is always parallel to its direction of travel. The difference between neutrinos and antineutrinos is that the sense of rotation of the spin when viewed in the direction of the motion of the particle is counterclockwise for one particle, clockwise for the other. Except for the fact that one particle is the mirror image of the other, their characteristics are presumably the same. In our discussion we shall primarily refer to the particle that obeys Eq. (41.1).

A neutrino has three translatory degrees of freedom, represented conveniently by the three constants of the motion p_k. It is not difficult to show that p_0 is then determined, except for its sign. Multiply Eq. (41.1) from the left by the spinor quantity $\tau^\alpha p_\alpha$. Because of the algebraic relationship between the spin matrices (17.7) it follows immediately that

$$\eta^{\mu\nu} p_\mu p_\nu = 0, \qquad p_0 = \pm\sqrt{\boldsymbol{p}^2}. \tag{41.3}$$

Depending on whether p_0 is positive or negative, Eq. (41.1) will be satisfied by a spinor u whose components are functions of the p_k, except for a common factor.

[1] See references given in Sect. 16.

Hence the spin of a neutrino is determined uniquely by the three components of its (three-dimensional) linear momentum and the sign of its relativistic mass (i.e. energy). For the usual choice of the components of the spin matrices the ratio between the two components of the spinor u is found to be

$$\frac{u^2}{u^1} = -\frac{p_1 + i p_2}{p_0 - p_3}. \tag{41.4}$$

We shall now construct a Hilbert space for the possible states of a single neutrino by two alternative (but equivalent) methods. In the first method we shall start out with the differential equation satisfied by the neutrino wave in configuration (i.e. coordinate) space, the second version of Eq. (41.1). According to this equation, we can construct an ordinary four-vector whose divergence vanishes,

$$v^\varrho = u^* \sigma^\varrho u, \qquad v^\varrho{}_{,\varrho} = 0. \tag{41.5}$$

Provided the wave function u satisfies the usual boundary conditions at infinity, it follows from GAUSS's theorem that the integral

$$J = \int v^\varrho \, d\Sigma_\varrho \tag{41.6}$$

is invariant with respect to Lorentz transformations and with respect to changes in the surface of integration which is space-like and extends in all directions to infinity (Sect. 15). Hence the integral (41.6) may be used as an invariant normalization integral for the wave function.

Within this formalism we shall also construct the operators representing linear and angular momentum, as generators of (infinitesimal) Lorentz transformations. Under the most general inhomogeneous Lorentz transformation,

$$\delta x^\varrho = P^\varrho{}_\sigma x^\sigma + D^\varrho \tag{41.7}$$

the wave function u transforms as follows:

$$\bar{\delta} u (x^\mu) = \delta a \, u - u_{,\varrho} \, \delta x^\varrho = \frac{i}{\hbar} Q u. \tag{41.8}$$

This equation defines the Hermitian operator Q, which in turn must be a function of the ten parameters that define an infinitesimal Lorentz transformation. The infinitesimal matrix δa is related to the coefficients of the infinitesimal homogeneous Lorentz transformation, the $P^\varrho{}_\sigma$, by the infinitesimal analog of Eq. (16.14). We may conveniently introduce two real three-vectors with the components α_k, β_k, respectively, in terms of which we write both the matrix δa and the coefficients $P^\varrho{}_\sigma$. These expressions are:

$$\left.\begin{aligned} &\delta a = \tfrac{1}{2}(\alpha_k + i \beta_k) \sigma^k, \\ &P^0{}_0 = 0, \qquad P^0{}_k = P^k{}_0 = \alpha_k, \\ &P^k{}_l = \delta^{klm} \beta_m. \end{aligned}\right\} \tag{41.9}$$

In terms of these two three-vectors and the displacement four-vector D^ϱ, the operator Q may be given the form

$$\left.\begin{aligned} Q = \sigma^k p_k D^0 - p_k D^k + [\tfrac{1}{2}(x^k p_l + p_l x^k) \sigma^l - p_k x^0] \alpha_k + \\ + \left[\frac{\hbar}{2} \sigma^k + \delta^{klm} x^l p_m\right] \beta_k, \qquad p_k = \frac{\hbar}{i} \frac{\partial}{\partial x^k}. \end{aligned}\right\} \tag{41.10}$$

The coefficients are all constants of the motion whose physical significance is that of linear momentum, angular momentum, and location at $x^0 = 0$, respectively.

That this construction is Lorentz-invariant may be inferred from the method of its derivation, and subsequently proved explicitly, by the demonstration that the Lie algebra of the ten constants of the motion that we have obtained is identical with the Lie algebra of the infinitesimal Lorentz group; hence one is a representation of the other.

We shall now construct our Hilbert space in a manner similar to that followed in Sect. 40. Instead of working in the four-dimensional coordinate space, we consider the four-dimensional momentum space, and in it the three-dimensional conical hypersurface obeying Eq. (41.3). If we let u be a function of the four arguments p_ϱ, it can be different from zero only on that hypersurface, which we might again represent parametrically, in terms of three parameters f_k. Now if we form on that hypersurface the four-vector $u^* \sigma^\varrho u$ [with arguments u obeying Eq. (41.4), it is easy to show that this vector is a null-vector (i.e. a vector with vanishing norm) and, in fact, parallel to the four-momentum p_ϱ. On the other hand, if we form the hypersurface element $d S^\varrho$,

$$d S^\varrho = \delta^{\varrho\alpha\beta\gamma} \frac{\partial p_\alpha}{\partial f_1} \frac{\partial p_\beta}{\partial f_2} \frac{\partial p_\gamma}{\partial f_2} d^3 f, \tag{41.11}$$

in analogy to the expression (15.8), that vector, also, turns out to be parallel to the momentum four-vector. Accordingly, the scalar product $u^* \sigma_\varrho u \, d S^\varrho$, which one might envisage as the analog of the integrand of Eq. (41.6), vanishes. We obtain a nonvanishing integrand by forming the constants of proportionality ν and $d S$, respectively, between these three parallel vectors, defined by the equations

$$u^* \sigma_\varrho u = \nu p_\varrho, \qquad d S^\varrho = p^\varrho d S. \tag{41.12}$$

The integral

$$J' = \int \nu \, d S \tag{41.13}$$

is then Lorentz-invariant. If we choose as our three parameters f_k the three momentum components p_k, the expressions (41.12) turn into

$$\nu = p_0^{-1} u^* u; \qquad d S = p_0^{-1} d^3 p. \tag{41.14}$$

The spinor $p_0^{-1} u$ of this expression is the three-dimensional Fourier transform of the x^ϱ-dependent spinor u of Eq. (41.6) if the hypersurface of integration of Eq. (41.6) is chosen as $x^0 = 0$. Three-dimensional Fourier transformation is, of course, not a Lorentz-covariant procedure. On the other hand, the four-dimensional Fourier transform of an x^ϱ-dependent spinor u that obeys Eq. (41.1) is not well-defined because the four-dimensional integral over $u^* u$ is necessarily infinite; if we form this Fourier transform formally, the resulting expression will contain a one-dimensional delta function. That is why the finite expressions in x-space and in p-space are related to each other somewhat indirectly.

The individual points on the conical hypersurface (41.3) correspond reversibly uniquely to all the different stationary states of the neutrino. In a Lorentz transformation this cone is mapped on itself. The normalization integral (41.13) is manifestly invariant. These facts are sufficient to prove the Lorentz invariance of our Hilbert space.

42. The spinning electron. Dirac was the first to succeed in formulating a relativistically invariant wave equation for a particle that is of the first order of differentiation and thus may be cast into the form of a Schrödinger-type wave equation,

$$\frac{\hbar}{i} \frac{\partial \psi}{\partial t} + H_{\mathrm{op}} \psi = 0. \tag{42.1}$$

DIRAC's wave equation describes a particle of non-zero rest mass with a spin of $\frac{1}{2}$. Moreover, the components of his wave function, which together form a four-component spinor, lend themselves to the formation of a time-like vector whose divergence vanishes and whose zero-component is positive-definite. In the one-particle formalism, this vector is treated as the probability density and probability flux.

In some respects these aspects of DIRAC's theory are today not as highly valued as they were in the twenties. Today a one-particle wave function is ordinarily viewed as the point of departure for hyperquantization, that is the transition to a formulation in which any number of similar particles may be present. In a hyperquantized theory, the original one-particle wave function goes over into a field operator, and the probability density into the density operator for particles. A Schrödinger equation of the type (42.1) need not hold for the field operators, but it will exist for the state functional of the n-particle system, whose norm is again equal to unity. In this section, we shall not concern ourselves with quantum field theories, which are treated later, but present DIRAC's original theory of a single electron, further simplified by the assumption that the external electromagnetic field vanishes.

In terms of four-component spinors DIRAC's equation for the free electron may be written in the form

$$\frac{\hbar}{i}\gamma^{\varrho}\psi_{,\varrho} + \mu\psi = 0 \quad \text{or} \quad (\gamma^{\varrho} p_{\varrho} + \mu)\psi = 0. \tag{42.2}$$

In terms of two-component spinors, the same equation is

$$\left.\begin{aligned} -\frac{\hbar}{i}\tau^{\varrho} v_{,\varrho} + \mu u &= 0, \\ +\frac{\hbar}{i}\sigma^{\varrho} u_{,\varrho} + \mu v &= 0, \\ \text{or} \qquad -\tau^{\varrho} p_{\varrho} v + \mu u &= 0, \\ +\sigma^{\varrho} p_{\varrho} u + \mu v &= 0. \end{aligned}\right\} \tag{42.3}$$

The 4×4 matrices γ^{μ} are defined by Eq. (18.7) and satisfy the algebraic relations (18.8). By very simple algebraic manipulations on either Eqs. (42.2) or (42.3) one can prove that

$$\eta^{\mu\nu} p_{\mu} p_{\nu} - \mu^2 = 0. \tag{42.4}$$

In the four-dimensional momentum space the permissible states are located on the same two-branched hyperboloid hypersurface that we encountered in Sect. 40.

We shall follow the same procedure as in the case of the neutrino and first construct an invariant integral of normalization, and the operators associated with infinitesimal Lorentz transformations, in configuration space. We shall form the four-vector ν^{ϱ},

$$\nu^{\varrho} = \psi^{\dagger}\gamma^{\varrho}\psi = u^{*}\sigma^{\varrho} u - v^{*}\tau^{\varrho} v. \tag{42.5}$$

Its divergence vanishes because of the wave equation (42.2) or (42.3). Hence the integral

$$J = \int \nu^{\varrho}\, d\Sigma_{\varrho} \tag{42.6}$$

over an infinitely extended space-like hypersurface is independent of the choice of that surface and invariant with respect to Lorentz transformations. For the special choice of the coordinate hypersurface $x^0 = 0$ this integral takes the form

$$J = \int \psi^{*}\psi\, d^3x = \int (u^{*} u + v^{*} v)\, d^3x. \tag{42.7}$$

The integrand is manifestly positive-definite.

Under an infinitesimal Lorentz transformation the transformation law for the wave function is

$$\left.\begin{aligned}\bar{\delta}\psi &= \delta A\,\psi - \psi,_{\varrho}\,\delta x^{\varrho},\\ \bar{\delta}u &= \delta a\,u - u,_{\varrho}\delta x^{\varrho},\\ \bar{\delta}v &= -\,\delta a^{*}\,v - v,_{\varrho}\,\delta x^{\varrho}, \qquad \delta x^{\varrho} = P^{\varrho}{}_{\sigma}\,x^{\sigma} + D^{\varrho}.\end{aligned}\right\}\tag{42.8}$$

Just as in Sect. 41 we shall again introduce the two real three-vectors α_k and β_k. With their help the generator of an infinitesimal Lorentz transformation comes out as

$$\left.\begin{aligned}Q = -D^0\begin{pmatrix}+\sigma^k p_k, & +\mu\\ +\mu, & -\sigma^k p_k\end{pmatrix} - D^k p_k + \boldsymbol{\beta}\cdot\left(\boldsymbol{x}\times\boldsymbol{p} + \frac{\hbar}{2}\boldsymbol{\sigma}\right) -\\ -\alpha_k\begin{pmatrix}x^0 p_k + \frac{1}{2}\sigma_l(x^k p_l + p_l x^k), & -\mu x^k\\ \mu x^k, & x^0 p_k - \frac{1}{2}\sigma_l(x^k p_l + p_l x^k)\end{pmatrix}.\end{aligned}\right\}\tag{42.9}$$

The p_k are the usual differential operators; the σ-symbols signify PAULI's three spin matrices. Q is manifestly Hermitian and defines the appropriate observables that we identify as the linear (three-) momentum and as the angular momentum. The operators that represent the energy and location at zero time are much harder to interpret, because of the off-diagonal elements in the matrices.

Physically, the Dirac particle differs from those discussed in the preceding two sections in that it possesses an internal degree of freedom, its spin, in addition to the translatory degrees of freedom common to all free particles with non-vanishing mass. In order to exhibit this spin, it is desirable to introduce linear combinations of the spinors u and v which belong uniquely to the two signs of energy belonging to every set of spatial linear momentum. Such linear combinations exist; they are determined by the wave equation (42.3). But the coefficients depend on the values of the p_k. That is to say, we can diagonalize the energy matrix; but the diagonalizing matrix elements themselves are not constants.

43. Energy separation in DIRAC's theory. To simplify the notation we shall introduce the symbols:

$$\left.\begin{aligned}\frac{1}{\mu}p_k = \varphi_k,\quad \sigma^k\varphi_k = \boldsymbol{\varphi},\quad p_k p_k = \mu^2\varphi^2,\quad 1+\varphi^2 = e_0^2,\quad e_0 > 0\\ \boldsymbol{\varphi}^2 = \varphi^2,\quad (e_0+\boldsymbol{\varphi})(e_0-\boldsymbol{\varphi}) = 1.\end{aligned}\right\}\tag{43.1}$$

With their help we define the two sets of operators which appear in the expression for the operator Q, (42.9): The coefficient of D^0, the energy,

$$E = \mu\begin{pmatrix}-\boldsymbol{\varphi}, & -1\\ -1, & \boldsymbol{\varphi}\end{pmatrix} \equiv \mu\,\varepsilon \tag{43.2}$$

and the coefficient of the vector α_k,

$$\left.\begin{aligned}\xi^k &= -\,x^0 p_k + \tfrac{1}{2}(E\,x^k + x^k E)\\ &\equiv \mu\left[-\,x^0\varphi_k + \tfrac{1}{2}(\varepsilon\,x^k + x^k\,\varepsilon)\right],\end{aligned}\right\}\tag{43.3}$$

which is to be interpreted as the location of the particle at time zero, multiplied by its relativistic mass. We shall diagonalize both of these matrices, at least partly, by means of a similarity transformation M,

$$M = \begin{pmatrix}1, & e_0-\boldsymbol{\varphi}\\ \boldsymbol{\varphi}-e_0, & 1\end{pmatrix},\qquad M^{-1} = \frac{e_0+\boldsymbol{\varphi}}{2e_0}M^{*}. \tag{43.4}$$

This matrix implies the transition from the two spinors u and v to two linear combinations,

$$\left.\begin{aligned} U &= u + (e_0 - \varphi)\, v, \\ V &= v - (e_0 - \varphi)\, u, \end{aligned}\right\} \tag{43.5}$$

which represent solutions of the wave equation (in momentum space) belonging to negative and to positive energy values, respectively. This similarity transformation diagonalizes the matrix ε outright. We find:

$$M\,\varepsilon\, M^{-1} = e_0 \begin{pmatrix} -1 & 0 \\ 0 & 1 \end{pmatrix}. \tag{43.6}$$

The (spatial) coordinates x_k are converted into matrices as follows:

$$M\, x^k\, M^{-1} = x^k + \frac{\hbar}{i\,\mu}\left(\sigma^k - \frac{\varphi_k}{e_0}\right) \frac{e_0 + \varphi}{2 e_0} \begin{pmatrix} 1, & \varphi - e_0 \\ e_0 - \varphi, & 1 \end{pmatrix}. \tag{43.7}$$

That is to say, if we multiply a single-energy wave function by x_k, then the result will be a wave function that is a linear combination of stationary solutions loated on both positive-energy and negative-energy branches of the hyperboloid hypersurface. Nevertheless, the operator (43.3) will be split; an infinitesimal Lorentz transformation will not mix the two branches. Indeed a brief calculation will show that $M\xi_k M^{-1}$ is

$$\left.\begin{aligned} M\,\xi^k\, M^{-1} = \mu \Bigl\{ &- x^0\,\varphi_k + \begin{pmatrix} -1, & 0 \\ 0, & 1 \end{pmatrix} \Bigl[\frac{1}{2}\,(e_0\, x^k + x^k\, e_0) + \\ &+ \frac{\hbar}{2 i\,\mu}\,(e_0 + \varphi)\left(\sigma^k - \frac{\varphi_k}{e_0}\right)\Bigr]\Bigr\}. \end{aligned}\right\} \tag{43.8}$$

Applied to both the linear and to the (total) angular momentum, the similarity transformation with M merely reproduces these expressions. In summary, then, this similarity transformation has the effect of bringing all ten of the usual constants of the motion into a form in which the states of positive and those of negative energy are completely uncoupled. Within each energy branch the stationary states are completely characterized, for instance by the values of the linear momentum components p_k and by whether the spin is in the direction of the motion or backwards. The eigenvalues of the operator φ, which is a constant of the motion, are $\pm\varphi$. At the one point $p_k = 0$ any one component of the spin is a suitable constant of the motion for characterizing the two possible states.

This characterization, at one point by the eigenstates of one matrix σ^k and elsewhere by the matrix φ, suffers from the fact that it is not covariant. Under a Lorentz transformation these variables are subject to relatively involved transformation laws. It is possible to show that the observable φ is actually one component of a four-vector, all of whose components are constants of the motion. In the four-component formalism we form the spinor[1]

$$\left.\begin{aligned} \sigma^{\mu\nu} &= \frac{1}{2i}\,(\gamma^\mu \gamma^\nu - \gamma^\nu \gamma^\mu) = \frac{1}{2i}\begin{pmatrix} \tau^\nu \sigma^\mu - \tau^\mu \sigma^\nu, & 0 \\ 0, & \sigma^\nu \tau^\mu - \sigma^\mu \tau^\nu \end{pmatrix}, \\ \sigma^{k l} &= \frac{1}{2i}\,[\sigma^k, \sigma^l] = \delta^{k l m}\,\sigma^m, \\ \sigma^{k 0} &= i \begin{pmatrix} 1 & 0 \\ 0 & -1 \end{pmatrix} \sigma^k, \end{aligned}\right\} \tag{43.9}$$

[1] The symbols δ^{klm} and $\delta^{\alpha\beta\mu\nu}$ have the same numerical values as those on p. 186.

and with its help the four-vector S^α,

$$S^\alpha = \tfrac{1}{2}\,\delta^{\alpha\beta\mu\nu}\,p_\beta\,\sigma_{\mu\nu}. \tag{43.10}$$

The component S^0 is simply equal to $\sigma^k\,p_k = \mu\,\varphi$. As for the remaining components there appears an uncertainty concerning the sequence of the two factors p_0 and σ^k, which do not commute. This uncertainty can be resolved easily, however, by subjecting S^0 to an infinitesimal Lorentz transformation. An infinitesimal Lorentz transformation is generated by the operator ξ^k, Eq. (43.3). We find for S^k the expression

$$\boldsymbol{S} = \frac{1}{2}\,(p_0\,\boldsymbol{\sigma} + \boldsymbol{\sigma}\,p_0) = -\frac{\mu}{2}\,(\boldsymbol{\varepsilon}\,\boldsymbol{\sigma} + \boldsymbol{\sigma}\,\boldsymbol{\varepsilon}). \tag{43.11}$$

If we perform our similarity transformation on this expression, we obtain the usual separation of positive and negative energy, as follows:

$$M\,\boldsymbol{S}\,M = \mu\,e_0\begin{pmatrix}-1 & 0\\ 0 & 1\end{pmatrix}\left(\boldsymbol{\sigma} + \frac{1}{e_0}\,\boldsymbol{\varphi}\times\boldsymbol{\sigma}\right). \tag{43.12}$$

On either energy branch, then, the separation of the two spin states for a given p_k may be accomplished by an arbitrary linear combination of the S^ϱ, which goes over into another linear combination of the same vector components under any Lorentz transformation.

II. Quantum field theory.

44. Hyperquantization of particle fields. For a quantum theory to be relativistically invariant it must satisfy the following requirements: (a) All solutions of the Schrödinger equation must be expandable with respect to an orthonormal complete system of such solutions, which are the relativistic analog of a set of complete orthonormal base vectors (at a time t_0) in the non-relativistic theory. (b) Under a Lorentz transformation the base system undergoes a unitary transformation; as a corollary of this requirement, the Schrödinger equation itself is invariant under Lorentz transformations; as a further corollary, the norm of the Hilbert space is invariant under Lorentz transformations. This last property may be non-trivial if the set of commuting operators that defines the original base system possesses a (partly) continuous spectrum.

In the past several sections, devoted to single-particle theories, we have demonstrated that various types of free single particles and their quantum states satisfy these requirements. Field theories must possess the same properties. Because of their greater complexity, the discussion is somewhat more involved. Quantum field theories originate in two different connections. Given a classical field theory, such as the theory of the electromagnetic field, we may set ourselves the task of replacing the classical dynamical variables by quantum-theoretical operators (observables). Or we may "hyperquantize" a single-particle theory, that is to say, construct a formalism in which we can describe arbitrary numbers of like particles, which are either bosons or fermions. It is well known that both approaches lead to very similar structures; this similarity may be considered one of the most striking contributions that quantum field theory has made to the conceptual foundations of physics. Because of this similarity, we may think of field quanta such as photons in terms of particles, and conversely a collection of like particles, such as electrons, possesses some of the attributes of a field. In this section we shall discuss hyperquantization of particle equations.

Suppose we describe the state of a single particle in terms of a complete set of commuting constants of the motion, as we did in Sects. 39 through 43, then we may describe a collection of like particles in terms of their distribution over this range of variables. Because at least some of these constants (such as the f_k) possess a continuous spectrum, the construction of a complete set of operators that would define the state of a many-particle system requires some elaboration. Generally speaking, the type of operator useful for our purpose is an occupation number, that is an operator that counts the number of particles in a particular state. Given a complete set of mutually orthogonal states of a single particle, we may obtain a complete set of commuting operators for the many-particle system by associating a separate occupation number with each one of the single-particle states. However, to this end it is desirable that the single-particle states belong to a denumerable, rather than a continuous, set.

To this end we shall introduce on the range of the f_k (or whatever continuous and unbounded parameters may have been used to identify single-particle states) a complete denumerable set of orthonormal functions, $G_s(f_k)$, $s=1, 2, \ldots$. With their help we may first set up a denumerable complete set of orthonormal states for a single particle,

$$|s\rangle = \int_{f'} \overline{G}_s(f') \, |f'\rangle \, d^3 f' \tag{44.1}$$

and

$$|f'\rangle = \sum_s G_s(f') \, |s\rangle . \tag{44.2}$$

If a single particle is in any one of these states, it is assuredly in none of the others. Hence for a collection of particles the statement that n_s particles are in the state "s", made separately for each one of the (infinitely many) states $s = 1, 2, \ldots$, is a complete description of the state of the system, provided we also know the statistics of the species of particles considered. Any set of occupation numbers $n_1, n_2, \ldots, n_s, \ldots$ represents a state of the system that excludes all states with different occupation numbers. Conversely, any state of the system whatsoever (that is compatible with the statistics) may be described as a linear superposition of states of the type $|(n_s)\rangle$. Hence these states represent a complete orthonormal set for the many-particle system. The operator whose eigenvalues are the possible integral values of n_s, N_s, commutes with all the $N_{s'}$ $(s' \neq s)$; accordingly the set of all N_s is a complete commuting set of operators for the many-particle system.

With respect to our chosen base system the matrix elements of the operators N_s are

$$\langle (n'_{s'}) | \, N_s \, | (n''_{s'}) \rangle = \prod_{s'=1}^{\infty} \delta_{n'_{s'} n''_{s'}} \, n_s . \tag{44.3}$$

If we were to go over to a different coordinate system in Hilbert space, based on some other complete set of orthonormal functions $F_r(f_k)$ on the domain f_k, the construction of the unitary matrix leading from one coordinate system to the other would not be overly difficult but at any rate cumbersome. Clearly, matrix elements leading from the coordinate system based on the K_s to that built on the F_r would differ from zero only for combinations in which the sum $\sum_{s=1}^{\infty} n_s$ equals the sum $\sum_{r=1}^{\infty} m_r$, either of these sums representing the total number of particles in the system. For such non-vanishing combinations each matrix element would be a sum of products of the numbers $\langle F_r | G_s \rangle$, where the number of times the index r recurs must equal m_r, and the number of times the index s recurs will be n_s.

These unitary transformations can be replaced very conveniently by the transformation laws of certain non-Hermitian operators, the creation and annihilation operators. The annihilation operator a_s is defined as follows. It has non-vanishing matrix elements only between such states where all occupation numbers on the left and on the right are the same except the occupation number n_s; n_s on the left must be one less than n_s on the right. Such a non-vanishing matrix element has the numerical value $\sqrt{n_s}$, where n_s is the index on the right. In other words,

$$\langle (n'_{s'}) | a_s | (n''_{s'}) \rangle = \prod_{s' \neq s} \delta_{n'_{s'} n''_{s'}} \, \delta_{(n'_s + 1) n''_s} \sqrt{n''_s} \,. \tag{44.4}$$

The creation operators a_s^* are simply defined as the Hermitian conjugates of the corresponding annihilation operators. These operators a_s, a_s^*, have the following properties that make them useful in the hyperquantization formalism:

(1) They possess very simple commutation (or anticommutation) rules. Every annihilation operator (anti-) commutes with every other annihilation operator, and every creation operator with every other creation operator. The (anti-) commutator between an annihilation operator and a creation operator is

$$[a_{s_1}, a_{s_2}^*] = \delta_{s_1 s_2} \,. \tag{44.5}$$

(2) All operators may be constructed as series in creation and annihilation operators. If an operator conserves the total number of particles in the system, then each term in the series has an equal number of creation and annihilation operators. In particular the operator N_s is

$$N_s = a_s^* \, a_s \,. \tag{44.6}$$

(3) Given two systems of orthonormal functions G_s and F_r as explained above. Then the transformation law leading from the a_s to the a_r, and vice versa, is simply

$$a_r = \sum_{s=1}^{\infty} \langle r | s \rangle \, a_s \,, \qquad \langle r | s \rangle \equiv \int_{f'} \overline{F}_r \, G_s \, d^3 f' \,. \tag{44.7}$$

In other words, the relationship between the annihilation operators belonging to different orthonormal systems of single-particle wave functions is the same as between the single-particle states themselves. Occupation number operators, for instance, do not possess this property.

If we are to investigate the behavior of hyperquantization formalisms under Lorentz transformations, then the transformation properties of annihilation operators (and, by implication, of creation operators) are of great value. Under a Lorentz transformation a coordinate system f_k (and possibly other, discrete observables needed to make up a complete set of commuting operators for a single-particle system) is subjected to a mapping of the type we have established in Sect. 40. That is to say, we obtain a representation of the Lorentz group if we adopt for the F_r above the Lorentz transforms of the G_s and hence obtain the unitary matrix with the elements $\langle r | s \rangle$, Eq. (44.7), as a representation of the Lorentz transformation that is being considered. In that case the a_s, which are operators in the many-particle formalism, transform with the same transformation coefficients as the base vectors of the single-particle system. And in view of the fact that all operators in the many-particle system may be considered as functions of the a_s, a_s^*, we may form the Lorentz transformation law of any and all observables in the many-particle formalism.

The transformation law of the G_s themselves is obtained as follows. Under an infinitesimal transformation of the parameters f_k the G_s transform as scalar densities,

$$\bar{\delta} G_s = -\frac{\partial G_s}{\partial f_k} \delta f_k + \frac{1}{2} G_s \frac{\partial \delta f_k}{\partial f_k}. \tag{44.8}$$

This law leaves their orthonormality intact. If we substitute for δf_k the expression (40.9), (40.8), we get

$$\bar{\delta} G_s = -\frac{i}{\hbar} P_{\mu\nu} \Lambda^{\mu\nu}{}_s{}^{s'} G_s', \tag{44.9}$$

where

$$\Lambda^{\mu\nu}{}_s{}^{s'} = \frac{\hbar}{4i} \frac{\partial}{\partial f_k} (F_k^\mu p^\nu - F_k^\nu p^\mu) \delta_s^{s'} + \frac{1}{2} (F_k^\mu p^\nu - F_k^\nu p^\mu) A^k{}_s{}^{s'}, \tag{44.10}$$

$$A^k{}_s{}^{s'} = \frac{\hbar}{i} \int \bar{G}_{s'} \frac{\partial G_s}{\partial f_k} d^3 f \equiv -\langle s' | X^k | s \rangle, \tag{44.11}$$

and otherwise the notation is the same as in Sect. 40. The matrices $\Lambda^{\mu\nu}{}_s{}^{s'}$ are a representation of the infinitesimal Lorentz group by matrices of infinite order. Whether the representation is irreducible or not depends on the particular choice of the functions G_s. At any rate, these matrices also determine the transformation law of the a_s, a_s^* under an infinitesimal Lorentz transformation.

The simple form of the transformation law (44.7) makes it possible to introduce operators in the hyperquantization formalism whose indices (s) are not discrete but continuous, and, more particularly, whose indices (or, more properly speaking, arguments) refer to points in the manifold whose coordinates are the f_k. We shall define

$$\alpha(f_k) = \sum_{s=1}^{\infty} G_s(f_k)\, a_s, \qquad a_s = \int \bar{G}_s(f_k)\, \alpha(f_k)\, d^3 f. \tag{44.12}$$

The corresponding operator

$$\nu(f_k) = \alpha^*(f_k)\, \alpha(f_k) \tag{44.13}$$

is not an occupation number but an occupation number density; its expectation value is the mean density of particles found in the vicinity of the point f_k. We shall now go on from the operator $\alpha(f_k)$ and introduce an annihilation operator whose arguments are the coordinates of the single-particle configuration space. We define this operator as follows:

$$\psi(x^0, x^k) = (2\pi)^{-\frac{3}{2}} \int \sqrt{\left| \frac{\partial(p_1 \cdots p_3)}{\partial(f_1 \cdots f_3)} \right|}\, e^{\frac{i}{\hbar} p_\varrho(f_k) x^\varrho}\, \alpha(f_k)\, d^3 f. \tag{44.14}$$

Its arguments are not operators but, in a manner of speaking, continuous indices. But as the domain of original continuous indices f_k is three-dimensional, one of the four indices x^0, x^k must be redundant; that is to say, there must exist a one-parametric relationship between annihilators $\psi(x^0, x^k)$ belonging to different numerical values of the arguments. We obtain this relationship immediately if we form the derivatives of ψ with respect to its arguments. Differentiation with respect to x^ϱ adds a factor $\frac{i}{\hbar} p_\varrho$ under the integral. There will exist some relationship between the momentum components of a single particle, the Hamiltonian constraint, which is also contained parametrically in the dependence of the p_ϱ on the f_k. To this relationship belongs a corresponding relationship between the derivatives of ψ, that is to say, ψ satisfies a partial differential equation. This equation has the same form as the single-particle differential equation for the wave equation. For instance, for spin-zero particles of rest mass μ this differential equation is

$$\hbar^2 \eta^{\mu\nu} \psi,_{\mu\nu} + \mu^2 \psi = 0. \tag{44.15}$$

The product $\psi^* \psi$ at a fixed world point represents the three-dimensional particle density at the time x^0.

For fermions the development of the formalism is very similar; the annihilation and creation operators satisfy the same transformation laws as they do for bosons, but they must be constructed so that instead of commutation laws they satisfy anticommutation laws. Again it is possible to define operators whose arguments are the configuration coordinates of a single particle, in such a manner that these operators satisfy the same differential equations as do the wave functions of a single particle. Because of this peculiar property of the annihilation and creation operators, which depends entirely on the simplicity of their transformation laws, the whole formalism has been dubbed hyperquantization. Formally, the wave functions are treated as if they were classical field variables, and subjected to a "second quantization". We shall quantize a true classical field in the next section.

45. Quantization of classical fields. Because the electromagnetic field possesses special features, connected with its gauge invariance, we shall in this section construct the quantum theory corresponding to a fictitious classical field, one that has only one component, u, and which satisfies the Lorentz-invariant field equation

$$\eta^{\mu\nu} u,_{\mu\nu} + \varkappa^2 u = 0. \tag{45.1}$$

We shall find that the quantization of this field leads to a theory of operators that have precisely the same properties as those obtained in the hyperquantization of scalar bosons in Sect. 44.

The Lagrangian of the field u has the form

$$L = \tfrac{1}{2} (\eta^{\mu\nu} u,_\mu u,_\nu - \varkappa^2 u^2), \qquad S = \int L \, d^4x. \tag{45.2}$$

In order to construct the commutators of this field, we shall consider those infinitesimal mappings in the function space consisting of functions $u(x^\varrho)$ that carry solutions of the field equation (45.1) over into solutions. These transformations are *invariant transformations*; their generators must be constants of the motion. The invariant transformations (i.e. those that preserve the form of the Lagrangian of the theory) form a group. Moreover, the Lorentz transformations are members of the group; hence the Lorentz transformations give rise to similarity transformations that map the group on itself. In other words, any commutator algebra based on a consideration of the group of invariant transformations of a Lorentz-invariant theory is itself Lorentz-invariant.

Instead of considering the most general mappings imaginable, we shall work with the relatively simple infinitesimal transformations in which δu itself satisfies the field equation (45.1). If added to a solution it will certainly lead to another solution. A solution may be represented as a superposition of plane waves, each separately satisfying Eq. (45.1). To each three-dimensional propagation vector k_s there belong two independent plane waves, one propagating with a positive and the other with a negative frequency. To describe this manifold of plane waves, we introduce three parameters f_k and let the four components of the wave propagation vector k_ϱ depend on them. We may then write δu in the form

$$\left.\begin{aligned} \delta u(x^\varrho) &= \int [e^{i k_\varrho x^\varrho} b(f) + e^{-i k_\varrho x^\varrho} \bar{b}(f)] \, d^3 f, \\ k^\varrho k_\varrho - \varkappa^2 &= 0, \qquad k_0 > 0, \end{aligned}\right\} \tag{45.3}$$

where b is an arbitrary (though integrable and square-integrable) function of the f_k. An infinitesimal change in u of the form (45.3) must induce a change in S which reduces to a surface integral and hence is without effect as far as the field equation is concerned. That is the meaning of an invariant transformation. Indeed, a brief calculation leads to the result

$$\left.\begin{aligned} \delta S &= -\oint_{\Sigma} C^{\mu}\, d\Sigma_{\mu}, \\ C^{\mu} &= \eta^{\mu\nu} \int_{f} \left[e^{i k_{\varrho} x^{\varrho}} (u_{,\nu} - i k_{\nu} u)\, b + e^{-i k_{\varrho} x^{\varrho}} (u_{,\nu} + i k_{\nu} u)\, \bar{b} \right] d^3 f \\ &= \eta^{\mu\nu} (u_{,\nu}\, \delta u - u\, \delta u_{,\nu}), \end{aligned}\right\} \tag{45.4}$$

which holds whether or not u itself is a solution of the field equation.

The vector field C^{μ} may be regarded as a generating density. The generator itself is the integral of C^{μ} over a surface which must be chosen so that the values of u and its normal derivative determine u throughout the Minkowski universe by virtue of the field equation. An infinitely extended surface whose normal is everywhere time-like will serve this purpose. If u satisfies the wave equation, and also standard boundary conditions at spatial infinity, then the value of this integral will be the same over any two surfaces, because the difference then equals a surface integral over an infinitely far surface. In this sense the integral is a constant of the motion. The field variable u, its normal derivative, which we shall denote by u_n, the variation δu, and its normal derivative δu_n, may all be chosen arbitrarily on the surface on which we construct our generating integral. In particular we may let the variation and its normal derivative be arbitrary functions (or even functionals) of u and u_n. Of course, any choice that we make will determine u and δu throughout the Minkowski universe. In any case, two such infinitesimal variations will then no longer commute with each other. We may form the commutator and the generator belonging to the commutator. In the classical (i.e. non-quantum) field theory, the relationship between the generators of the two non-commuting variations and the generator of the commutator serves to define uniquely the Poisson bracket between the first two integrals. If we are to form the Poisson bracket between two physical variables off the chosen hypersurface, the two variables must first be expressed as functionals on the hypersurface, whereupon the bracket can be formed by standard procedure.

If we adopt for our defining surface the hypersurface $x^0 = 0$, then u_n reduces to $\dot{u}$, $d\Sigma$ to d^3x, and the standard Poisson bracket to

$$\big(u(\boldsymbol{x}_1),\ \dot{u}(\boldsymbol{x}_2)\big) = \delta_3(\boldsymbol{x}_1, \boldsymbol{x}_2). \tag{45.5}$$

Poisson brackets between u at any two points vanish, and so do brackets between two $\dot{u}$. These relationships can, of course, be obtained also by simple analogy. But the derivation by means of infinitesimal invariant transformations guarantees the Lorentz covariance of the results from the outset.

It is also possible to set up Poisson brackets exclusively between constants of the motion. Given a solution of the field equation, u, the following integrals are constants of the motion:

$$\left.\begin{aligned} v(k_{\varrho}) &= \int_{\Sigma} \eta^{\mu\nu} e^{-i k_{\varrho} x^{\varrho}} (u_{,\mu} + i k_{\mu} u)\, d\Sigma_{\nu}, \\ \bar{v}(k_{\varrho}) &= \int_{\Sigma} \eta^{\mu\nu} e^{i k_{\varrho} x^{\varrho}} (u_{,\mu} - i k_{\mu} u)\, d\Sigma_{\nu}, \\ k^{\varrho} k_{\varrho} &= \varkappa^2. \end{aligned}\right\} \tag{45.6}$$

We shall not compute the details. Suffice it to say that the Poisson brackets between any two v, or between any two $\bar{v}$, vanish, and that the Poisson bracket between a v and a $\bar{v}$ is proportional to a delta function between its arguments, as which we may adopt again a three-dimensional f_k-manifold.

It is well known that there exists no one-to-one correspondence between the Lie algebra of all the dynamical variables of a classical theory and the observables of its quantum counterpart. That is because in the quantum theory we can always form many observables that possess no classical analog. At any rate, there are generally infinitely many quantum observables that correspond to one classical dynamical variable. Accordingly it is somewhat arbitrary as to which particular Poisson brackets we take over into the quantized theory. For this purpose we shall consider a subgroup of the invariant canonical transformations which includes the Lorentz transformations and which, under a Lorentz transformation, therefore is mapped on itself. This subgroup consists of those transformations in which the δu, δu_n are linear functions (or functionals) of u, u_n. The generators of this subgroup are quadratic in u, u_n and may, therefore, be represented as double integrals over the hypersurface on which all generators are defined. This subgroup in the quantum theory may be brought into one-to-one correspondence with the corresponding classical transformation group, by the simple requirement that in the quantum theory all quadratic expressions in the observables (u and u_n) be symmetrized. It then turns out that the desired quantum commutation relations must be the precise analoges of the classical relations (45.5),

$$\left.\begin{aligned} &[\dot{u}(\boldsymbol{x}_1), u(\boldsymbol{x}_2)] = \frac{\hbar}{i}\,\delta_3(\boldsymbol{x}_1, \boldsymbol{x}_2), \\ &[u(\boldsymbol{x}_1), u(\boldsymbol{x}_2)] = 0, \qquad [\dot{u}(\boldsymbol{x}_1), \dot{u}(\boldsymbol{x}_2)] = 0. \end{aligned}\right\} \tag{45.7}$$

Again, these commutation relations may be converted into commutation relations between constants of the motion which depend on parameters f_k and which may be treated as independent of the choice of Lorentz frame, the observables v and $\bar{v}$.

The classical variables v and $\bar{v}$ are square-integrable over the range of the f_k if the total energy of the field, its total linear momentum, and similar over-all characteristics are to be finite. We may therefore represent them in terms of some complete orthonormal function system G_s defined on the f_k-manifold. With the help of such a system we may finally construct from the $v(f)$, $\bar{v}(f)$ a set of operators a_s, a_s^* which satisfy the same commutation relations (44.5) as the annihilation and creation operators of the hyperquantization formalism. From the commutation relations alone one can show that the eigenvalue spectrum of the product $a_s^*\, a_s$ consists of the non-negative integers and that with respect to the joint eigenstates of these products the operators a_s, a_s^* have the matrix elements (44.4). These calculations are not peculiar to relativistic quantum theories; accordingly we shall refer the reader for details to the articles on quantum field theory. The intent of the present section was to show that the quantization of a classical field theory may be carried out in a manner which makes the Lorentz covariance of the procedures manifest at every stage.

46. The electromagnetic field. We shall now quantize the electromagnetic field by the same methods as used in the preceding section. As mentioned previously, the gauge covariance of the electromagnetic field gives rise to complications. If we consider again infinitesimal transformations that add to a field a solution of the (vacuum) Maxwell equations, the generating hypersurface integral turns out to be

$$C = \int_{\Sigma} (\varphi_\varrho\, \delta\varphi^{\varrho\nu} - \varphi^{\varrho\nu}\, \delta\varphi_\varrho)\, d\Sigma_\nu. \tag{46.1}$$

If we choose as our hypersurface the surface $x^0 = 0$, this generating integral assumes the form

$$C = \int (\boldsymbol{A} \cdot \delta \boldsymbol{E} - \boldsymbol{E} \cdot \delta \boldsymbol{A})\, d^3x. \tag{46.2}$$

Because of the structure of MAXWELL'S equations, we cannot choose $\delta \boldsymbol{E}$ completely arbitrarily, but only subject to the requirement that its divergence vanish. Further, because the divergence of $\boldsymbol{E}$ vanishes, the addition of a gradient to $\delta \boldsymbol{A}$ does not change the value of the integral C; this integral depends only on the curl of $\boldsymbol{A}$, that is to say $\boldsymbol{B}$. In fact, the integral (46.1) or (46.2) is gauge-invariant, though the integrand is not. If we restrict ourselves to permissible transformations (in which $\delta \varphi_\mu$ satisfies the Maxwell equations), knowledge C of as a functional of the field variables φ_μ (assumed to be solutions, as well) and their normal derivatives on the hypersurface Σ determines the infinitesimal transformation involved only up to a gauge transformation; a pure gauge transformation has the generator zero.

Within our program, which interprets Poisson brackets in terms of commutators of invariant transformations, it is, therefore, impossible to define a Poisson bracket for quantities that are not gauge-invariant. Moreover, the Poisson bracket between any quantity and one that vanishes if the field equations are satisfied must be zero. In fact, from the generator (46.2) we obtain, after a somewhat lengthy but straightforward calculation, the following relations:

$$\left.\begin{aligned}\big(E_k(\boldsymbol{x}_1), B_l(\boldsymbol{x}_2)\big) &= \delta_{kls} \frac{\partial \delta_3(\boldsymbol{x}_1, \boldsymbol{x}_2)}{\partial x_2^s} = -\delta_{kls} \frac{\partial \delta_3(\boldsymbol{x}_1, \boldsymbol{x}_2)}{\partial x_1^s}, \\ (E_k, E_l) &= 0, \ (B_k, B_l) = 0.\end{aligned}\right\} \tag{46.3}$$

The corresponding quantum rules are:

$$\left.\begin{aligned}[E_k(\boldsymbol{x}_1), B_l(\boldsymbol{x}_2)] &= \frac{\hbar}{i} \delta_{kls} \frac{\partial}{\partial x_1^s} \delta_3(\boldsymbol{x}_1, \boldsymbol{x}_2), \\ [E_k(\boldsymbol{x}_1), E_l(\boldsymbol{x}_2)] &= 0, \quad [B_k(\boldsymbol{x}_1), B_l(\boldsymbol{x}_2)] = 0.\end{aligned}\right\} \tag{46.4}$$

It is to be noted that all commutators with div $\boldsymbol{E}$ or with div $\boldsymbol{B}$ vanish automatically.

To obtain the concept of the photon, it is convenient to go over to a normal-mode description of the electromagnetic field. With some similarity to the scalar case, we may, for instance, introduce a surface integral

$$\left.\begin{aligned}\Gamma(k^\varrho, \alpha^\varrho) &= \int_\Sigma e^{i k_\varrho x^\varrho} [i(k^\nu \alpha^\varrho - k^\varrho \alpha^\nu)\varphi_\varrho - \alpha_\varrho \varphi^{\varrho\nu}]\, d\Sigma_\nu, \\ k^\varrho k_\varrho &= 0, \quad k_\varrho \alpha^\varrho = 0, \quad k_0 > 0,\end{aligned}\right\} \tag{46.5}$$

and its conjugate complex. The permissible world vectors k_ϱ form a three-dimensional manifold, which may be parametrized as before. The four-vector α_ϱ is required to be normal to k_ϱ and is thus reduced to a three-dimensional manifold (with k_ϱ fixed). However, the value of Γ remains unaffected if we add to α_ϱ a multiple of k_ϱ (an operation that is compatible with the first requirement on α_ϱ, as the wave propagation vector k_ϱ is normal to itself). Accordingly, we can characterize an electromagnetic field completely if for each null vector k_ϱ we evaluate the integral Γ for just two properly chosen vectors α_ϱ. We shall now fix these two vectors α_ϱ by additional invariant requirements. We shall permit the vectors α_ϱ to be complex, but we shall require that each of them be a null vector, that its product by its own conjugate complex be -1, and that the two vectors be each other's conjugate complex; or, to summarize:

$$k_\varrho \alpha^\varrho = 0, \quad \alpha_\varrho \alpha^\varrho = 0, \quad \alpha_\varrho \bar{\alpha}^\varrho = -1. \tag{46.6}$$

If we separate into real and imaginary parts,

$$\alpha^{\varrho} = \beta^{\varrho} + i\gamma^{\varrho}, \quad \bar{\alpha}^{\varrho} = \beta^{\varrho} - i\gamma^{\varrho}, \tag{46.7}$$

then these will satisfy the real conditions

$$k_{\varrho}\beta^{\varrho} = 0, \quad k_{\varrho}\gamma^{\varrho} = 0, \quad \beta_{\varrho}\gamma^{\varrho} = 0, \quad \beta_{\varrho}\beta^{\varrho} = \gamma_{\varrho}\gamma^{\varrho} = -\tfrac{1}{2}. \tag{46.8}$$

These conditions determine α^{ϱ} except for the addition of a (possibly complex) multiple of k_{ϱ} and for the multiplication by a constant of magnitude 1. There is also an ambiguity in that we can interchange the two vectors α_{ϱ} and $\bar{\alpha}_{\varrho}$. The first operation corresponds to a gauge transformation, the second to the (arbitrary) choice of the origin of the phase angle. The last ambiguity may be eliminated by the convention that the spatial part of k_{ϱ}, and the (spatial) vectors β_{ϱ} and γ_{ϱ}, in that sequence, are to form a screw with the same sense as the three spatial coordinates x^1, x^2, x^3. With this convention α_{ϱ} and $\bar{\alpha}_{\varrho}$ will be interchanged by Lorentz transformations that involve a space reflection.

With all these requirements and conventions we have accomplished, for a given (null) propagation vector k_{ϱ}, a decomposition into left- and right-circularly polarized waves. This decomposition is invariant with respect to all proper Lorentz transformations, whereas a decomposition into plane-polarized normal modes is not. We find, in conclusion, that we may characterize an electromagnetic field completely by means of two functions, $\Gamma_L(f_k)$ and $\Gamma_R(f_k)$. The only non-vanishing Poisson brackets and hence also quantum commutators, are between Γ_L and $\bar{\Gamma}_L$, and between Γ_R and $\bar{\Gamma}_R$, and these are proportional to delta functions of the arguments. Just as in the scalar case we may now proceed to normalize the amplitudes so that the quantum commutators become equal to delta functions, and finally introduce annihilation and creation operators with a discrete index. The corresponding occupation number will then indicate how many photons are present in a given state.

47. Concluding remarks. In the last two sections we have treated field theories that were strictly linear, theories whose Lagrangian was no more than quadratic in the field variables. In physics we are, of course, principally interested in fields and particles that interact with each other. For such interaction to be meaningful it is necessary that the Lagrangian contain at least some terms that are of higher than second degree in the field variables (or particle wave amplitudes). If these higher terms are multiplied by small coefficients, we speak of weak interactions, otherwise of intermediate or strong interactions. In any case the occupation numbers for the separate particles are then no longer constants of the motion or, in the terminology of quantum theory, "good quantum numbers".

Such theories have been approached in the so-called interaction representation Such a representation is based on a complete system of orthonormal base vectors that would be stationary states in the absence of interaction. The Lorentz covariance of the commutation laws and of the dynamical law (i.e. the Hamiltonian) are assured by describing the system (both particles and field) on an arbitrary space-like hypersurface and by formulating the dynamical law in terms of transition to an infinitesimally close neighboring hypersurface. Because the hypersurfaces are not coordinate planes, the proof of Lorentz covariance involves only the local transformation properties of the observables and their commutation relations, and requires proof that the dynamical law is "integrable", i.e. that the transition operator between two hypersurfaces separated by a finite interstice is independent of the choice of the infinitesimal intermediate steps. More recent approaches have tended to replace interaction representations by Heisenberg representations, in which the base system of Hilbert vectors adopted is stationary under the full dynamical law, including the interaction terms. To what extent these theoretical efforts have led to physically and mathematically satisfying theories is an exceedingly complex question, which is treated in detail in the articles on quantum electrodynamics and general quantum field theory.

It should also be mentioned that the identification of observables with infinitesimal invariant transformations leads to a group-theoretical interpretation of commutators, but never to anticommutators. It is true that anticommuting quantities are generally not observables, and that observables formed from them, such as occupation numbers, energy-momentum densities, and the like, satisfy again commutation relations. But anticommuting operators are needed to describe fermion fields; one would like to be able to interpret their anticommutation relations in terms of intuitive invariant operations. Such an intuitive interpretation is not readily available within the compass of hyperquantization of particles subject to PAULI's exclusion principle. There is, on the other hand, no classical counterpart to a fermion field. Very possibly, a better understanding of the mappings generated by anticommuting field operators will depend on the further clarification of such concepts as particle conjugation, isotopic spin, and the new quantum numbers associated with strange particles. At the time of this writing (1961) we find that some symmetry properties of nature which we have hitherto taken for granted are satisfied only by a restricted class of interactions (such as parity), while at the same time we have discovered other symmetry properties (such as charge independence of nuclear forces) whose range of validity is as yet uncertain. It appears not unlikely that the operations associated with fermion fields resemble symmetry operations (with but two possible states of the property considered) rather than the continuous groups of which the canonical transformations are but the most important example.

The bibliography will be found on p. 272 of this volume, at the end of the article on the general theory of relativity, which follows.

The General Theory of Relativity.

By

PETER G. BERGMANN.

A. The original version of the theory.

I. Foundations.

1. Origin of the theory. The general theory of relativity grew out of EINSTEIN'S attempts to apply his earlier theory, the special theory of relativity, to the gravitational field. These efforts came to fruition in 1916, eleven years after the first publication of the fundamentals of the special theory. As it turned out, the inclusion of gravitation into the scope of the newly developed space-time concepts was by no means straight-forward but required a further modification of these concepts away from the Galileo-Newtonian ideas. As we shall see later, because the general theory of relativity enriches the geometric structure of MINKOWSKI still further, it requires a new revision of all physical theories so as to achieve consistency with the underlying space-time continuum. This revision is still proceeding. In particular we do not possess today a fully worked-out general relativistical quantum theory.

Why is it that there exists no satisfactory Lorentz-covariant theory of the gravitational field? The principal reason is that the sources of the gravitational field consist of all ponderable matter in the universe, and the strength of a source is simply its mass.

In Newtonian mechanics the mass is defined as the ratio between the external force experienced by a body and the resulting acceleration; this mass is found to be a permanent attribute of the body, and it is additive. These two properties of the mass may be summarized for continuous mass distributions in a single differential equation, the equation of continuity,

$$\frac{\partial \varrho}{\partial t} + \operatorname{div}(\varrho \boldsymbol{u}) = 0, \tag{1.1}$$

in which ϱ is the mass density and $\boldsymbol{u}$ the local velocity of the streaming matter. As a result of the astronomical observations of TYCHO DE BRAHE and their summarization by KEPLER as well as on the basis of elementary terrestrial experiments (some of which go back to GALILEO), NEWTON established that the same inertial mass that appears as the ratio between force and acceleration (regardless of the physical nature of the force) in the Second Law of mechanics is numerically equal to the gravitational mass which appears, so to speak, as the gravitational charge in NEWTON'S inverse-square law of gravitation. Though it must have appeared remarkable to NEWTON that a *universal* attribute of matter should do double duty as source strength of a *particular* type of force, this circumstance gave rise neither to further development of the foundations of gravitational theory, nor did it lead to any special mathematical difficulties.

In the special theory of relativity the situation is complicated in that there appear several different types of mass within the realm of relativistic mechanics.

There is the invariant rest mass of the mass point (or of the isolated mechanical system), which may be defined as the magnitude (square root of the norm) of the energy-linear momentum vector. There is the relativistic mass, defined as the ratio between three-dimensional linear momentum and three-dimensional velocity. There are also, in the older literature, the "longitudinal" and the "transverse" mass, defined as the ratio between (three-dimensional) force and acceleration, which depends on the angle between force and velocity. Of these various quantities, only the rest mass is invariant with respect to Lorentz transformations, but it is not additive. The relativistic mass is additive, being the fourth component of the energy-momentum four-vector, which is subject to a universal conservation law. But being only one component of a vector, the relativistic mass can serve as the source of a field only if the remaining components of the vector play a similar role; otherwise the resulting set of laws will not be Lorentz-covariant. Longitudinal and transverse mass may be dismissed as candidates for the source strength, because they represent only two extreme cases of a continuous set of "masses", depending on the angle between force and velocity in a three-dimensional formalism.

For continuously distributed matter the relativistic mass density forms one component of a world tensor of rank 2 that is symmetric in its two indices, the stress-energy tensor (SR[1], Sects. 29 and 34). If this tensor is to act as a source of the gravitational field, the gravitational potentials must form a tensor field of rank 2. By analogy the source of the electromagnetic field (the electric charge) is an invariant for an isolated assembly of particles, its density forms a world vector (the charge-current density), and the electromagnetic potentials themselves form a world vector as well.

To summarize the argument up to this point, the only available Lorentz-covariant source field for the gravitational field appeared to be a symmetric tensor field. If this result was to be accepted, the gravitational field, which in NEWTON's theory has but one (scalar) potential, would be required to possess a whole potential tensor with 10 components, and there would presumably be 40 components of the gravitational field strength.

Such a field might appear to the observer to possess but one potential as long as all interacting bodies have velocities small compared to the speed of light. Similarly, as long as we restrict ourselves to experiments in electrostatics, the electric field would seem to possess but one (scalar) potential. However, the task of guessing at possible field laws to be satisfied by the gravitational potentials and field strengths, and of developing corresponding ponderomotive laws, is beset with so much arbitrariness (granted that we are to restrict ourselves to Lorentz-covariant schemes), and the possibilities for observing relativistic (i.e. non-Newtonian) effects so slight, that one would have little faith in a theory based only on these premises.

In order to create a theory that would combine logical simplicity with a compelling physical appeal, EINSTEIN went back to the long-known equality of inertial and gravitational mass.

2. The principle of equivalence. Just as in electromagnetic interactions, the gravitational mass appears in NEWTON's law of gravitation in a double role: It acts as the source strength of the gravitational field, and it determines the magnitude of the force that a given body experiences in a gravitational field of fixed strength. This double role is required by NEWTON's Third Law, which

[1] References to the preceding article on Special Relativity are throughout denoted by the symbol SR.

requires that the forces that two bodies exert on each other be equal in magnitude (and in opposite directions). If we wish we may speak of the *active* and the *passive* roles of the gravitational mass. Careful experiments have been performed to establish that the passive gravitational mass and the inertial mass are numerically equal for a wide variety of materials, including radioactive substances[1]. These experiments have had the expected outcome.

The result implies that in the presence of a gravitational field all test particles that are in the same state of motion will undergo the same acceleration with respect to an inertial frame of reference. We cannot eliminate the effects of gravitation on test particles by choosing a particular type or substance. Hence we have no means of establishing an inertial frame of reference in the presence of a gravitational field by means of an experiment that employs only properties of the immediate neighborhood. In practice an inertial frame of reference is established, or at least approximated, by means of astronomical observations. These observations enable us first to eliminate the daily rotation of the Earth about its axis, then its orbital motion about the Sun, then the motion of the whole solar system with respect to the galaxy, etc. But as far as *local* observations are concerned, the effects of a gravitational field resemble those caused by the adoption of an accelerated frame of reference. The acceleration of test particles is independent of their masses. This resemblance between so-called inertial forces and the gravitational force is called the *principle of equivalence.*

The principle of equivalence is the corner stone of the general theory of relativity. Because it has been misunderstood frequently, we shall delimit it with some care. The principle may be formulated concisely in terms of the equality of gravitational and inertial mass. But for heuristic purposes it is preferable to express the principle in terms of the similarity between inertial forces and gravitational forces. We may then state that according to the principle of equivalence *we cannot distinguish between gravitational and inertial forces by means of experiments that employ the kinetic effects of these forces on test particles at a fixed location in space and time.* The term "inertial forces" includes all kinetic effects that are the result of referring the motion of test particles to a frame of reference that is not inertial; typical examples are centrifugal and Coriolis forces. For both inertial and gravitational forces the observed kinematic effects (that is to say, the accelerations, rather than the forces) are independent of the mass or substance of the particles that we use for the experiment.

Clearly, there are experiments and observations that permit us to distinguish between these two kinds of "forces". As mentioned before, there are, first of all, astronomical observations that relate our terrestrial observations to landmarks outside the scope of the gravitational field of the Earth, the Sun, our galaxy. But we may also establish the existence of a gravitational field (as contrasted to a pure inertial field) by means of kinetic experiments: The lines of force of a gravitational field converge in a manner in which those of an inertial field could not converge. Hence the presence of a gravitational field would become apparent if we were to plot the lines of force over a somewhat extended domain. In the case of the gravitational field on the surface of the Earth, we should need to perform experiments throughout a region sufficiently extended that the convergence of the vertical directions toward the center of the Earth is observable.

Hence it is not true, and a proper statement of the principle of equivalence does not claim, that inertial and gravitational fields are mechanically indistinguish-

[1] R. Eötvös: Math. naturw. Ber. Ung. **8**, 65 (1890). — L. Southerns: Proc. Roy. Soc. Lond. A **84**, 325 (1910). — R. Eötvös, D. Pekár and E. Fekete: Ann. Physik **68**, 11 (1922). — R.H. Dicke: Science **129**, 621 (1959).

able. The principle claims that they cannot be distinguished by mechanical experiments confined to a very small region in space (and time). In other words, the actions of gravitational and inertial fields on test particles are alike; the two kinds of field are indistinguishable as regards their ponderomotive effects. But there is a difference between the two kinds of fields that may be expressed mathematically as a set of relationships between the gradients (first partial derivatives) of the field intensities, equivalent to second-order differential relationships between the potentials. What these relationships are in detail we shall develop later. But right now we may infer from the physical argument that there are certain differential expressions which vanish for purely inertial fields but do not vanish for gravitational or mixed gravitational-inertial fields.

Because these relations are differential, they will not enable us to *separate* gravitational from inertial fields in the event that both are present. In fact, they enable us to do no more than to establish the existence or non-existence of a gravitational field. Take the instance of the gravitational field on the surface of the Earth. Suppose an experimenter were sealed in a large enclosure and could perform experiments only within the box but that he could not view, or otherwise communicate with, the surroundings. If the box rests on the ground, he would observe the acceleration of all test particles toward the ground relative to his enclosure; from this fact alone he could not decide whether this acceleration is caused by a gravitational field or whether the whole box (in the absence of a gravitational field) is accelerated upwards. However, if the box is large enough, the observer inside will find that the directions of acceleration are not strictly parallel but slightly convergent toward the bottom. From this fact alone he will be able to deduce the presence of a gravitational field, but not the absence of an inertial field as well. The degree of convergence is rigidly related to the location of the source (the Earth) only if the source possesses spherical symmetry but not, for instance, in the region between the two components of a double star system.

It follows that in the presence of a gravitational field it is impossible to establish by mechanical experiments alone an inertial frame of reference. EINSTEIN conjectured that this impossibility was fundamental, that the usual astronomical determinations could only ameliorate but not entirely cure the elusiveness of the inertial frames of reference. Briefly, EINSTEIN contended that in general the inertial frames of reference do not play a special role among all the conceivable frames of reference but that such a privileged set of frames exists strictly only in a universe entirely devoid of matter and hence free of gravitational fields. That we actually can construct inertial frames with considerable accuracy is, from this point of view, to be explained by the extremely low average density of ponderable matter in the universe.

This conjecture, just as all physical hypotheses, cannot be "derived" by a purely logical argument. It is based on certain experimental evidence, and on an interpretation of that evidence. It can be tested in terms of the predictions of a completed physical theory based on, or at least incorporating, the hypothesis. But this hypothesis, of the fundamental equivalence of all frames of reference, known as *the principle of (general) covariance*, forms the cornerstone of EINSTEIN's theory of gravitation, which we know under the name of the general theory of relativity.

3. The principle of covariance. Like so many other concepts in theoretical physics, that of the "frame of reference" needs to be defined, and redefined, with any thoroughgoing revision of our foundations. In Newtonian mechanics the frame of reference was a Cartesian, three-dimensional coordinate system which was determined uniquely at every instant in time. Ideally at least, the coordinate system could be realized as a material scaffolding extending through space and moving as a rigid body. Time was universal and absolute, hence common to all conceivable frames of reference. Among the totality of all frames of reference, the inertial frames were those with respect to which a test particle would remain in unaccelerated motion if freed of the influence of all external forces.

The special theory of relativity does away with universal time and with the concept of a rigid body. However, Minkowski geometry still provides for such concepts as straight (world) lines, plane hypersurfaces, etc., so that we may construct a Cartesian coordinate system in unaccelerated motion. The idea of an accelerated (non-inertial) Cartesian coordinate system, however, loses all meaning. Hence we may introduce the four-dimensional equivalent of a Cartesian coordinate system, and that is what we call a Lorentz frame. The Lorentz frame of reference has all the essential attributes of an inertial frame of reference of Newtonian physics; without straining the terminology we may refer to it as an inertial frame as well. But we cannot make any fundamental distinction between an accelerated Cartesian and an accelerated non-Cartesian coordinate system. A non-inertial frame of reference in the special theory of relativity, if the concept is to be introduced at all, will include both rectilinear and curvilinear coordinate systems engaged in arbitrary motion. But the need for such non-inertial frames rarely arises. The normal frame of reference in special relativity is the Lorentz frame. The frequent employment of spherical or cylindrical coordinate systems in relativistic calculations does not contradict this statement. These coordinate systems are related in an obvious way to Cartesian coordinate systems. They are introduced because work with such special functions as spherical harmonics or Bessel cylinder functions is thereby facilitated.

If we are to drop the special role of Lorentz frames of reference in accordance with EINSTEIN'S hypothesis, of course without abandoning the accomplishments of special relativity, then we must search for a theory which is based on the fundamental equality of all curvilinear four-dimensional coordinate systems. In all that follows we shall use the terms "curvilinear four-dimensional coordinate system" and "frame of reference" interchangeably. A frame of reference is, then, any convention that associates with every world point ("elementary event", a particular location at an instant in time) a set of four real numbers, in such a manner that these four numbers identify the world point uniquely, and so that if the coordinates of two world points differ by small amounts then the two world points lie close together. It is also assumed that given one acceptable frame of reference the coordinate transformations leading to another acceptable frame are continuous functions that possess at least a finite number of derivatives. Analyticity is usually not assumed, because we operate only with real functions of real variables.

A theory that embodies EINSTEIN'S hypothesis must involve field equations and ponderomotive laws that retain their form unchanged under curvilinear coordinate transformations. Moreover, it must be impossible to introduce a special class of frames, connected with each other by a restricted group of coordinate transformations, with respect to which the laws of the theory are simplified. Otherwise these frames would in effect represent a privileged set, resembling the inertial frames of Newtonian physics and of the special theory of relativity.

The mathematical requirement that the laws of nature are to be invariant with respect to the group of curvilinear coordinate transformations (with some restrictions on continuity and differentiability), reinforced by the further requirement that within the theory no privileged set of frames can be constructed, is the technical form of the principle of covariance.

The principle of covariance does not preclude the possibility that particular solutions of the field equations may possess special symmetries. In fact, there exists what might be called a correspondence principle, according to which there must exist one solution that corresponds to the Minkowski universe and which, physically, represents the complete absence of gravitational fields. This solution will admit a special set of coordinate frames, with a restricted transformation group. But the existence of a restricted transformation group is confined to the one particular solution, or to the class of particular solutions (e.g. static fields), and is not an intrinsic feature of the field equations.

At least in the usual development of general relativity the principle of covariance plays a local, not a global role. One is usually interested in solutions which at spatial infinity go over into the Minkowski universe, or into a global solution corresponding to some particular cosmological model (such as an "expanding" universe). Topological questions have been explored, in the large in connection with cosmology, and in the small as a possible approach to the structure of elementary particles. However, by and large, work in general relativity has tended to assume simple boundary conditions at spatial infinity, and the set of acceptable frames of reference is restricted to such coordinate systems that give expression to the assumed boundary conditions at infinity. The group of coordinate transformations is thereby considerably restricted. Whether this asymptotic restriction is inherent in the present theory, or merely a historic accident, is at present not known.

4. Riemannian spaces. The principle of covariance rules out Minkowski geometry as a basis for a theory of gravitation. By itself it tells us nothing about what particular geometry should take its place. Historically the choice was made on the basis of a correspondence argument. In view of the fact that the ponderomotive effects of gravitational fields are indistinguishable from those of inertial fields, EINSTEIN examined the form that the law of inertia would assume in a non-inertial frame of reference. The whole experience of mechanics teaches that such an examination is performed to advantage with the help of a variational principle. Accordingly we shall first cast the law of inertia, in a Lorentz frame, into a variational form. Then we shall express the variational principle in terms of curvilinear coordinates.

Let us designate, for an instant, the Lorentzian coordinates of a Minkowski universe by $y^\varkappa$ ($\varkappa = 0, 1, 2, 3$). Then the law of inertia may be stated to the effect that the world lines corresponding to the motion of force-free test particles are straight lines. As such they satisfy the Lorentz-invariant principle

$$\delta S = 0, \qquad S = + \int \sqrt{\eta_{\mu\lambda}\, dy^\varkappa\, dy^\lambda} \tag{4.1}$$

(cf. SR, Sect. 23). Let us now transform this variational principle to a different coordinate system,

$$x^\varrho = x^\varrho(y^\varkappa). \tag{4.2}$$

Then, evidently,

$$S = \int \sqrt{g_{\varrho\sigma}\, dx^\varrho\, dx^\sigma}, \qquad g_{\varrho\sigma} = \eta_{\mu\nu} \frac{\partial y^\mu}{\partial x^\varrho} \frac{\partial y^\nu}{\partial x^\sigma}. \tag{4.3}$$

The new set of 10 quantities $g_{\varrho\sigma}$ are related to the proper time differential,

$$g_{\varrho\sigma}\, dx^\varrho\, dx^\sigma = d\tau^2. \tag{4.4}$$

In this respect they play a similar role in the presence of curvilinear coordinates as the coefficients $\eta_{\mu\nu}$ do in Lorentz frames. By their definition (4.3) they transform according to the law

$$g'_{\alpha\beta} = \frac{\partial x^{\varrho}}{\partial x^{\alpha'}} \frac{\partial x^{\sigma}}{\partial x^{\beta'}} g_{\varrho\sigma} \tag{4.5}$$

if we go over from one curvilinear frame of reference to another. They satisfy the definition of a tensor[1]. In contrast to $\eta_{\mu\nu}$ they are not constants but are functions of the coordinates. The $g_{\varrho\sigma}$ form a *tensor field*; this tensor is called the *(covariant) metric tensor*, in obvious analogy to the terminology that we have already introduced in the Minkowski universe.

We shall now determine the differential equations obeyed by the straight world lines, in other words the Euler-Lagrange equations belonging to the variational principle (4.3). These differential equations are obtained by standard methods. We have, on introducing an arbitrary parameter ϑ, first

$$\left.\begin{aligned} L &= \sqrt{g_{\varrho\sigma}\frac{dx^{\varrho}}{d\vartheta}\frac{dx^{\sigma}}{d\vartheta}}, \qquad \frac{\partial L}{\partial\left(\frac{dx^{\mu}}{d\vartheta}\right)} = \left(g_{\varrho\sigma}\frac{dx^{\varrho}}{d\vartheta}\frac{dx^{\sigma}}{d\vartheta}\right)^{-\frac{1}{2}} g_{\mu\nu}\frac{dx^{\nu}}{d\vartheta} \\ &= g_{\mu\nu}\frac{dx^{\nu}}{d\tau}, \\ \frac{\partial L}{\partial x^{\mu}} &= \left(g_{\varrho\sigma}\frac{dx^{\varrho}}{d\vartheta}\frac{dx^{\sigma}}{d\vartheta}\right)^{-\frac{1}{2}} g_{\alpha\beta,\mu}\frac{dx^{\alpha}}{d\vartheta}\frac{dx^{\beta}}{d\vartheta} = \frac{d\tau}{d\vartheta} g_{\alpha\beta,\mu}\frac{dx^{\alpha}}{d\tau}\frac{dx^{\beta}}{d\tau}. \end{aligned}\right\} \tag{4.6}$$

If we substitute these expressions into the usual formula for the variational equations, we obtain the following:

$$\left.\begin{aligned} \frac{\partial L}{\partial x^{\mu}} - \frac{d}{d\vartheta}\left(\frac{\partial L}{\partial\left(\frac{dx^{\mu}}{d\vartheta}\right)}\right) &= \frac{d\tau}{d\vartheta}\left[g_{\alpha\beta,\mu}\dot{x}^{\alpha}\dot{x}^{\beta} - \frac{d}{d\tau}(g_{\mu\nu}\dot{x}^{\nu})\right] \\ &= -\frac{d\tau}{d\vartheta}\left[g_{\mu\nu}\ddot{x}^{\nu} + \frac{1}{2}(g_{\mu\alpha,\beta} + g_{\mu\beta,\alpha} - g_{\alpha\beta,\mu})\dot{x}^{\alpha}\dot{x}^{\beta}\right] \equiv -\frac{d\tau}{d\vartheta}k_{\mu} = 0, \\ \dot{x}^{\mu} \equiv \frac{dx^{\mu}}{d\tau}, \quad \ddot{x}^{\mu} &\equiv \frac{d^2x^{\mu}}{d\tau^2}. \end{aligned}\right\} \tag{4.7}$$

The contents of the square bracket, which we have denoted by k_{μ}, is usually referred to as the *curvature vector* of a curve. Its transformation law is indeed that of a covariant vector, and it is obviously parameter-invariant. The integral over the scalar product $k_{\mu}\,\delta x^{\mu}$ along an arc whose endpoints are being held fixed tells how much the proper time interval of the arc is diminished by the variation δx^{μ}. Naturally, the curvature of a straight line vanishes.

Physically, Eq. (4.7) may be interpreted as follows: The second proper time derivatives of the coordinates depend on a field

$$[\mu, \alpha\beta] \equiv \tfrac{1}{2}(g_{\mu\alpha,\beta} + g_{\mu\beta,\alpha} - g_{\alpha\beta,\mu}) \tag{4.8}$$

and, quadratically, on the instantaneous velocity $\dot{x}^{\alpha}$, in rough analogy to the ponderomotive law of electrodynamics (SR, Sect. 32). In that theory, however, the dependence on the velocity is linear. The expressions (4.8), 40 in number, may be interpreted as the components of the field intensity of the inertial forces, arising because we have abandoned Lorentz frames. They do not form a tensor, but a so-called affine connection. In any event, these field intensities are combinations of the first derivatives of the components of the metric tensor. EINSTEIN hypothesized, accordingly, that *the components of the metric tensor form the*

[1] See H. TIETZ: Geometrie, Vol. II of this Encyclopedia.

potentials both for inertial and for gravitational fields. Accordingly he constructed his theory in a space with a fundamental quadratic form of the type (4.4), a metric. Spaces with a quadratic metric were first investigated by B. RIEMANN. They are called *Riemannian spaces.* They are by no means the most general metric spaces that we can conceive of, nor are spaces necessarily metric; in theoretical physics we encounter non-metric spaces. For instance, the phase space of classical analytical mechanics is not a metric space, though it possesses an invariant volume (a metric assigns an invariant distance to pairs of points). Hilbert space possesses a metric in a complex space that is bilinear in the coordinate differences and their conjugate complexes. Hence, the choice of Riemannian spaces for the theory of gravitation is no more logically necessary than the choice of Minkowski spaces for electrodynamics. It suggests itself as the least radical departure from Minkowski spaces that offers us the possibility to abandon the special and privileged role of inertial frames among all conceivable frames of reference. Like many other constructs of theoretical physics, it may well require modification in the future.

In the construction (4.3) we have assumed that there are Lorentz frames and that the introduction of non-inertial frames is gratuitous. The idea of general relativity is that the presence of a gravitational field is mathematically equivalent to the non-existence of Lorentz frames. In other words, it is assumed that there is no coordinate system in which the metric tensor assumes the values $\eta_{\mu\nu}$ throughout space-time unless the universe is free of gravitational fields. Before we can proceed with the development of the general theory of relativity as such, we shall first summarize briefly those aspects of Riemannian theory needed for our purposes.

5. Affine connection, curvature. No effort will be made here to derive or prove properties of Riemannian spaces. For details we refer to TIETZ's article in Vol. II or to textbooks. The purpose of this section is to remind the reader who has studied the subject in the past, and to indicate to the novice which parts of Riemannian geometry he will have to study as a bare minimum in order to be able to comprehend the techniques used in general relativity.

Vectors, tensors. Just as in the Minkowski universe, we must define contravariant and covariant quantities. The prototype of contravariant vectors is the differential coordinate increment or displacement vector dx^ϱ, the prototype of the covariant vector the gradient of a scalar field, $V_{,\varrho}$. Hence the basic definitions of these two types of quantity are:

$$V^{\varrho'} = \frac{\partial x^{\varrho'}}{\partial x^\sigma} V^\sigma, \qquad W'_\varrho = \frac{\partial x^\sigma}{\partial x^{\varrho'}} W_\sigma. \tag{5.1}$$

In curvilinear coordinates the transformation coefficients are themselves fields, i.e. functions of the coordinates. The definitions of covariant, contravariant, and mixed tensors are analogous to Eqs. (5.1). Linear combination, multiplication, and contraction are subject to the same rules as in Minkowski geometry. The following are special tensors: $g_{\mu\nu}$ has already been introduced in Sect. 4. It is called the (covariant) metric tensor. With the help of the equations

$$g^{\mu\nu} = \frac{\partial x^\mu}{\partial y^\alpha} \frac{\partial x^\nu}{\partial y^\beta} \eta^{\alpha\beta} \tag{5.2}$$

we define likewise a contravariant metric tensor. It satisfies the set of equations

$$g_{\mu\varrho} g^{\varrho\nu} = \delta^\nu_\mu. \tag{5.3}$$

The Kronecker symbol is a mixed tensor with invariant components, just as in Minkowski geometry.

The analogues to the pseudotensors of Minkowski geometry are here the *tensor densities*. They are distinguished from tensors in that they multiply also by some power of the Jacobian of the transformation, the power being called the *weight* of the density. For instance, a scalar density of weight K will obey the transformation law:

$$\sigma' = \left| \frac{\partial x}{\partial x'} \right|^K \sigma. \tag{5.4}$$

The contravariant Levi-Civita symbol $\delta^{\mu\nu\varrho\sigma}$ (in a four-dimensional space) is a tensor density of weight 1; the covariant Levi-Civita symbol $\delta_{\mu\nu\varrho\sigma}$, which has the same numerical values for all of its components as the contravariant symbol, possesses the weight -1. Weights add under multiplication.

If we choose a single world point P, then we may perform a coordinate transformation so that at that point the metric tensor takes the values $\eta_{\mu\nu}$. In fact, one can do more: One can even achieve that all the first derivatives of the metric tensor vanish as well. This accomplishment, however, does not permit us to conclude that our Riemannian space is in fact a Minkowski space. A Riemannian space is a Minkowski space only if we can render $g_{\mu\nu}=\eta_{\mu\nu}$ in a whole region of space. Actually, the second derivatives of the metric tensor already in general cannot be reduced to zero merely by the choice of suitable coordinates. Conversely, if we can make all first and second derivatives of $g_{\mu\nu}$ zero at one point, then at that point the Riemannian space is flat. Whether it is flat in a whole region (whether it is Minkowskian) we can decide by examining a tensor that incorporates second derivatives of the metric tensor in *any* coordinate system. This tensor, the *Riemann-Christoffel curvature tensor*, plays a central role in general relativity. We shall lead up to it in two stages.

Affine connection. Because the transformation law of vectors and tensors depends on local values of the partial derivatives $\frac{\partial x^\alpha}{\partial x^{\mu'}}$, $\frac{\partial x^{\mu'}}{\partial x^\alpha}$, we cannot obtain vectors by forming linear combinations of like vectors or tensors at different world points (in Minkowski geometry this is possible, because the Lorentz coefficients are constants). That is why not even the first derivatives of a tensor field form a tensor field themselves. We can partly cure this situation by introducing into our space a structure that permits us, while proceeding along a curve in space from one point to the next, to state what we mean by "the same vector" or "the parallel vector" to a given vector at one point of the curve. The required structure consists of a number of field quantities $\Gamma^\gamma_{\alpha\beta}$ which assign to a vector A^γ at a world point with the coordinates x^ϱ the vector $A^\gamma+\delta A^\gamma$ at the point $x^\varrho+\delta x^\varrho$ (δx^ϱ being a displacement along the curve) as the parallel vector by means of the law of parallel displacement:

$$\delta A^\gamma = -\Gamma^\gamma_{\alpha\beta}\, A^\alpha\, \delta x^\beta. \tag{5.5}$$

The coefficients $\Gamma^\gamma_{\alpha\beta}$ are called the components of the affine connection. They do not form a tensor but obey the transformation law

$$\Gamma^{\gamma'}_{\alpha\beta} = \frac{\partial x^{\gamma'}}{\partial x^\lambda}\left(\frac{\partial x^\iota}{\partial x^{\alpha'}}\,\frac{\partial x^\varkappa}{\partial x^{\beta'}}\,\Gamma^\lambda_{\iota\varkappa} + \frac{\partial^2 x^\lambda}{\partial x^{\alpha'}\,\partial x^{\beta'}}\right). \tag{5.6}$$

In most geometries it is assumed that the affine connection is symmetric in the two subscripts, a property that is preserved under coordinate transformations.

With the help of the affine connection we define a *covariant differentiation*,

$$V^\mu_{;\varrho} \equiv V^\mu_{,\varrho} + \Gamma^\mu_{\alpha\varrho}\, V^\alpha, \qquad U_{\mu;\varrho} \equiv U_{\mu,\varrho} - \Gamma^\alpha_{\mu\varrho}\, U_\alpha. \tag{5.7}$$

The covariant derivatives indicate to what extent a given field deviates from being parallel to itself (according to the chosen definition). A relation between the (symmetric) affine connection and the metric tensor is established, uniquely, by the requirement that all covariant derivatives of the metric tensor are to vanish,

$$g_{\mu\nu;\varrho} = 0, \qquad g^{\mu\nu}{}_{;\varrho} = 0. \tag{5.8}$$

This requirement leads to a special form of the affine connection, known as Christoffel symbols (of the second kind),

$$\Gamma^{\gamma}_{\alpha\beta} = \left\{ {\gamma \atop \alpha\beta} \right\} \equiv \tfrac{1}{2} g^{\gamma\varepsilon} (g_{\alpha\varepsilon,\beta} + g_{\beta\varepsilon,\alpha} - g_{\alpha\beta,\varepsilon}). \tag{5.9}$$

The identification (5.9) is an integral part of Riemannian geometry.

The covariant derivatives of a scalar density are:

$$\sigma_{;\varrho} \equiv \sigma_{,\varrho} - K \left\{ {\alpha \atop \alpha\varrho} \right\} \sigma. \tag{5.10}$$

The covariant derivatives of vector and tensor densities are defined accordingly.

We shall interpolate two remarks: First, there are certain covariant derivatives, and combinations of derivatives, which reduce to ordinary derivatives. Among the most important are the antisymmetrized derivatives of a covariant vector field and of a covariant skewsymmetric tensor field,

$$\left.\begin{aligned} &A_{\mu;\nu} - A_{\nu;\mu} \equiv A_{\mu,\nu} - A_{\nu,\mu}, \\ &A_{[\varrho\sigma];\tau} + A_{[\sigma\tau];\varrho} + A_{[\tau\varrho];\sigma} \equiv A_{[\sigma\varrho],\tau} + A_{[\sigma\tau],\varrho} + A_{[\tau\varrho],\sigma}, \end{aligned}\right\} \tag{5.11}$$

and the divergence of a contravariant vector density of weight 1,

$$\mathfrak{A}^{\varrho}{}_{;\varrho} \equiv \mathfrak{A}^{\varrho}{}_{,\varrho}. \tag{5.12}$$

The other remark is concerned with the expression (5.9) for the Christoffel symbols. Evidently the affine connection vanishes at a world point if, and only if, at that point the first derivatives of the $g_{\mu\nu}$ vanish. If we choose a coordinate system which has this property at the point P, then at that point the "field intensities" (4.8) (which are called the Christoffel symbols of the first kind) vanish, and a vector field at that point is "parallel to itself" if the first partial derivatives of its components all vanish. Finally, a straight line passing through P will have vanishing second derivatives of the coordinates with respect to the proper time at P.

The curvature tensor. According to Eq. (5.5) the parallel displacement of a vector along a curve is a linear operation on that vector. Suppose the curve $\mathfrak{C}$ connects two points P_1 and P_2 with each other. Then the law (5.5) integrated along $\mathfrak{C}$ maps the totality of all imaginable vectors at the point P_1 (this totality is sometimes referred to as the *tangent vector space* at P_1) on the similar totality at point P_2 in an invariant manner. We shall now ask whether this law depends only on the choice of the two points P_1 and P_2 or also on the choice of the connecting curve $\mathfrak{C}$.

In order to obtain an answer we vary the curve $\mathfrak{C}$, but so that its two end points are not shifted. As a result the mapping law is (or is not) changed as well. It depends on the exact form of the affine connection whether this change vanishes for all permissible variations. The actual calculation is carried out most conveniently with the help of a notation which to the theoretical physicist is known as the *time-ordered product*. We shall consider the infinitesimal law (5.5) as an infinitesimal linear operation on the vector space A and rewrite the law in the form

$$\frac{dA}{d\vartheta} = -\Gamma_{\beta} \frac{dx^{\beta}}{d\vartheta} A, \tag{5.13}$$

where ϑ is a parameter that we introduce in order to describe the curve $\mathfrak{C}$ parametrically. The formal integral of the differential law (5.13) is then

$$A_2 = T\left\{e^{-\int_{\mathfrak{C}} \Gamma_\beta dx^\beta}\right\} A_1. \tag{5.14}$$

This is the expression we shall vary by changing the x^β as functions of ϑ. The calculation is a little lengthy but straightforward. Its result is, in the same abbreviating notation,

$$\left.\begin{aligned} \delta A_2 &= \int_{\vartheta=\vartheta_1}^{\vartheta_2} T\left\{\exp\left[-\int_{\vartheta}^{\vartheta_2} \Gamma_\beta dx^\beta\right]\right\} R_{\iota\varkappa}(\vartheta)\, T\left\{\exp\left[-\int_{\vartheta_1}^{\vartheta} \Gamma_\beta dx^\beta\right]\right\} \delta x^\iota(\vartheta)\, dx^\varkappa(\vartheta),\\ R_{\iota\varkappa} &= \Gamma_{\iota,\varkappa} - \Gamma_{\varkappa,\iota} - \Gamma_\iota \Gamma_\varkappa + \Gamma_\varkappa \Gamma_\iota. \end{aligned}\right\} \tag{5.15}$$

$R_{\iota\varkappa}$ is itself a linear differential operator on the A-space, and evidently in a certain sense the commutator of the parallel displacements along the curve $\mathfrak{C}$ and along its variation δx^ι. If, and only if, the operator $R_{\iota\varkappa}$ vanishes, then the linear mapping of the vectors at P_1 on the vectors at P_2 is independent of the connecting curve. The operator $R_{\iota\varkappa}$ therefore measures an intrinsic property of the affine connection, its *vorticity*. If $R_{\iota\varkappa}$ is non-zero, then the parallel displacement of a vector about a closed curve from P back to P will not reproduce the original vector.

If $R_{\iota\varkappa}$ vanishes, then the affine connection is integrable, and we can construct to any given vector at a point P a whole vector field that is everywhere parallel to the original chosen vector. We then say that the space is (affinely) flat. In a flat space we can construct a coordinate system of self-parallel curves, in terms of which the affine connection vanishes everywhere. If the affine connection is related to the metric tensor by Eq. (5.9), as it is in Riemannian spaces, then in such a coordinate system the metric tensor is constant, and it is a trivial matter to introduce a Lorentz frame. The vanishing of $R_{\iota\varkappa}$ is then the necessary and sufficient condition for the existence of Lorentz frames. In the less abbreviated notation of tensor calculus, this critical tensor, the *curvature tensor*, has the form

$$R_{\iota\varkappa\lambda\cdot}^{\ \ \ \ \nu} \equiv \Gamma^\nu_{\lambda\iota,\varkappa} - \Gamma^\nu_{\lambda\varkappa,\iota} - \Gamma^\nu_{\varrho\iota}\Gamma^\varrho_{\lambda\varkappa} + \Gamma^\nu_{\varrho\varkappa}\Gamma^\varrho_{\lambda\iota}. \tag{5.16}$$

It has a number of analytic and algebraic properties which we shall enumerate in the next section.

6. Properties of the Riemann-Christoffel tensor. In enumerating properties of the curvature tensor, we shall proceed from the properties possessed by any curvature tensor to those that presuppose a symmetric affine connection, and finally to those properties that are peculiar to a curvature tensor that is formed with the help of Christoffel symbols. For the first set we shall use the abbreviating linear operator notation of Eqs. (5.13) through (5.15). These properties are in fact independent of whether the affine connection is defined with respect to vectors of either kind, densities, tensors, or spinors, as long as the quantities in question obey a *linear* transformation law.

Because the curvature tensor is a commutator, it is skewsymmetric in its two subscripts,

$$R_{\iota\varkappa} + R_{\varkappa\iota} \equiv 0. \tag{6.1}$$

It also satisfies the Jacobi relations, which may be written in the form

$$R_{\iota\varkappa,\lambda} + R_{\varkappa\lambda,\iota} + R_{\lambda\iota,\varkappa} + [\Gamma_\lambda, R_{\iota\varkappa}] + [\Gamma_\iota, R_{\varkappa\lambda}] + [\Gamma_\varkappa, R_{\lambda\iota}] \equiv 0. \tag{6.2}$$

These are closely related to a set of somewhat more elaborate differential identities, the Bianchi identities, which hold when the affine connection is symmetric. They are:

$$R_{\iota\varkappa\varrho\cdot;\lambda}^{\ \ \ \sigma} + R_{\varkappa\lambda\varrho\cdot;\iota}^{\ \ \ \sigma} + R_{\lambda\iota\varrho\cdot;\varkappa}^{\ \ \ \sigma} \equiv 0. \tag{6.3}$$

In addition we obtain a set of algebraic identities,

$$R_{\iota\varkappa\lambda\cdot}^{\ \ \ \nu} + R_{\varkappa\lambda\iota\cdot}^{\ \ \ \nu} + R_{\lambda\iota\varkappa\cdot}^{\ \ \ \nu} \equiv 0. \tag{6.4}$$

Finally, if the affine connection consists of Christoffel symbols, the curvature tensor has two additional symmetry properties, which are

$$R_{\iota\varkappa\lambda\mu} + R_{\iota\varkappa\mu\lambda} \equiv 0, \tag{6.5}$$

where $R_{\iota\varkappa\lambda\mu}$ stands for

$$R_{\iota\varkappa\lambda\mu} \equiv g_{\mu\nu} R_{\iota\varkappa\lambda\cdot}^{\ \ \ \nu}, \tag{6.6}$$

and

$$R_{\iota\varkappa\lambda\mu} - R_{\lambda\mu\iota\varkappa} \equiv 0. \tag{6.7}$$

Contracted forms. If we contract the tensor $R_{\iota\varkappa\lambda\cdot}^{\ \ \ \nu}$ on the first and on the last index, we obtain the so-called *Ricci tensor*,

$$R_{\varkappa\lambda} \equiv R_{\sigma\varkappa\lambda\cdot}^{\ \ \ \sigma}, \qquad R_{\varkappa\lambda} \equiv R_{\lambda\varkappa}. \tag{6.8}$$

If we contract once more, we get the curvature *scalar*, R:

$$R \equiv g^{\varkappa\lambda} R_{\varkappa\lambda}. \tag{6.9}$$

The combination

$$G_{\varkappa\lambda} \equiv R_{\varkappa\lambda} - \tfrac{1}{2} g_{\varkappa\lambda} R \tag{6.10}$$

is called the *Einstein tensor*. The Ricci tensor and the Einstein tensor are both symmetric tensors. The Einstein tensor has the further property that its covariant divergence vanishes identically,

$$G^{\varkappa\varrho}{}_{;\varrho} \equiv 0, \tag{6.11}$$

a fact that can be readily derived from the Bianchi identities (6.3) by contracting them on two separate index pairs. Hence the identities (6.11) are often called the *contracted Bianchi identities*.

Number of algebraically independent components. Because of the many symmetry relations connecting the various components of the curvature tensor with each other, the number of algebraically independent components is quite small. In an n-dimensional Riemannian space this number is $\frac{1}{12} n^2 (n^2 - 1)$; in a two-dimensional space it amounts to only one component, in a three-dimensional space to six, and in a four-dimensional space to 20.

The number of components of the Ricci tensor, or of the Einstein tensor, is 6 in a three-dimensional space, and 10 in a four-dimensional space. Four is, accordingly, the lowest number of dimensions of a space in which the full curvature tensor contains a greater amount of information about the structure of the space than either contracted form.

Commutator of covariant differentiation. For symmetric affine connections the commutator of two covariant derivatives is

$$\left.\begin{aligned} V^{\nu}{}_{;\iota\varkappa} - V^{\nu}{}_{;\varkappa\iota} &\equiv R_{\iota\varkappa\lambda\cdot}^{\ \ \ \nu} V^{\lambda}, \\ V_{\lambda;\iota\varkappa} - V_{\lambda;\varkappa\iota} &\equiv - R_{\iota\varkappa\lambda\cdot}^{\ \ \ \nu} V_{\nu}. \end{aligned}\right\} \tag{6.12}$$

II. Formulation of the theory.

7. The action principle. We shall now develop the general theory of relativity essentially in the form in which it was originally conceived by A. EINSTEIN in 1916. The complete theory consists of the field equations, which determine the possible fields of gravitational potentials $g_{\mu\nu}$ in empty space or in the presence of matter, and of the ponderomotive law that tells us how matter behaves in the presence of the field. Later on we shall show in some detail that the ponderomotive law is actually determined to a large extent by the form of the field equations, but these logical connections were not known at the time the theory was first put forward.

We adopt an attitude in which the geometer's point of view, who regards the $g_{\mu\nu}$ as the components of the metric, and the physicist's, who regards them as the potentials of the gravitational field, are both valid. By analogy with the electromagnetic field, the only classical field theory that had been completely formulated previously, and also in analogy to the "field law" of the Newtonian gravitational potential V,

$$\nabla^2 V = 4\pi\varkappa\varrho, \tag{7.1}$$

we expect the field equations to be of the second differential order and to equal in number the potentials, ten. Because of the principle of general covariance, the field equations must be covariant with respect to curvilinear coordinate transformations. We may either conjecture the field equations directly, or we may attempt to obtain the field equations from an invariant action principle. For this presentation we shall adopt the latter procedure, which recommends itself because the number of available invariants is extremely limited. For the free field the only scalar of the required differential order is the curvature scalar R, Eq. (6.9). If we adopt as the source field of gravitation a tensor $P^*_{\mu\nu}$, which is to be related to the energy-mass density, the linear momentum density, and the stress, then the only source scalar that we can construct for our proposed action integral is $g^{\mu\nu} P^*_{\mu\nu}$. Accordingly we shall try an action principle of the form

$$\delta S = 0, \qquad S = \int L\, d^4x, \qquad L = \mathfrak{g}^{\mu\nu}(\alpha R_{\mu\nu} + \beta P^*_{\mu\nu}), \tag{7.2}$$

where we use the notation

$$R_{\mu\nu} \equiv \Gamma^{\varrho}_{\mu\varrho,\nu} - \Gamma^{\varrho}_{\mu\nu,\varrho} - \Gamma^{\sigma}_{\sigma\varrho}\Gamma^{\varrho}_{\mu\nu} - \Gamma^{\varrho}_{\mu\sigma}\Gamma^{\sigma}_{\nu\varrho}, \qquad \mathfrak{g}^{\mu\nu} \equiv \sqrt{-|g|}\, g^{\mu\nu}, \tag{7.3}$$

and where α, β are two universal constants which serve together as a conversion factor between the units for length and for mass, as we shall find later.

In the variation of this action principle we shall employ a technique invented by PALATINI[1]. Though eventually the components of the affine connection must be the Christoffel symbols, i.e. specified differential expressions of the metric tensor, we shall treat them as if they were independent variables. We do assume from the beginning that they are symmetric in their subscripts. PALATINI's procedure bears some analogy to the introduction of the momenta as independent variables in classical mechanics.

When we perform the independent variations, we find the following:

$$\delta L = (\alpha R_{\mu\nu} + \beta P^*_{\mu\nu})\, \delta\mathfrak{g}^{\mu\nu} + \alpha\, \mathfrak{g}^{\mu\nu}\, \delta R_{\mu\nu}. \tag{7.4}$$

One set of field equations will, therefore, surely be:

$$\alpha R_{\mu\nu} + \beta P^*_{\mu\nu} = 0. \tag{7.5}$$

[1] A. PALATINI: Rend Circ. Mat. Palermo **43**, 203 (1919).

To obtain the other set, we perform the following calculation:

$$\left.\begin{aligned}\mathfrak{g}^{\mu\nu}\,\delta R_{\mu\nu} = \mathfrak{g}^{\mu\nu}\big[(\delta\Gamma^{\varrho}_{\mu\varrho,\nu} - \delta\Gamma^{\varrho}_{\mu\nu,\varrho}) - (\Gamma^{\sigma}_{\sigma\varrho}\,\delta\Gamma^{\varrho}_{\mu\nu} + \Gamma^{\varrho}_{\mu\nu}\,\delta\Gamma^{\sigma}_{\sigma\varrho}) +\\ + (\Gamma^{\varrho}_{\mu\sigma}\,\delta\Gamma^{\sigma}_{\nu\varrho} + \Gamma^{\sigma}_{\nu\varrho}\,\delta\Gamma^{\varrho}_{\mu\sigma})\big].\end{aligned}\right\}\tag{7.6}$$

In the first parenthesis we perform a differentiation by parts, so that

$$\mathfrak{g}^{\mu\nu}(\delta\Gamma^{\varrho}_{\mu\varrho,\nu} - \delta\Gamma^{\varrho}_{\mu\nu,\varrho}) = (\mathfrak{g}^{\mu\nu}\,\delta\Gamma^{\varrho}_{\mu\varrho})_{,\nu} - (\mathfrak{g}^{\mu\nu}\,\delta\Gamma^{\varrho}_{\mu\nu})_{,\varrho} + \mathfrak{g}^{\mu\nu}{}_{,\varrho}\,\delta\Gamma^{\varrho}_{\mu\nu} - \mathfrak{g}^{\mu\nu}{}_{,\nu}\,\delta\Gamma^{\varrho}_{\mu\varrho}. \tag{7.7}$$

Now we may rename indices and collect terms. In doing so we remark that, although $\Gamma^{\varrho}_{\mu\nu}$ is not a tensor, its variation $\delta\Gamma^{\varrho}_{\mu\nu}$ is, as shown by the transformation law (5.6). It is, therefore, not surprising that its coefficients arrange themselves into a tensor expression:

$$\mathfrak{g}^{\mu\nu}\,\delta R_{\mu\nu} = (\mathfrak{g}^{\mu\varrho}\,\delta\Gamma^{\sigma}_{\varrho\sigma} - \mathfrak{g}^{\varrho\sigma}\,\delta\Gamma^{\mu}_{\varrho\sigma})_{;\mu} + \big[\mathfrak{g}^{\mu\nu}{}_{;\varrho} - \tfrac{1}{2}\,(\delta^{\mu}_{\varrho}\,\mathfrak{g}^{\nu\sigma}{}_{;\sigma} + \delta^{\nu}_{\varrho}\,\mathfrak{g}^{\mu\sigma}{}_{;\sigma})\big]\,\delta\Gamma^{\varrho}_{\mu\nu}. \tag{7.8}$$

The first parenthesis is both a covariant and an ordinary divergence, because of Eq. (5.12). We may convert it into a surface integral. If we confine our variations to the interior of the domain of integration on which S is defined, the contribution of the surface integral vanishes. As for the remainder, the field equation resulting from the variation of the affine connection is quite apparently to the effect that the square bracket in Eq. (7.8) must vanish. And this requirement is equivalent to the one that the affine connections are in fact the Christoffel symbols (5.9). The temporary omission of this requirement from the variation has simply given rise to an extra set of field equations which reintroduce the connection between metric and affine connection. The actual field equations are Eqs. (7.5). Our next task is to study the properties of the field equations and to relate them to the known properties of the gravitational field.

8. Correspondence with NEWTON's theory. Field equations are susceptible to a physical interpretation in connection with their ponderomotive equations, which we have already set up in Sect. 4. We shall give the ponderomotive equations (4.7) the convenient form

$$\frac{dU^{\mu}}{d\tau} + \left\{{\mu \atop \alpha\beta}\right\} U^{\alpha}\,U^{\beta} = 0, \qquad U^{\mu} \equiv \frac{dx^{\mu}}{d\tau}, \qquad g_{\mu\nu}\,U^{\mu}\,U^{\nu} \equiv 1. \tag{8.1}$$

The symbol $\left\{{\mu \atop \alpha\beta}\right\}$ denotes, as it will from now on, the Christoffel symbol (5.9). On the strength of the principle of equivalence we assume that these equations of motion are valid not only in purely inertial fields but in the presence of a truly gravitational field as well. If we interpret the Christoffel symbols as components of the field strength, we find that these field strengths must be multiplied bilinearly by components of the velocity. Somewhat similarly, in the relativistic form of the equations of motion of electrodynamics, the field strength must be multiplied linearly by the velocity. That the relativistic acceleration (that is, the change in velocity with proper time) must depend in some fashion on the velocity may also be concluded from the formal but unavoidable requirement that U^{ϱ} can only change in such a manner that it remains a unit vector. In a Riemannian space, this requirement implies that

$$0 = \frac{1}{2}\,\frac{d}{d\tau}\,(g_{\mu\nu}\,U^{\mu}\,U^{\nu}) = \frac{1}{2}\,g_{\mu\nu,\varrho}\,U^{\mu}\,U^{\nu}\,U^{\varrho} + g_{\mu\nu}\,U^{\mu}\,\dot{U}^{\nu}. \tag{8.2}$$

The law (8.1) satisfies this requirement.

In order to examine the relationship of the ponderomotive law (8.1) with any law in Newtonian physics, we must consider a double limiting case in which (a) the deviations of the metric from the Minkowski values are small and in which (b) the three-dimensional velocities are small compared to c, and the derivatives of a potential with respect to x^0 are small compared to its derivative with respect to a space coordinate. We call the first approximation a *linearization* of the theory (because it casts the laws into a linear form), and the second assumption a "*slow*" approximation[1]. An approximation which linearizes the theory without being restricted to motions small compared to c is also called a "*fast*" approximation (without an implication that convergence is rapid).

With these two assumptions the components of U^ϱ reduce to (1, 0, 0, 0), and Eq. (8.1) takes the approximate form

$$\frac{d\boldsymbol{u}}{d\tau} + \frac{1}{2}\nabla^2 g_{00} = 0. \tag{8.3}$$

If the motion is non-relativistic and the gravitational-inertial field not too large, then $\frac{1}{2}g_{00}$ corresponds to NEWTON's single scalar potential. In most applications of celestial mechanics, orbital relative velocities are of the order of $10^{-4}\,c$, and that is a measure of the appropriateness of the "slow" assumption for this type of application. We can also make a very rough approximation for the difference of g_{00} on the surface of the Earth and at infinity by considering the magnitude of the so-called escape velocity (the least velocity a body would have to possess on Earth to escape its gravitational field entirely). This escape velocity is roughly 7.5×10^5 cm/sec, or $2.5\times 10^{-5}\,c$. Now if we integrate Eq. (8.3) in the customary manner, we find the energy integral, according to which the difference between g_{00} on the Earth's surface and at infinity must be equal to the difference in kinetic energies per unit mass. By definition a body that has precisely the escape velocity will have zero kinetic energy at infinity. Hence the value of g_{00} on the Earth's surface comes out as roughly $1-(3\times 10^{-10})$. Quantitatively, the linearizing approximation is extremely well justified at such locations as the surface of any member of our solar system. Even on the surface of a white dwarf, the deviation of g_{00} from unity does not exceed a value of the order of 10^{-5}. Thus we are justified in our hope that the full theory by EINSTEIN may be subjected to both the linearizing and the slow approximations in those situations in which NEWTON's theory, with which it is to be compared, has been confirmed by the observations.

Before we proceed with the reduction of the field equations, we shall call attention to one point. If we form the covariant divergence of the field equations (7.5), then we obtain, because of the identity (6.11), the result

$$\tfrac{1}{2}\alpha\, g^{\mu\varrho} R_{;\varrho} + \beta\, P^{*\mu\varrho}{}_{;\varrho} = 0. \tag{8.4}$$

If we form the scalar of Eqs. (7.5), by contracting on $g^{\mu\nu}$, we have further

$$\alpha R + \beta\, g^{\mu\nu} P^{*}_{\mu\nu} = 0, \tag{8.5}$$

and hence, by combining Eqs. (8.4) with (8.5),

$$(P^{*\mu\varrho} - \tfrac{1}{2} g^{\mu\varrho} g_{\alpha\beta} P^{*\alpha\beta})_{;\varrho} = 0. \tag{8.6}$$

This last divergence relation imposes on the source tensor $P^{*\mu\nu}$ a condition which must be satisfied if the field equations (7.5) are to possess any solutions. In the

[1] This terminology does, of course, not refer to the slow convergence of the approximation procedure, but to the physical assumption that all motion is slow.

special theory of relativity we had constructed an energy-momentum tensor whose (ordinary) divergence vanishes, because this divergence relation expresses the law of conservation of energy (or mass) and linear momentum, as well as the conservation of angular momentum after some manipulation, if matter and its interaction are to be described in terms of a continuous distribution. From Eq. (8.6) we gather that it is not the tensor $P^{*\mu\nu}$ itself that must play the role of our original energy-stress tensor, but the expression $P^{\mu\nu}$,

$$P^{\mu\nu} = P^{*\mu\nu} - \tfrac{1}{2} g^{\mu\nu} g_{\alpha\beta} P^{*\alpha\beta}, \tag{8.7}$$

inasmuch as in a Minkowski universe and in terms of a Lorentz frame the covariant and the ordinary partial differentiation are identical. Using this new tensor $P^{\mu\nu}$ we may write our field equations (7.5) in the alternative form:

$$\alpha G_{\mu\nu} + \beta P_{\mu\nu} = 0. \tag{8.8}$$

This form, by the way, is obtained immediately if in the variation we consider $g^{\mu\nu}$, rather than $\mathfrak{g}^{\mu\nu}$, as the independent variables.

In order to obtain an approximate expression for $G_{\mu\nu}$ in terms of the $\mathfrak{g}^{\mu\nu}$, we may use to advantage a computation which is based on the fact that a variation of the affine connection is a tensor and that for this reason such a variation must be expressible fully in terms of the covariant derivatives of the variation of the underlying metric. By means of a straightforward calculation, which first expresses the variation of $G_{\mu\nu}$ in terms of the variations of $\left\{ {\varrho \atop \mu\nu} \right\}$, and then the latter in terms of the $\delta \mathfrak{g}^{\mu\nu}{}_{;\varrho}$ we obtain the following rigorous expression:

$$\delta G_{\mu\nu} = \tfrac{1}{2} (\mathfrak{g}_{\mu\alpha} \delta \mathfrak{g}^{\alpha\varrho}{}_{;\nu\varrho} + \mathfrak{g}_{\nu\alpha} \delta \mathfrak{g}^{\alpha\varrho}{}_{;\mu\varrho} - \mathfrak{g}_{\mu\alpha} \mathfrak{g}_{\nu\beta} \mathfrak{g}^{\varrho\tau} \delta \mathfrak{g}^{\alpha\beta}{}_{;\varrho\tau} - \mathfrak{g}_{\mu\nu} \delta \mathfrak{g}^{\varrho\sigma}{}_{;\varrho\sigma}). \tag{8.9}$$

For our present purpose we draw on the linearizing assumption and remark that in a Minkowski geometry $G_{\mu\nu}$ vanishes. If we now consider a metric that deviates from the Minkowski metric by small amounts, then in any expression in which differentiated components of the metric are multiplied by undifferentiated components, the leading terms are those in which the undifferentiated factors are replaced by the Minkowski values. Furthermore, covariant derivatives are to first order identical with partial derivatives. Accordingly we conclude from our expression (8.9) that in the linearizing approximation we may set:

$$G_{\mu\nu} \approx \tfrac{1}{2} (\eta_{\mu\alpha} \gamma^{\alpha\varrho}{}_{,\varrho\nu} + \eta_{\mu\alpha} \gamma^{\alpha\varrho}{}_{,\varrho\nu} - \eta^{\varrho\sigma} \gamma_{\mu\nu,\varrho\sigma} - \eta_{\mu\nu} \gamma^{\varrho\sigma}{}_{,\varrho\sigma}), \tag{8.10}$$

where we have introduced the notation

$$\left. \begin{aligned} \gamma^{\mu\nu} &= \mathfrak{g}^{\mu\nu} - \eta^{\mu\nu}, \\ \gamma_{\mu\nu} &= \eta_{\mu\alpha} \eta_{\nu\beta} \gamma^{\alpha\beta} = \eta_{\mu\nu} - \mathfrak{g}_{\mu\nu}. \end{aligned} \right\} \tag{8.11}$$

By substituting the approximate expression (8.10) into the field equations (8.8) we obtain a set of 10 linear field equations for the 10 variables $\gamma_{\mu\nu}$, but these equations are not separated. We can separate them, however, by adopting a specialized coordinate frame, one in which the four expressions

$$\sigma^{\mu} \equiv \mathfrak{g}^{\mu\varrho}{}_{,\varrho} \approx \gamma^{\mu\varrho}{}_{,\varrho} = 0 \tag{8.12}$$

vanish (DE DONDER's coordinate conditions[1]). In general such a frame can always be constructed, a fact that we shall at least make plausible by showing that in the linearized approximation this construction involves no more than the inte-

[1] T. DE DONDER: La gravifique einsteinienne. Paris: Gauthier-Villars 1921.

gration of four inhomogeneous d'Alembert equations; this integration can be accomplished by means of a suitable GREEN's function, for instance the Liénard-Wiechert potentials.

Under an infinitesimal coordinate transformation

$$x^{\varrho\prime} = x^{\eta} + \xi^{\varrho}(x) \tag{8.13}$$

the transformation law of the σ^{μ} as a function of the coordinates is

$$\bar{\delta}\sigma^{\mu} = \sigma^{\varrho}\,\xi^{\mu}{}_{,\varrho} - (\sigma^{\mu}\,\xi^{\varrho})_{,\varrho} + \mathfrak{g}^{\varrho\sigma}\,\xi^{\mu}{}_{,\varrho\sigma}. \tag{8.14}$$

In the linearized approximation the terms that contain σ^{μ} on the right vanish,

$$\bar{\delta}\sigma^{\mu} \approx \eta^{\varrho\sigma}\,\xi^{\mu}{}_{,\varrho\sigma} = \Box\,\xi^{\mu}. \tag{8.15}$$

Hence a coordinate transformation which is described by ξ^{μ} such that

$$\Box\,\xi^{\mu} + \sigma^{\mu} = 0 \tag{8.16}$$

leads to the desired result that the σ^{μ} vanish. If we go over to such a coordinate frame, three of the four terms in the expression (8.10) vanish, and Eq. (8.8) reduces to the separated system

$$-\frac{\alpha}{2}\,\Box\,\gamma^{\mu\nu} + \beta\,P^{\mu\nu} = 0. \tag{8.17}$$

This system of equations is, of course, susceptible to direct integration, again with the help of the Liénard-Wiechert potentials or some similar GREEN's function. However, we shall again add the "slow" to the "linearizing" approximation, so that the d'Alembert operator reduces to a Laplace operator. Using the identification suggested by Eq. (8.3), we find in particular

$$\frac{\alpha}{2}\,\nabla^2\,\gamma_{00} + \beta\,P_{00} = 0. \tag{8.18}$$

If P_{00} is considered to be the (relativistic) mass density, it follows that the ratio between α and β must be

$$\frac{\beta}{2} = 8\pi\varkappa \tag{8.19}$$

if our theory is to lead to NEWTON's theory in the slow, linear approximation. This determination we shall accept in all that follows.

9. WEYL's tensor and the Petrov-Pirani notation. In the next several sections we shall discuss some of the known solutions of EINSTEIN's field equations; some of these are closed-form solutions of the rigorous field equations (8.8), some belong to the linearized equations (8.10) or (8.17) with (8.12). The interpretation of these solutions is hampered by the arbitrariness of the choice of coordinate systems implied by the principle of covariance. One gravitational field may appear in several forms which appear entirely different. Even the trivial (flat) space-time metric may appear in disguise. The general problem of identifying physical situations irrespective of the coordinate system chosen for their presentation is known as the *equivalence problem*. In Sect. 23 seq. we shall present a complete solution of the equivalence problem by means of *intrinsic* coordinates, a relatively recent development in general relativity. For the time being, we shall direct our attention to certain properties of the curvature tensor that are relatively independent of the choice of coordinates.

All twenty components of the curvature tensor vanish in a flat space-time manifold, and this property is certainly invariant. In the non-trivial cases, ten components of the curvature tensor, those combined into the Ricci tensor, or into the Einstein tensor, are algebraically tied by the field equations to the source distribution, the components of the matter tensor $P_{\mu\nu}$. The remaining components of the full curvature tensor, ten in number, are not algebraically restricted by the field equations, though they are subject to certain differential conditions. We may combine these remaining ten field components into a tensor of their own, known as *Weyl's tensor*, which is constructed so as to possess all algebraic symmetry properties of the Riemann-Christoffel tensor and to obey the further requirement that its contractions vanish. WEYL's tensor has the form:

$$\left.\begin{aligned} C_{\iota\varkappa\lambda\mu} &= R_{\iota\varkappa\lambda\mu} + \tfrac{1}{2}\,(g_{\iota\lambda} R_{\varkappa\mu} + g_{\varkappa\mu} R_{\iota\lambda} - g_{\varkappa\lambda} R_{\iota\mu} - g_{\iota\mu} R_{\varkappa\lambda}) + \\ &\quad + \tfrac{1}{6}\,(g_{\varkappa\lambda} g_{\iota\mu} - g_{\iota\lambda} g_{\varkappa\mu})\,R, \\ C_{\varkappa\lambda} &\equiv g^{\iota\mu} C_{\iota\varkappa\lambda\mu} \equiv 0. \end{aligned}\right\} \tag{9.1}$$

Although this tensor is of rank 4, because of its many algebraic symmetries it has only 10 algebraically independent components.

In the absence of sources of the gravitational field, WEYL's tensor and the Riemann-Christoffel tensor are equal. Both of them depend on two pairs of indices that are each skewsymmetric. Accordingly, we may consider WEYL's tensor as an algebraic form in a linear vector space whose base consists of the six linearly independent bivectors (= skewsymmetric tensors of rank 2) whose index pairs are:

Bivector index $(A, B, \ldots)$	*Index pair*
1	(23)
2	(31)
3	(12)
4	(10)
5	(20)
6	(30)

(9.2)

Thus, for purposes of algebraic manipulation we shall transcribe WEYL's tensor as a 6×6 matrix, which may be interpreted as a linear operator in the space of bivectors. We have, for instance, the following components of the new matrix C:

$$C_{41} = -C_{0123}, \quad C_{31} = C_{1223}, \quad C_{45} = C_{0102}. \tag{9.3}$$

The matrix C is symmetric with respect to its subscripts, but it has other simple algebraic properties as well, which together reduce the total number of independent components to ten. In order to formulate these properties, we shall define two additional matrices, of which one will play the role of a metric tensor and will permit us to raise and lower bivector indices, whereas the other will be the Levi-Civita tensor of the four-dimensional tensor calculus. We define:

$$g^{AB} = g^{[\alpha\beta][\iota\varkappa]} = g^{\alpha\iota} g^{\beta\varkappa} - g^{\beta\iota} g^{\alpha\varkappa} \tag{9.4}$$

and

$$\varepsilon^{A}{}_{B} = \varepsilon^{[\alpha\beta]}{}_{[\iota\varkappa]} = \frac{1}{\sqrt{-g}}\, g_{\varrho\iota}\, g_{\sigma\varkappa}\, \delta^{\alpha\beta\varrho\sigma}. \tag{9.5}$$

These new formations satisfy the relations

$$\varepsilon^2 = -I \tag{9.6}$$

and

$$\varepsilon g = (\varepsilon g)^T = -g\,\varepsilon^T, \tag{9.7}$$

where the superscript T indicates transition to the transpose matrix. The totality of all algebraic properties of C may now be summarized as follows:

$$\left.\begin{aligned} &C = C^T, \quad C\,\varepsilon \equiv C^* = C^{*T} \equiv \varepsilon^T C, \\ &\operatorname{tr} C = 0, \quad \operatorname{tr} C^* = 0. \end{aligned}\right\} \tag{9.8}$$

The symbol tr denotes the formation of the trace.

The technique of passing over to the six-dimensional space of bivectors was initiated by PETROV[1].

We can illuminate the algebraic relations (9.8) by changing our notation in the sense that we introduce at one world point a tetrad of mutually orthogonal unit vectors and separate the first three bivector indices from the second triplet. If we retain matrix notation for each of these two triplets, the 6×6 matrices (9.4), (9.5) assume the form

$$g = \begin{pmatrix} I & 0 \\ 0 & -I \end{pmatrix}, \quad \varepsilon = \begin{pmatrix} 0 & -I \\ I & 0 \end{pmatrix}. \tag{9.9}$$

In these expressions I denotes the 3×3 unit matrix. With their help, we may combine the relations (9.8) into the following set, which deals with the two 3×3 matrices A and B:

$$C = \begin{pmatrix} A & B \\ B & -A \end{pmatrix}, \quad \begin{aligned} A &= A^T, & B &= B^T, \\ \operatorname{tr} A &= 0, & \operatorname{tr} B &= 0. \end{aligned} \tag{9.10}$$

Both A and B are symmetric and have vanishing traces. Thus each of these two matrices has five independent matrix elements, and we recover once more the result that the total number of algebraically independent elements must equal ten. In our rather specialized tetrad frame the elements of A are those components of WEYL's tensor with an even number of indices 0 (none or two), whereas the elements of the matrix B correspond to those components in which the index 0 appears exactly once.

The transformation of the matrix C, or the matrices A and B, under Lorentz transformations of the underlying tetrad of unit vectors may be treated conveniently by transition to a complex notation. In view of the fact that the matrices A and B have the same algebraic properties, we combine them into one symmetric, complex, and trace-less matrix Γ,

$$\Gamma = A + i\,B. \tag{9.11}$$

If we now combine the two three-vectors that characterize a Lorentz transformation (SR, Sects. 11 and 12) into one complex vector and use it to define complex orthogonal transformations in three-space, the matrix Γ will transform as a tensor of rank 2 under these orthogonal transformations. With this technique one may investigate the problem of orthogonalization of the matrix Γ.

In the real domain any symmetric matrix may be diagonalized by orthogonal transformations; likewise a complex Hermitian matrix can be diagonalized by unitary transformations. But it turns out that a complex symmetric matrix cannot always be diagonalized by complex orthogonal matrices. An attempt to do so in formal analogy to the procedure in the real domain leads to expressions whose denominators are no longer bounded away from zero. Thus we find that there are a number of special algebraic situations, which lend themselves to

[1] A. S. PETROV: Sci. Nat. Kazan State Univ. **114**, 55 (1954). — F. A. E. PIRANI: Phys. Rev. **105**, 1089 (1957). — P. JORDAN, J. EHLERS and W. KUNDT: Akad. Wiss. Lit. Mainz, math.-nat. Kl. **1960**, Nr. 2, 21—105 [*21*].

classification. This classification has been accomplished by a number of different methods. There are three main classes, usually referred to as Petrov classes I, II, and III. Class I consists of those matrices Γ which can be fully diagonalized. In class II Γ may be separated into two diagonal boxes, of which the larger (the 2×2 box) cannot be diagonalized. In class III even partial diagonalization is impossible. Each of these classes possesses subclassifications, which correspond to varying degrees of degeneracy of the characteristic values of Γ. For details we refer to the article by JORDAN, EHLERS and KUNDT, cited above.

The Petrov classification has proven useful for obtaining closed-form solutions of the field equations, in that each class, or subclass, with the exception of the non-degenerate class I, permits the introduction of a coordinate system in which the field equations are considerably reduced in complexity. It has been suggested that classes II and III somehow represent "pure gravitational radiation"; certainly one subclass of II has those properties that in the linearized case are represented by plane gravitational waves (Sect. 11). The physical significance of class III, if any, has so far remained elusive.

10. Linearized static solutions. In this and the following three sections we shall discuss specific types of solutions of the field equations that lend themselves readily to physical interpretation. A discussion of all solutions is made difficult by the complexity of the field equations. At present we know only a limited number of solutions of the rigorous field equations, not nearly enough to satisfy us that we understand all the possible physical implications of the theory. The linearized field equations, together with the coordinate conditions, Eqs. (8.17), (8.12), may be integrated without difficulty, and the solutions interpreted in terms of concepts familiar from other field theories (foremost among them electrodynamics). There are, furthermore, methods available for correcting these solutions for second-order non-linear terms in the field equations, by a systematic expansion in powers of deviation from the Minkowski metric. But we may not trust such expansion approaches uncritically, as there are no mathematical theorems that would assure us that our expansions converge, or even that to every type of linearized solution there exists a corresponding rigorous solution toward which the power series could converge. Accordingly, the value of the linearized solutions is largely conjectural. We shall nevertheless describe some of them, in the hope that mathematical justification will be forthcoming eventually.

Beginning with the simplest problem of all, we shall consider the static fields associated with point sources. In view of the fact that $P_{\mu\nu}$ in the linear approximation will have its ordinary divergence vanish, we may choose a distribution of matter in which only P_{00} is different from zero and possesses a delta-function-like distribution at the origin of the spatial coordinates, vanishing everywhere else in three-space. If this distribution is centrosymmetric, the corresponding solution of the field equations (8.17) will be

$$\gamma_{00} = \frac{2\varkappa\mu}{r}, \qquad \gamma_{\varrho s} = 0. \tag{10.1}$$

The parameter μ is to be interpreted as the total mass of the source distribution,

$$\mu = \int P_{00}\, d^3x. \tag{10.2}$$

As we shall see later, there exists a solution of the rigorous field equations to which the expression (10.1) represents the simplest non-trivial approximation; that is the famous Schwarzschild solution. The centrosymmetric solution is to be considered as the prototype of a field surrounding a compact celestial body.

Its existence makes it at all possible to establish a connection between the formalism of general relativity and the body of observational experience that we possess about the properties of gravitation.

Just as in electrostatics, additional solutions of the static field equations may be constructed by the simple expedient of replacing the monopole solution (10.1) of LAPLACE's equation by higher spherical harmonics, or by linear combinations of higher spherical harmonics. None of these solutions have any known physical significance, nor is it known whether there are rigorous solutions corresponding to all of them; there are rigorous solutions known that correspond to the spherical harmonics with cylindrical symmetry, i.e. to the harmonics having the form $P_l(\cos\vartheta)\,r^{-(l+1)}$. We can also easily envisage the physical realizability of all those solutions whose corresponding mass distributions are nowhere negative. For instance, the Earth, and likewise the Sun and the planets, all have non-vanishing quadrupole moments, because of their oblate shapes.

Given a harmonic function $f(x, y, z)$, we can construct the corresponding curvature tensor. If we form the two 3×3 matrices A and B introduced in Sect. 9, we find that B vanishes, whereas A turns out to be simply

$$A_{kl} = C_{kl} = -\tfrac{1}{4} f_{,kl}, \qquad k, l = 1, 2, 3. \tag{10.3}$$

Additional static solutions of the field equations with point-like sources correspond to sources whose physical meaning is considerably more obscure than that of the distributions described so far. There are solutions that are produced by sources whose components P_{0k} $(k = 1, 2, 3)$ are non-zero. In order to satisfy the divergence relationship, these sources must be solenoidal (i.e. the curls of some three-dimensional vector field) and so must be the resulting field γ_{0k}. Accordingly, we set

$$\gamma_{0k} = \delta_{krs} f_{s,r}, \qquad \nabla^2 f_s = 0. \tag{10.4}$$

In this manner we can, for instance, produce solutions that look like the solutions of a mass current loop with zero mass density. Strictly speaking, the solution (10.4) is not unique, as the De Donder coordinate conditions need not be satisfied at the singular point, the coordinate origin. Accordingly, the most general solution would contain, in addition to the expression (10.4), the gradient of the function $(1/r)$. However, it is not difficult to show that such an additional gradient can be transformed away by a coordinate transformation in which δx^0 is proportional to r^{-1}.

In nature we cannot produce mass distributions of this kind, simply because we possess no negative masses. The general theory of relativity does not exclude such masses; as far as that theory is concerned, the discovery of particles with negative mass would serve neither to confirm nor to refute the theory. If a continued search for such particles should fail to lead to their discovery, their non-existence would probably add to our doubts that the theory in its present form is complete.

If we calculate again the curvature tensor corresponding to a field of type (10.4), we find that of the two matrices A and B the first one vanishes, whereas B depends solely on the divergence of the vector field f_s,

$$A = 0 \qquad B_{ik} = C_{i,k+3} = -\tfrac{1}{2}(\operatorname{div} \boldsymbol{f})_{,ik}, \qquad i, k = 1, 2, 3. \tag{10.5}$$

The lowest multipole character of $\operatorname{div} \boldsymbol{f}$ is not centrosymmetry but that of a dipole. To the extent that one can speak of an analogy between the electromagnetic and the gravitational fields, the type (10.5) corresponds to the field of a magnetostatic dipole or higher multipole.

If we now proceed systematically and investigate the solutions whose γ_{ij} $(i, j=1, 2, 3)$ are different from zero, we find that such solutions do indeed exist and that the lowest multipole of this kind is a quadrupole. Generally speaking, such solutions may be given the form

$$\gamma_{ij} = f_{is,sj} + f_{js,si} - \delta_{ij} f_{sr,rs} - f_{ss,ij}, \qquad f_{rs} = f_{sr}, \qquad \nabla^2 f_{rs} = 0. \tag{10.6}$$

Further examination shows, however, that this type of solution may be transformed into the first kind by means of a linearized coordinate transformation,

$$\delta h_{ij} = -(\xi_{i,j} + \xi_{j,i}), \qquad \delta\gamma_{ij} = \xi_{i,j} + \xi_{j,i} - \delta_{ij}\xi_{s,s}, \tag{10.7}$$

so that actually no new solutions are obtained in this manner[1]. Thus the most general linearized solutions whose sources are concentrated at one point may be thought of as generated by static mass distributions and by stationary solenoidal mass current distributions.

In contrast to the full field equations, the linearized field equations permit the linear superposition of solutions. Hence we may add, for instance, the solutions that correspond to a mass monopole, a mass quadrupole, and a solenoidal mass current loop. This particular combination, with suitable coefficients, may be a fairly realistic approximation to the field caused by a celestial body that rotates about its axis and is thereby oblate. The magnetic-type field caused by the rotation of the mass will give rise to forces in the vicinity of the planet that resemble Coriolis forces, in that they will have a ponderomotive effect only on test particles that are in motion. For a possible experimental test cf. Sect. 15.

11. Gravitational waves. Next we shall consider solutions of the system of Eqs. (8.12), (8.17) that represent waves. We begin with the coordinate conditions (8.12) and consider them as integrability conditions for a set of functions $U^{\mu[\nu\varrho]}$, from which the gravitational potentials $\gamma^{\mu\nu}$ are obtained as divergences,

$$\gamma^{\mu\nu} = U^{\mu[\nu\varrho]}{}_{,\varrho}, \qquad U^{\mu[\varrho\sigma]} + U^{\mu[\sigma\varrho]} \equiv 0. \tag{11.1}$$

The existence of the U-quantities assures that the conditions (8.12) are in fact satisfied, but it does not assure the symmetry of the $\gamma^{\mu\nu}$ with respect to their two superscripts. Hence we must require separately that

$$(U^{\mu[\nu\varrho]} - U^{\nu[\mu\varrho]})_{,\varrho} = 0, \tag{11.2}$$

a divergence relationship, which in turn is an integrability condition and insures the existence of a field $V^{[\mu\nu][\varrho\sigma]}$, antisymmetric with respect to the first pair of indices as well as with respect to the second, which is related to the U-field thus:

$$U^{\mu[\nu\varrho]} - U^{\nu[\mu\varrho]} = V^{[\mu\nu][\varrho\sigma]}{}_{,\sigma}, \qquad V^{[\mu\nu][\varrho\sigma]} = -V^{[\nu\mu][\varrho\sigma]} = -V^{[\mu\nu][\sigma\varrho]}. \tag{11.3}$$

We can solve this equation for $U^{\mu[\varrho\sigma]}$, that is to say, we can express the U-field in terms of the V-field, by permuting the indices three times. The result of this operation is

$$U^{\mu[\nu\varrho]} = \tfrac{1}{2}(V^{[\mu\nu][\varrho\sigma]} + V^{[\varrho\nu][\mu\sigma]} + V^{[\varrho\mu][\nu\sigma]})_{,\sigma}. \tag{11.4}$$

Substitution of this expression into Eq. (11.1) yields the following expression:

$$\gamma^{\mu\nu} = \Psi^{\mu\varrho,\nu\sigma}{}_{,\varrho\sigma}, \tag{11.5}$$

where $\Psi^{\mu\varrho,\nu\sigma}$ is a linear combination of V-quantities,

$$\left.\begin{aligned}\Psi^{\mu\varrho,\nu\sigma} = -\tfrac{1}{4}(&V^{[\mu\nu][\varrho\sigma]} + V^{[\varrho\sigma][\mu\nu]} + V^{[\mu\varrho][\nu\sigma]} + \\ &+ V^{[\nu\varrho][\mu\sigma]} + V^{[\mu\sigma][\nu\varrho]} + V^{[\nu\sigma][\mu\varrho]}),\end{aligned}\right\} \tag{11.6}$$

[1] R. SACHS and P. G. BERGMANN: Phys. Rev. **112**, 674 (1958).

and possesses the following symmetry properties:

$$\Psi^{\mu\varrho,\nu\sigma} = -\Psi^{\varrho\mu,\nu\sigma} = -\Psi^{\mu\varrho,\sigma\nu} = +\Psi^{\nu\sigma,\mu\varrho}, \tag{11.7}$$

that is to say, it is antisymmetric in each index pair, and symmetric with respect to an interchange of the two index pairs. From now on, we shall work with the new Ψ-field, which is subject to the symmetry conditions (11.7) but may be otherwise chosen at will to guarantee that the $\gamma^{\mu\nu}$, Eq. (11.5), will satisfy the coordinate conditions (8.12).

Now we come to the field equations (8.17). In the absence of sources, the gravitational potentials will satisfy D'ALEMBERT's equation; they will certainly do so if the components of the Ψ-field satisfy the same wave equation. Though there are obviously solutions of the field equations whose corresponding Ψ-field does not satisfy D'ALEMBERT's equation, we shall in this section consider Ψ-fields that do. There is, first of all, the plane wave solution, which we can construct by assuming that all components of the Ψ-field depend on the single argument

$$A = k_\varrho x^\varrho, \qquad \eta^{\varrho\sigma} k_\varrho k_\sigma = 0. \tag{11.8}$$

k_ϱ represents the propagation vector. We have, very obviously,

$$\gamma^{\mu\nu} = k_\varrho k_\sigma \Psi^{\mu\varrho,\nu\sigma\,\prime\prime}. \tag{11.9}$$

With these expressions we can form the (linearized) components of the uncontracted curvature tensor and thus establish the various types of gravitational radiation that exist in this approximation. However, to perform such a calculation would be extremely tedious, and the results somewhat disappointing. Given a certain direction of propagation, the formalism just developed would yield not only true gravitational waves (i.e. waves that cannot be transformed away) but also all those deviations from a Minkowski metric that would be produced from a flat metric by a coordinate transformation in which the ξ^ϱ, Eq. (8.13), are solutions of D'ALEMBERT's equation, according to Eq. (8.15). Furthermore, there exist Ψ-fields whose divergences vanish and which, though non-zero themselves, lead to vanishing gravitational potentials. As a matter of fact, calculations that are considerably less cumbersome than those carried out with the help of a Ψ-field lead to the result that for a given direction of propagation there are only two linearly independent true gravitational waves, and these two correspond to two different states of polarization. They can be carried over into each other by a rotation of the spatial coordinates about the direction of propagation through an angle of $\pi/4$.

We may obtain non-trivial solutions quite easily by orienting the propagation vector parallel to a coordinate axis, say the x^1-axis, so that the argument A turns into $(x^0 - x^1)$, and by letting only one component of Ψ be different from zero, for instance the component $\Psi^{21,31}$. We have then only one non-vanishing gravitational potential,

$$\gamma^{23} = \Psi^{21,31}{}_{,11} \equiv 2f(x^0 - x^1) \equiv h_{23} = h_{32}. \tag{11.10}$$

If we form the matrices A and B, Eq. (9.10), we obtain the following expressions:

$$A = \begin{pmatrix} 0 & 0 & 0 \\ 0 & 0 & f'' \\ 0 & f'' & 0 \end{pmatrix}, \qquad B = \begin{pmatrix} 0 & 0 & 0 \\ 0 & +f'' & 0 \\ 0 & 0 & -f'' \end{pmatrix}. \tag{11.11}$$

Under a rotation through $\pi/4$ about the x^1-axis, the matrices A and B interchange their forms, except for a change in sign. The matrices (11.11) belong to Petrov class II.

Though the Ψ-formalism is unnecessarily cumbersome for the construction of plane-wave solutions, it is the only method known by which we can construct (linearized) spherical waves. Even then, there is no doubt that the number of linearly independent solutions is much smaller than the number of possible Ψ-components, which is 21. In a complete evaluation, there will appear again both non-trivial solutions and those that represent a flat continuum in a curvilinear coordinate system. The details have never been worked out, but it is apparent that the lowest form of symmetry of a spherical wave must be quadrupole radiation[1]. The nonvanishing components of the Ψ-field would then have the form of an S-wave. If they are to represent an out-going wave, their functional dependence would be $r^{-1}F(r-t)$. If the wave is to be completely free of singularities, the functional dependence would be $r^{-1}[F(t+r)-F(t-r)]$. Of course, there is no difficulty about constructing spherical waves with more complex symmetry than quadrupole waves.

Though none of these considerations prove anything about the existence of spherical waves in the rigorous theory, we may at least conjecture that if such solutions do exist, the linear approximation ought to be fairly realistic at spatial infinity. Even this conjecture must be qualified. To the extent that a gravitational progressive wave carries energy, a steady-state progressive wave (e.g. an infinite sinusoidal out-going wave) is inconsistent with the field equations, and a second-order approximation would show this fact. A standing wave in all likelihood possesses an energy density that drops off at great distances from the center as r^{-2}; the integral over this energy density thus diverges, and we must assume that it is impossible to satisfy the usual boundary conditions at infinity. Hence we must assume that if spherical waves with standard boundary conditions at infinity exist at all, such waves must be pulses of finite total energy transport. If they are purely outgoing, then the effective mass connected with the center of radiation must be smaller after the pulse has left than it was before the pulse was initiated[2].

It would be most desirable to possess at least one rigorous solution of the gravitational field equations that resembles a spherical wave pattern; but this task has not yet been solved. As a preliminary, it might be remarked that the existence of a Ψ-field is a consequence of the De Donder coordinate conditions; hence it may be inferred in the rigorous case as well. However, if we were to rewrite the field equations of general relativity in terms of the Ψ-field, we should obtain non-linear partial differential equations of the fourth order. Whether the problem is amenable to treatment in this formulation remains to be seen.

12. Rigorous static solutions. In this section and the next we shall consider a few cases in which EINSTEIN's field equations have been solved in closed form. We shall begin with *static* solutions, that is with cases in which it is possible to introduce a coordinate system in which none of the gravitational potentials depends on x^0 and in which also the potentials g_{0s} vanish. Invariantly, the existence of such a coordinate system implies the existence of a vector field τ^ϱ with two properties: (1) A mapping of the manifold on itself in which each world point is displaced along a curve to which τ^ϱ is tangential, and by a distance proportional to the magnitude of the vector τ^ϱ, leaves the length of all curves unchanged. (2) The vorticity of this vector field τ^ϱ, i.e. the tensor $(\tau_{\varrho,\sigma}-\tau_{\sigma,\varrho})$, has zero projections in the hypersurface at right angles to the τ-direction. In terms of covariant differential equations, these two conditions read as follows:

$$\tau_{\mu;\nu}+\tau_{\nu;\mu}=0, \tag{12.1}$$

[1] J. BOARDMAN and P. G. BERGMANN: Phys. Rev. **115**, 1318 (1959).

[2] A. TRAUTMAN: To be published.

KILLING's equation, provides for the existence of the invariant mapping. And the condition on the vorticity is:

$$(\delta_\alpha^\mu - \tau^{-2}\tau_\alpha\tau^\mu)(\tau_{\mu;\nu} - \tau_{\nu;\mu})(\delta_\beta^\nu - \tau^{-2}\tau_\beta\tau^\nu) = 0, \qquad \tau^2 \equiv g_{\varrho\sigma}\tau^\varrho\tau^\sigma. \tag{12.2}$$

Manifolds that satisfy KILLING's equation but not the requirement on vorticity are often called *stationary* (rather than static).

The intuitive significance of either of these conditions is brought out if we do introduce a special coordinate system in which the contravariant components τ^ϱ take the values (0, 0, 0, 1). In this special coordinate system KILLING's equation takes the simple form

$$g_{\mu\nu,0} = 0. \tag{12.3}$$

By a suitable further restriction of the coordinate system, condition (12.2) turns into

$$g_{0m} = 0. \tag{12.4}$$

The condition (12.4) implies that a particle that is nearly at rest (momentarily) with respect to the chosen coordinate system will experience an acceleration only as a result of the gradient of the "scalar" gravitational potential g_{00}, but not a Coriolis type (i.e. velocity-dependent) acceleration. In a stationary manifold it is not possible to eliminate the velocity-dependent forces (for non-relativistic speeds) everywhere by the choice of a coordinate system in which the potentials are time-dependent.

Once we restrict ourselves to static manifolds and describe them in terms of a suitably chosen coordinate system, then the Christoffel symbols and the components of the curvature tensor are somewhat simplified. The components of the contracted curvature tensor $R_{\mu\nu}$ take the form:

$$\left.\begin{aligned} R_{mn} &= \overline{R}_{mn} + \tfrac{1}{2}\lambda_{|mn} + \tfrac{1}{4}\lambda_{|m}\lambda_{|n}, \qquad R_{0m} = 0,\\ R_{00} &= g_{00}\,g^{rs}(\tfrac{1}{2}\lambda_{|rs} + \tfrac{1}{4}\lambda_{|r}\lambda_{|s}), \end{aligned}\right\} \tag{12.5}$$

where the symbol $\overline{R}$ denotes the three-dimensional curvature tensor formed from the 3×3 metric g_{mn}, λ denotes $\ln g_{00}$, and the symbol $_{|m}$, etc., denotes covariant differentiation in three-space formed with the help of the Christoffel symbols formed from the 3×3 metric g_{mn}.

Let us consider briefly the group of coordinate transformations that remain at our disposal. The time coordinate is completely determined except for an additive constant ($t' = t + c$) and for a reflection, $t' = -t$; but the three remaining coordinates may still be transformed among themselves without any restrictions. Under this group of transformations g_{00} is a scalar field. This group is, however, insufficient to disentangle the field equations (12.5) sufficiently to permit their solution. To this end we must restrict the manifold by a second symmetry condition quite analogous to the first.

We shall now assume that there exists, in the static domain, a field of three-vectors v^k with exactly the same properties as the vector field τ^ϱ in the four-dimensional manifold[1]. That is to say, a mapping along the v-directions leaves the length of all curves unchanged, and so does a reflection from a (two-dimensional) surface at right angles to the v-curves. With this assumption, we may introduce a coordinate system in which all components of the metric tensor depend only

[1] H. WEYL: Ann. Phys. **54**, 117 (1917). — T. LEVI-CIVITA: R. C. Accad. Lincei **26**, 458 (1917).

on two of the four coordinates and in which the components $g_{01}, g_{02}, g_{03}, g_{13}$, and g_{23} all vanish. The metric tensor will now have the form

$$g_{\mu\nu} = \begin{pmatrix} g_{11} & g_{12} & 0 & 0 \\ g_{12} & g_{22} & 0 & 0 \\ 0 & 0 & g_{33} & 0 \\ 0 & 0 & 0 & g_{00} \end{pmatrix}. \tag{12.6}$$

All non-vanishing components are functions of x^1 and x^2 only. By means of a coordinate transformation that affects only these same two coordinates, we may further introduce conformal coordinates, which reduce the form (12.6) to

$$g_{\mu\nu} = \begin{pmatrix} A & 0 & 0 & 0 \\ 0 & A & 0 & 0 \\ 0 & 0 & g_{33} & 0 \\ 0 & 0 & 0 & g_{00} \end{pmatrix}. \tag{12.7}$$

We shall now introduce the following notation: We shall denote g_{33} by C, and g_{00} by B. The natural logarithms of the functions A, B, and C shall be denoted by α, β, and γ, respectively. Furthermore, we shall let Latin indices take only the values 1 and 2, and summations are to run only over these two values. With these conventions the field equations become:

$$\left.\begin{aligned} \frac{2A}{B} R_{00} &\equiv \beta_{,ss} + \frac{1}{2}\beta_{,s}(\beta+\gamma)_{,s} = 0 \\ \frac{2A}{C} R_{33} &= \gamma_{,ss} + \frac{1}{2}\gamma_{,s}(\beta+\gamma)_{,s} = 0 \\ R_{03} &\equiv 0, \quad R_{0k} \equiv 0, \quad R_{3k} \equiv 0, \\ 2R_{kl} &\equiv (\beta+\gamma)_{,kl} + \delta_{kl}\alpha_{,ss} + \frac{1}{2}\delta_{kl}\alpha_{,ss}(\beta+\gamma)_{,s} + \\ &\quad + \frac{1}{2}[\beta_{,k}\beta_{,l} + \gamma_{,k}\gamma_{,l} - \alpha_{,k}(\beta+\gamma)_{,l} - \alpha_{,l}(\beta+\gamma)_{,k}] = 0. \end{aligned}\right\} \tag{12.8}$$

These are altogether five equations for the three unknowns α, β, and γ. There are, however, a number of identities between them, which represent BIANCHI's identities in this highly symmetric situation. A detailed calculation shows that the number of these identities is two, and that they involve derivatives of the set of equations denoted by R_{kl}, and the remainder algebraically.

By adding the first two Eqs. (12.8), we obtain a single equation for the sum of β and γ,

$$\zeta_{,ss} + \tfrac{1}{2}\zeta_{,s}\zeta_{,s} = 0, \quad \zeta \equiv \beta + \gamma. \tag{12.9}$$

This equation may be converted into the two-dimensional Laplace equation by the substitution

$$Z_{,ss} = 0, \quad Z \equiv e^{\frac{1}{2}\zeta} = \sqrt{BC}. \tag{12.10}$$

This last equation is satisfied by the real (or the imaginary) part of any analytic function. We shall discuss below what kinds of analytic functions are consistent with reasonable boundary conditions.

Given a solution of Eq. (12.10), we may substitute into either the first or the second of Eqs. (12.8) or, indeed into their difference,

$$(Z\xi_{,s})_{,s} = 0, \quad \xi \equiv \beta - \gamma, \tag{12.11}$$

yielding another linear homogeneous equation.

Solutions of this equation, along with those of Eq. (12.10), determine the two fields β and γ. The remaining variable, α, is subject to the three equations R_{11}, R_{12}, and R_{22}. These three equations satisfy two Bianchi differential identities,

$$\left.\begin{aligned}&\left(R_{11}-R_{22}-\frac{A}{B}R_{00}-\frac{A}{C}R_{33}\right)_{,1}+2R_{12,2}+(\beta+\gamma)_{,1}R_{11}+\\&\quad+(\beta+\gamma)_{,2}R_{12}+(\alpha-\beta)_{,1}\frac{A}{B}R_{00}+(\alpha-\gamma)_{,1}\frac{A}{C}R_{33}\equiv 0,\\&\left(R_{22}-R_{11}-\frac{A}{B}R_{00}-\frac{A}{C}R_{33}\right)_{,2}+2R_{12,1}+(\beta+\gamma)_{,2}R_{22}+\\&\quad+(\beta+\gamma)_{,1}R_{12}+(\alpha-\beta)_{,2}\frac{A}{B}R_{00}+(\alpha-\gamma)_{,2}\frac{A}{C}R_{33}\equiv 0.\end{aligned}\right\}\tag{12.12}$$

The equations themselves may be conveniently considered in the following combinations:

$$R_{ss}-\left(\frac{A}{B}R_{00}+\frac{A}{C}R_{33}\right)\equiv\alpha_{,ss}-\frac{1}{2}\beta_{,s}\gamma_{,s}=0,\tag{12.13}$$

$$2R_{12}\equiv(\beta+\gamma)_{,12}+\frac{1}{2}(\beta_{,1}\beta_{,2}+\gamma_{,1}\gamma_{,2})-\frac{1}{2}[(\beta+\gamma)_{,1}\alpha_{,2}+(\beta+\gamma)_{,2}\alpha_{,1}]=0,\tag{12.14}$$

and either

$$\left.\begin{aligned}R_{11}-R_{22}+\frac{A}{B}R_{00}+\frac{A}{C}R_{33}\equiv(\beta+\gamma)_{,11}+\frac{1}{2}[\beta_{,1}\gamma_{,1}+\beta_{,2}\gamma_{,2}+\\+(\beta_{,1})^2+(\gamma_{,1})^2+(\beta+\gamma)_{,2}\alpha_{,2}-(\beta+\gamma)_{,1}\alpha_{,1}]=0\end{aligned}\right\}\tag{12.15}$$

or

$$\left.\begin{aligned}2(R_{11}-R_{22})\equiv\zeta_{,11}-\zeta_{,22}+\frac{1}{2}[(\beta_{,1})^2-(\beta_{,2})^2+(\gamma_{,1})^2-(\gamma_{,2})^2]\\+(\beta+\gamma)_{,2}\alpha_{,2}-(\beta+\gamma)_{,1}\alpha_{,1}=0.\end{aligned}\right\}\tag{12.16}$$

An examination of the identities (12.12) shows that of the three Eqs. (12.13), (12.14), and (12.16) the first one will be satisfied identically if the other two are satisfied (assuming, of course, that the two equations R_{00} and R_{33} have been solved previously). The two "off-diagonal" equations (12.14), (12.16) are first-order equations for α. These two equations are related to each other by the differential identities, which represent the integrability conditions necessary so that both may be satisfied. Accordingly, our system of equations may be solved locally by means of quadratures only. Any restrictions will come about by requirements of non-singularity of the solutions and by boundary conditions.

For rotational symmetry the coordinates that we have introduced must be interpreted as cylinder coordinates-plus-time; the coordinate x^3 is the azimuthal angle, and we may, for instance, adopt the line $x^1=0$ as the axis of symmetry. If we do, then on the axis g_{33} must vanish as $(x^1)^2 g_{11}$, that is to say, we must have:

$$\lim_{x^1\to 0}\left\{\frac{(x^1)^2 A}{C}\right\}=1.\tag{12.17}$$

This requirement must be met on the axis of symmetry whereever the metric is not singular. Hence, if we place isolated singularities on the axis, a path of integration that continues the α-field on the axis from one side of the singularity to the other must be compatible with Eq. (12.17). Because a closed path in a singly connected domain will not give rise to multiply-valued α-fields, the integration about an isolated singularity will place but one restriction on the form of that singularity. This restriction was recognized early as a forerunner of a more complete theory of motion. Because there are no solutions of the two-dimensional

Laplace equation that are free of singularities everywhere and which satisfy reasonable boundary conditions at infinity, any non-empty solution must have some singularities somewhere. Off-axis singularities represent singular rings, which have no physical interest. Hence we must place some singularities on the axis of symmetry. Now it turns out that these singularities are not capable of existence unless they satisfy certain conditions. We shall examine these conditions in the next section. Before doing so we shall briefly treat the case of spherical symmetry.

The metric (12.7) of course includes the solution with spherical symmetry as a special case. Nevertheless, because it singles out the axis of rotational symmetry, this is not the best form for exhibiting spherical symmetry. The spherically symmetric metric was first obtained by K. SCHWARZSCHILD and is named after him[1]. In one of the several forms to be found in the literature, SCHWARZSCHILD'S line element takes the form

$$g_{\mu\nu} = \begin{pmatrix} -\left(\delta_{mn} + \frac{R}{r}\,\frac{x^m x^n/r^2}{1-R/r}\right), & 0 \\ 0 & , \; 1 - \frac{R}{r} \end{pmatrix} \tag{12.18}$$

$$r^2 \equiv (x^1)^2 + (x^2)^2 + (x^3)^2, \qquad R \equiv 2\varkappa\mu.$$

The constant R is known as the Schwarzschild radius of the mass μ. All line elements, static or otherwise, that have spherical symmetry are coordinate transforms of this one[2]. Incidentally, this theorem implies that a spherically oscillating source of the gravitational field will not produce outgoing gravitational radiation.

13. Singularities on the axis. Let us return to the solutions having rotational symmetry. We shall in particular consider the situation in which there are point-like singularities on the axis of symmetry. This case represents a gravitational field that would be generated by several mass points on a straight line and at rest. We specialize the situation by choosing as the solution of Eq. (12.9) the function $\zeta = 2\ln x^1$. This choice includes the possibility of the trivial (flat) solution. The relationship between β and γ will then be:

$$\gamma = 2\ln\varrho - \beta, \tag{13.1}$$

where we write ϱ for x^1. We enter this expression for γ into the remaining field equations. Eq. (12.11) assumes the form

$$\beta_{,11} + \frac{1}{\varrho}\beta_{,1} + \beta_{,22} = 0, \tag{13.2}$$

whereas Eqs. (12.4) and (12.15) turn, respectively, into:

$$\bar{\alpha}_{,1} = \frac{\varrho}{2}\left[(\beta_{,1})^2 - (\beta_{,2})^2\right], \quad \bar{\alpha}_{,2} = \varrho\,\beta_{,1}\beta_{,2}, \quad \bar{\alpha} \equiv \alpha + \beta. \tag{13.3}$$

Eq. (12.13) is identical in Eqs. (13.2), (13.3) and need not be considered further.

The new variable $\bar{\alpha}$ has been chosen so that it must vanish on the axis, in accordance with the condition (12.17). Considering now an isolated singularity

[1] Sitzgsber. preuss. Akad. Wiss. **1916**, 189, 424.

[2] G. D. BIRKHOFF: Relativity and Modern Physics, p. 253. Cambridge, Mass.: Harvard University Press.

on the axis, we come to the conclusion that an integral, taken over a semicircular path of integration that bypasses the singularity, must vanish:

$$I = \oint \varrho \{\tfrac{1}{2}[(\beta,_1)^2 - (\beta,_2)^2]\, d\varrho + \beta,_1 \beta,_2\, dz\} = 0, \qquad z \equiv x^2. \tag{13.4}$$

Because the two expressions (13.3) satisfy the local integrability conditions, we are assured beforehand that the value of the integral I will not depend on the radius of the semicircular path of integration. Accordingly, we obtain just one condition. We shall now think of the solution of the differential equation (13.2) as being expanded into Legendre polynomials in this fashion:

$$\beta(\varrho, z) = \sum_{l=0}^{\infty} [a_l r^{-(l+1)} + b_l r^l]\, P_l\left(\frac{z}{r}\right), \qquad \varrho \equiv x^1, \quad z \equiv x^2, \quad r = \sqrt{\varrho^2 + z^2}. \tag{13.5}$$

In this expansion about the singularity, we have permitted both negative and positive powers of r, the "distance" from the singularity. There is no implication that this expansion is valid for large distances from the singularities. Generally, the radius of convergence is determined by the presence of additional singularities elsewhere on the axis. If we were to consider the case of but one singularity, then all the terms with positive powers of r should have to be omitted. Right now we are concerned with an expansion that will be useful in the immediate vicinity of the singularity under consideration, on the assumption that it may not be the only one present.

The integrand of I is quadratic in the derivatives of β, which are multiplied by the factors $\varrho\, d\varrho$, $\varrho\, dz$, both of which are proportional to r^2. From the fact that I must be independent of r it follows that because of the well-known orthogonality properties of the Legendre polynomials the only non-vanishing terms are those in which each a_l is multiplied by b_{l+1}. We shall denote the integral over the appropriate bilinear expression in Legendre polynomials and their derivatives by c_l. These c_l are definite numerics (though we shall not trouble here to calculate their values). Our condition (13.4) then takes the final form:

$$I \equiv \sum_{l=0}^{\infty} c_l a_l b_{l+1} = 0. \tag{13.6}$$

This result may be interpreted quite intuitively as follows: Our singularity may possess a mass monopole, along with a dipole, linear quadrupole, etc. The condition that a static solution exist is that the sum of the forces exerted on the monopole by the gradient of the potential, on the dipole by the gradient of the field strength, and so forth, be zero.

The calculation just performed intimates (though of course it does not prove) that the field equations of general relativity contain within themselves the ponderomotive laws; it appears that a point-like singularity of the field cannot exist statically unless the forces (in the Newtonian sense of that term) acting on that "particle" cancel each other. We shall have occasion to elaborate on that topic later on (Sect. 19), but the instance just treated represents historically the first encounter with "automatic" ponderomotive conditions in general relativity.

14. Rigorous plane-wave solutions. We now return to the topic of gravitational waves, which we treated in the linearized theory in Sect. 11. The existence of gravitational radiation, and its connection with moving masses, has been a source of controversy, which is not completely settled at this time[1]. We shall not attempt

[1] Cf. L. INFELD and J. PLEBAŃSKI: Motion and Relativity. London: Pergamon Press and Warszawa: Państwowe Wydawnictwo Naukowe 1960 [3].

to treat the relationship between gravitational radiation and its sources here; this is a question which is, apparently, still beset with both conceptual and mathematical difficulties. What has been settled is the existence of solutions of EINSTEIN'S equations in empty space that can reasonably be considered plane waves[1].

Early attempts to construct solutions of the field equations in which all components of the metric tensor would depend only on one argument (say $x^0 - x^1$) and be non-singular everywhere were unsuccessful. In 1937 EINSTEIN and ROSEN[2] constructed a metric that they interpreted as gravitational radiation emanating from a spatially infinitely extended line source. They gave reasons why they considered a plane wave as physically impossible. EINSTEIN and ROSEN obtained their solutions by formally interchanging in WEYL'S and LEVI-CIVITA'S static solutions with rotational symmetry the coordinates x^0 and x^3 (the latter being the axis of rotational symmetry) and making the necessary changes in signs.

BONDI and coworkers succeeded in constructing plane waves by first redefining the meaning of that concept. Instead of requiring that a plane-wave solution should have a specified coordinate dependence everywhere, they proceeded by considering the group of "motions" required for this type of solution. A "motion" in the language of differential geometry is a one-parametric continuous group of coordinate transformations that leave the precise functional dependence of the metric on the coordinates unchanged. The existence of a motion in general relativity is equivalent to the existence of a vector field that satisfies KILLING'S equation, (12.1). A plane electromagnetic wave is characterized by five separate motions. Three of these consist of translation in the wave front, whereas the two remaining motions represent Lorentz transformations involving a combination of rotation in space with a transition to a new inertial frame of reference, so that the form of the functions $\boldsymbol{E}(\boldsymbol{x}, t)$, $\boldsymbol{B}(\boldsymbol{x}, t)$ remains unchanged. If we define a plane gravitational wave as a metric field which admits a five-parametric group of motions, we find, as a particularly simple metric, the following example:

$$g_{\mu\nu} = \begin{pmatrix} -e^{\varphi(u)} & 0 & 0 & 0 \\ 0 & -u^2 e^{2\beta(u)} & 0 & 0 \\ 0 & 0 & -u^2 e^{-2\beta(u)} & 0 \\ 0 & 0 & 0 & e^{\varphi(u)} \end{pmatrix}, \qquad u \equiv x^0 - x', \tag{14.1}$$

where β is an arbitrary function of the argument u and $\varphi(u)$ obeys the single equation:

$$\frac{d\varphi}{du} = u\left(\frac{d\beta}{du}\right)^2. \tag{14.2}$$

The solution will be flat if

$$\frac{d^2\beta}{du^2} + \frac{2}{u}\frac{d\beta}{du} - u\left(\frac{d\beta}{du}\right)^3 = 0. \tag{14.3}$$

The metric (14.1) obviously depends only on one argument, and the direction of propagation is a null vector. It suffers from the apparent defect that the metric determinant vanishes for $u=0$. However, the coordinates chosen are not even approximately Cartesian. If we consider the special case in which the metric is

---

[1] H. BONDI: Nature, Lond. **179**, 1072 (1957). — H. BONDI, F.A.E. PIRANI and I. ROBINSON: Proc. Roy. Soc. Lond. A **251**, 519 (1959).

[2] A. EINSTEIN and N. ROSEN: J. Franklin Inst. **223**, 43 (1937). — N. ROSEN: Phys. Z. Sowjet. **12**, 366 (1937).

flat and both β and φ vanish, the transition from BONDI'S to Lorentz coordinates is accomplished by the following set of equations:

$$\left.\begin{aligned} x^0 &= y^0 - \frac{1}{2}\,\frac{(y^2)^2 + (y^3)^2}{y^0 - y^1}, \\ x^1 &= y^1 - \frac{1}{2}\,\frac{(y^2)^2 + (y^3)^2}{y^0 - y^1}, \\ x^2 &= \frac{y^2}{y^0 - y^1}, \quad x^3 = \frac{y^3}{y^0 - y^1}. \end{aligned}\right\} \tag{14.4}$$

In these equations y^ϱ constitute a Lorentz coordinate system. The wave fronts $x^0 - x^1 = \text{const}$ are identical with the planes $y^0 - y^1 = \text{const}$. Within a wave front, however, the transformations are highly nonlinear. In particular, Eqs. (14.4) show that the plane $x^0 - x^1 = 0 = y^0 - y^1$ represents a domain in which the transformation equations are singular. We may displace this plane in which the coordinate transformation is singular by adding to $y^0 - y^1$ a constant, i.e. by moving to a neighboring wave front.

The choice of nonlinear coordinates within the wave fronts is unavoidable. Any attempt to pass over to a more nearly linear coordinate system in the presence of a real wave [when the flatness condition (14.3) is not satisfied] leads to a dependence of the metric field on the coordinates x^2, x^3. Intuitively, the need for an apparently artificial type of coordinate system can be explained as follows. If we consider a plane wave pulse, in which the flatness condition holds on both sides of a bounded range of values of $x^0 - x^1$, the pulse separates two flat space-time regions, "before" and "after". If we examine geodesic congruences which on one side of the wave pulse consist of parallel lines, these lines will converge, or diverge, on the other side of the pulse. In terms of the behavior of test particles, we may say that an assembly of test particles that were at rest relative to each other before the passage of the wave pulse will exhibit non-zero velocities relative to each other after the pulse has passed by. Thus, even for a plane wave pulse we cannot connect the Minkowski (flat) universes on both sides of the pulse with each other across the pulse in such a manner that the direction of a vector is integrable under our definition of parallel displacement.

The extension of the results of BONDI and coworkers to spherical waves is at present conjectural, because it is not yet known that spherical waves exist in any physically acceptable sense[1]. However, if we assume that there are spherical gravitational waves, associated perhaps with an oscillating quadrupole, then we must conclude that at large distances from the source their amplitude decreases as r^{-1} and that the appearance of such a wave in a small domain (small in the sense that its linear dimensions are small compared to r) resembles that of a plane wave. For such a wave the angular divergence of two vectors on one side of the wave pulse which are parallel on the other will be proportional to r^{-1}, and hence if observed over a region of constant angle as viewed from the source, the deviation from parallelity of a whole vector field that is parallel on the other side of the wave pulse will be asymptotically independent of the distance r.

Thus it would seem that there is no unique way of connecting a tetrad of directions on one side of a wave pulse with the tetrad on the other, but that there is an uncertainty which depends on the amplitude of the wave, and on the wave length, We can estimate the order of magnitude of this uncertainty in radians for a radiating double star (if indeed such a system radiates). Consider two stars separated by a distance $\bar{r}$, whose masses lead to a Schwarzschild radius of the

[1] I. ROBINSON and A. TRAUTMAN: Phys. Rev. Letters 4, 431 (1960).

order of R. The period of such a double star system, according to KEPLER's third law, will be of the order of $c^{-1}\,\bar{r}^{\frac{3}{2}}\,R^{-\frac{1}{2}}$, and hence the wave length of the radiated wave $\bar{r}^{\frac{3}{2}}\,R^{-\frac{1}{2}}$. If we assume that the potentials are of the order $(v/c)^2\,(R/r)$ in dimensionless units, then the WEYL's tensor at a considerable distance will be a potential divided by the square of the wave length, or

$$C \approx \left(\frac{v}{c}\right)^2 \frac{R^2}{r\,\bar{r}^3} \approx \frac{R^3}{r\,\bar{r}^4}\,. \tag{14.5}$$

We now ask for the change in the angle of a vector that is carried around a closed curve that penetrates twice through the wave front and whose linear dimension in the wave front is of the order r. The depth of penetration should be of the order of the wave length, because once the phase of the wave exceeds π, the contributions from different phases of the wave tend to cancel each other. Hence the effective area of the two-dimensional surface bounded by the curve will be of the order

$$A \approx r\,\bar{r}^{\frac{3}{2}}\,R^{-\frac{1}{2}}. \tag{14.6}$$

The change in angle will be of the order $C A$, or

$$\Delta\psi \approx \left(\frac{R}{\bar{r}}\right)^{\frac{5}{2}}. \tag{14.7}$$

Consider an extreme case, in which the masses are those of fixed stars ($R \approx 10^5$ cm) and the distance of the order of the distance of the moon from the earth ($\bar{r} \approx 10^{11}$ cm). Then the order of magnitude of the expression (14.7) would come out as 10^{-15}, which amounts to 10^{-10} seconds of arc. For some time to come, this order of magnitude is not accessible to astronomical observation, but it is noteworthy that the notion of a uniquely determined frame of directions far from our galaxy may not be compatible with the presence of radiating systems of gravitating masses.

15. Experimental tests. When it comes to the verification of the general theory of relativity, the number of actual experiments and astronomical observations that bear on the theory and its predictions is still disappointingly meager. In the last few years, some new results have been added, and a number of tests have been proposed.

a) The principal underlying observation of any general-relativistic theory is the principle of equivalence, i.e. the equality (or proportionality) of inertial and (passive) gravitational mass[1]. This proportionality is at present believed to be confirmed to an accuracy of at least 10^{-9}, and there is hope that this figure can be improved on further[2].

b) From the beginning of general relativity, three effects predicted by the theory were known to be within the realm of observations. These are the precession of the ellipses of planetary orbits, especially that of Mercury, which is by far the largest and amounts to 43″/century; the bending of light rays by the field of gravity of the Sun, which is 1.75″ at the limb of the Sun; and the gravitational red shift of spectral lines that are observed at a different gravitational potential than where they are produced. As far as astronomical observations are concerned, the consensus of opinion appears to be that these three effects have been verified as closely as is feasible with the best techniques now available[3].

---

[1] Cf. Refs. given in Sect. 2.

[2] Private communication by Prof. R.H. DICKE.

[3] R.J. TRUMPLER: Helv. phys. Acta Suppl. **4**, 106 (1956) [*17*]. — W. PAULI: Theory of Relativity, p. 218 (Note 16). London: Pergamon Press 1958 [*9*].

Recently the Mössbauer effect has been used successfully to determine the gravitational red shift by a laboratory experiment[1]. The reliability of that experiment is probably higher than that of the astronomical observations of the gravitational red shift. Its numerical accuracy will undoubtedly be improved further, as poor statistics was one of the principal sources of error in the first published papers.

c) In recent years, a number of experiments have been suggested, but not yet performed, that would utilize instrumental satellites. Originally, it was suggested that the gravitational red shift should be measured in a satellite experiment[2], but this particular experiment has lost much of its attraction, because of the successful performance of the same determination in the laboratory with the help of the Mössbauer effect. More recently, PUGH and SCHIFF have suggested the determination of gyroscopic precession of stabilized satellites[3]. There are two different precession effects, one caused by the gravitating mass of the earth and independent of its rotation, the other a direct effect of the earth's rotation about its axis[4]. The ratio of these two effects is of the order of the length of the day divided by the period of the satellite. Also, the first of the two precession effects is independent of the orientation of the satellite orbit, whereas the second effect depends on the angle between the equatorial plane of the earth and the plane of the satellite orbit. It is questionable whether these effects can be discovered with instrumentation that will be available in the near future. The product of the frequency of precession of the dominant precession effect by the period of the satellite orbit equals the ratio between the earth's Schwarzschild radius and the radius of the satellite, which is of the order 10^{-9}.

d) Search for gravitational waves. J. WEBER (unpublished) has suggested a search for gravitational waves, perhaps produced by plasma oscillations, or, alternatively, for their production and detection in the laboratory by piezoelectric transducers. A cursory calculation shows that requirements on the power dissipation of the transmitter, the sensitivity of the detector, and the shielding against other means of transmission are extreme.

III. Inclusion of non-gravitational fields.

16. Expanded variational principle. We combine the gravitational with other fields in a general-relativistic (i.e. covariant) manner most easily by adding appropriate terms to the variational principle that leads to the field equations. The field equations of all fields are universally derived from action integrals that lend themselves readily to a general-relativistic form. Let us take, as the prime example of a classical (i.e. non-quantum) field, the electromagnetic field. The action principle of the electromagnetic field is the four-dimensional space-time integral over the world scalar that equals the difference between the squares of the electric and the magnetic field strengths. In a Lorentz-covariant notation this action integral is:

$$S = -\tfrac{1}{2}\int \varphi^{\mu\nu}\varphi_{\mu\nu}\,d^4x, \qquad \varphi_{\mu\nu} \equiv \varphi_{\mu,\nu} - \varphi_{\nu,\mu}. \tag{16.1}$$

This expression will become invariant with respect to curvilinear coordinate transformations if we raise the indices in the first factor by means of the variable

[1] T.E. CRANSHAW, J.P. SCHIFFER and A.B. WHITEHEAD: Phys. Rev. Letters **4**, 163 (1960). — R.V. POUND and G.A. REBKA jr.: Phys. Rev. Letters **4**, 337 (1960).

[2] S.F. SINGER: Phys. Rev. **104**, 11 (1956).

[3] G.E. PUGH: Unpublished 1959. — L.I. SCHIFF: Phys. Rev. Letters **4**, 215 (1960).

[4] W. DE SITTER: Proc. Acad. Sci. Amsterd. **19**, 367 (1916). — J. LENSE and H. THIRRING: Phys. Z. **19**, 156 (1918).

metric tensor $g^{\mu\nu}$ of general relativity instead of the Minkowski metric $\eta^{\mu\nu}$, and if we multiply the integrand by $(-g)^{\frac{1}{2}}$, in order to give it the proper density weight; we thus obtain the new action integral:

$$S_M = -\tfrac{1}{2}\int \sqrt{-g}\, g^{\mu\varrho} g^{\nu\sigma} \varphi_{\mu\nu} \varphi_{\varrho\sigma}\, d^4x, \tag{16.2}$$

whose variation with respect to *all* field variables is:

$$\left.\begin{aligned} \delta S_M = &-2\int (\sqrt{-g}\,\varphi^{\mu\varrho})_{,\varrho}\, \delta\varphi_\mu\, d^4x - \\ &-\int \sqrt{-g}\,(\varphi^{\varrho}_{\mu}\varphi_{\nu\varrho} - \tfrac{1}{4} g_{\mu\nu}\varphi^{\varrho\sigma}\varphi_{\varrho\sigma})\, \delta g^{\mu\nu}\, d^4x. \end{aligned}\right\} \tag{16.3}$$

As indicated in Eqs. (7.2), (7.5), (8.7) and (8.8), we may combine the gravitational action with the electromagnetic action and obtain the expanded action principle:

$$S = \int L\, d^4x, \qquad L = \sqrt{-g}\left(\frac{1}{8\pi\varkappa} R - \frac{1}{2}\varphi^{\mu\nu}\varphi_{\mu\nu}\right). \tag{16.4}$$

The field variables with respect to which the variation is to be carried out are the gravitational potentials $g_{\mu\nu}$ (or $\mathfrak{g}^{\mu\nu}$) and the electromagnetic potentials φ_μ. The resulting field equations fall into two sets. Four of the field equations are MAXWELL'S equations of the electromagnetic field. In these equations the metric tensor appears undifferentiated, or at most differentiated once. The other set of equations represents EINSTEIN'S equations of the gravitational field, with MAXWELL'S energy-stress tensor forming the right-hand sides, the "sources" of the gravitational field.

This tensor of MAXWELL, which is symmetric and possesses a zero trace, is composed of three distinct prerelativistic portions. With respect to a locally Minkowskian coordinate system, six components of this tensor form a 3×3 square, whose components are:

$$t_{kl} = \tfrac{1}{2}\delta_{kl}(\boldsymbol{E}^2 + \boldsymbol{H}^2) - (E_k E_l + H_k H_l). \tag{16.5}$$

Three components turn out to be:

$$t_{0k} = (\boldsymbol{E}\times\boldsymbol{H})_k. \tag{16.6}$$

The remaining component is:

$$t_{00} = \tfrac{1}{2}(\boldsymbol{E}^2 + \boldsymbol{H}^2). \tag{16.7}$$

These expressions constitute the stress, the energy flux, and the energy density, respectively, of the electromagnetic field. In the theory of the combined gravitational and electromagnetic field, the components of the four-tensor $t_{\varkappa\lambda}$ appear thus as the ten source intensities of the gravitational field. The electromagnetic field energy accordingly has ponderable mass just to the extent called for by the principle of equivalence.

We shall now generalize Eq. (16.4) by introducing, in addition to the metric field tensor $g_{\mu\nu}$, one or several non-gravitational fields, which we shall designate by the symbol ψ_A, and by adding to the Lagrangian of general relativity a term $\mathfrak{M}$, having the transformation properties of a scalar density of weight 1, so that the total Lagrangian becomes:

$$L = \frac{\sqrt{-g}}{8\pi\varkappa} R + \mathfrak{M}\,[\psi_A, \psi_{A,\varrho}, g^{\mu\nu}, g^{\mu\nu}{}_{,\varrho}]. \tag{16.8}$$

This Lagrangian will lead to two sets of field equations,

$$\Psi^A \equiv \frac{\delta\mathfrak{M}}{\delta\psi_A} = 0, \tag{16.9}$$

and

$$(8\pi\varkappa)^{-1} G_{\mu\nu} + \frac{1}{2} P_{\mu\nu} = 0, \qquad P_{\mu\nu} \equiv \frac{2}{\sqrt{g}}\,\frac{\delta\mathfrak{M}}{\delta g^{\mu\nu}}. \tag{16.10}$$

We shall call the tensor $P_{\mu\nu}$ the source density of the gravitational field. Because of the field equations (16.10), the covariant divergence of the source density must vanish. It is not difficult to show that this requirement will be satisfied automatically, because of the transformation properties of $\mathfrak{M}$, if the field equations (16.9) are satisfied.

In Lorentz-covariant theories the source density is identical with the energy-stress density. Three-dimensional integrals over four of its components are constants of the motion, and they are called, respectively, the total energy and the total linear momentum of the system. In a general-relativistic theory the source density itself leads to no constant integrals, but it is intimately related to a set of quantities that do. We shall consider the general-relativistic energy in the next section.

17. The energy-stress complex. In classical mechanics the energy, or Hamiltonian, is defined as the generator of the evolution of the physical system in the course of time. The relationship between the conservation law of energy and the equations of motion may be summarized in the identity:

$$\frac{d}{dt}\left(\sum_k \frac{\partial L}{\partial \dot{q}_k}\dot{q}_k - L\right) + \sum_k \dot{q}_k \frac{\delta L}{\delta q_k} + \frac{\partial L}{\partial t} \equiv 0. \tag{17.1}$$

Hence, if the system is conservative ($\partial L/\partial t = 0$), and provided the equations of motion are satisfied ($\delta L/\delta q_k = 0$), the energy is a constant of the motion. Similarly, if there are no external forces, the linear momentum of a system will be constant. In general relativity it is also possible to develop an energy-stress complex by reference to the coordinate transformations relative to which the theory is invariant. In view of the fact that the theory is invariant with respect to *all* (sufficiently differentiable, etc.) curvilinear coordinate transformations, this approach is more powerful than in classical mechanics, or even in Lorentz-covariant field theories. There is, however, a certain measure of ambiguity about the development of an energy-stress complex, which we shall have to discuss later on. In this section we shall construct *one* expression that leads to a conservation law, without claiming any uniqueness properties.

Consider our field variables, $g^{\mu\nu}$ and ψ_A. In order to carry out coordinate transformations we must describe their infinitesimal transformation laws. We shall add to the notation first introduced in Sect. 8 by distinguishing the change of a quantity under an infinitesimal coordinate transformation at a fixed world point from the change of the same quantity with fixed values of the coordinates. The first kind of change we shall denote by the symbol δ_C, the second by the symbol $\bar{\delta}$. These two kinds of change are related to each other by the identity

$$\bar{\delta} K \equiv \delta_C K - K_{,\varrho}\, \xi^\varrho. \tag{17.2}$$

A few examples of infinitesimal coordinate transformation laws may suffice. For the coordinates themselves we have:

$$\delta_C x^\varrho = \xi^\varrho, \qquad \bar{\delta} x^\varrho \equiv 0. \tag{17.3}$$

For a scalar U, and for a scalar density $\mathfrak{B}$, respectively, we have the laws:

$$\left.\begin{aligned} \delta_C U &= 0, & \bar{\delta} U &= -U_{,\varrho}\,\xi^\varrho, \\ \delta_C \mathfrak{B} &= -\mathfrak{B}\,\xi^\varrho{}_{,\varrho}, & \bar{\delta}\mathfrak{B} &= -(\mathfrak{B}\,\xi^\varrho)_{,\varrho}. \end{aligned}\right\} \tag{17.4}$$

The $\bar{\delta}$-change, but not the change δ_C, satisfies the identity

$$\bar{\delta}(K_{,\mu}) \equiv (\bar{\delta} K)_{,\mu} = \bar{\delta} K_{,\mu}. \tag{17.5}$$

It also has the convenient property that with respect to a finite coordinate transformation the $\bar\delta$-change of a tensor, and of an affine connection, transform as tensors. For instance, the infinitesimal transformation laws of the contravariant metric tensor and of the Christoffel symbols, respectively, are:

$$\bar\delta g^{\mu\nu} = g^{\mu\alpha} g^{\nu\beta} (\xi_{\alpha;\beta} + \xi_{\beta;\alpha}), \tag{17.6}$$

and

$$\begin{aligned} \bar\delta \left\{ {\varrho \atop \mu\nu} \right\} &= \tfrac{1}{2} [(R_{\tau\mu\nu}{}^{\varrho} + R_{\tau\nu\mu}{}^{\varrho}) \xi^\tau - (\xi^\varrho{}_{;\mu\nu} + \xi^\varrho{}_{;\nu\mu})] \\ &\equiv \left\{ {\tau \atop \mu\nu} \right\} \xi^\varrho{}_{,\tau} - \left\{ {\varrho \atop \tau\nu} \right\} \xi^\tau{}_{,\mu} - \left\{ {\varrho \atop \mu\tau} \right\} \xi^\tau{}_{,\nu} - \xi^\varrho{}_{,\mu\nu} - \left\{ {\varrho \atop \mu\nu} \right\}_{,\tau} \xi^\tau . \end{aligned} \tag{17.7}$$

In the calculations that follow we shall assume that the expression $\delta_C \psi_A$ is linear in the first derivatives of ξ^ϱ, so that we may write

$$\bar\delta \psi_A = c_{A\varrho}{}^{\sigma} \xi^\varrho{}_{,\sigma} - \psi_{A,\varrho} \xi^\varrho . \tag{17.8}$$

This assumption is not essential to the existence of an energy-stress complex, but the calculations require some modification if ψ_A obeys a transformation law of a different type. We are now ready to begin.

In order to derive an identity similar to (17.1), we shall write down the infinitesimal transformation law of L, Eq. (16.8), in two different ways. We have, in accordance with the law (17.4),

$$\bar\delta L = -(L \xi^\mu)_{,\mu}, \tag{17.9}$$

but also

$$\begin{aligned} \bar\delta L = \frac{\partial L}{\partial g^{\mu\nu}} \bar\delta g^{\mu\nu} + \frac{\partial L}{\partial \left\{ {\varrho \atop \mu\nu} \right\}} \bar\delta \left\{ {\varrho \atop \mu\nu} \right\} + \frac{\partial L}{\partial \left\{ {\varrho \atop \mu\nu} \right\}_{,\sigma}} \bar\delta \left\{ {\varrho \atop \mu\nu} \right\}_{,\sigma} + \\ + \frac{\partial L}{\partial \psi_A} \bar\delta \psi_A + \frac{\partial L}{\partial \psi_{A,\varrho}} \bar\delta \psi_{A,\varrho} . \end{aligned} \tag{17.10}$$

Combination of Eqs. (17.9) and (17.10) yields the identity

$$\vartheta^\mu{}_{,\mu} - \frac{\partial L}{\partial g^{\mu\nu}} \bar\delta g^{\mu\nu} - \frac{\delta L}{\delta \left\{ {\varrho \atop \mu\nu} \right\}} \bar\delta \left\{ {\varrho \atop \mu\nu} \right\} - \frac{\delta L}{\delta \psi_A} \bar\delta \psi_A \equiv 0, \tag{17.11}$$

where ϑ^μ is defined as the expression

$$\vartheta^\mu \equiv -L \xi^\mu - \frac{\partial L}{\partial \left\{ {\gamma \atop \alpha\beta} \right\}_{,\mu}} \bar\delta \left\{ {\gamma \atop \alpha\beta} \right\} - \frac{\partial L}{\partial \psi_{A,\mu}} \bar\delta \psi_A . \tag{17.12}$$

According to PALATINI's method, which was presented in Sect. 7, the last three terms in Eq. (17.11) contain the left-hand sides of the field equations, and hence these terms will vanish if the field equations are satisfied. In that case, then, an integral of ϑ^0, extended over a three-dimensional domain of the coordinates $x^1 \dots x^3$, will change in the course of time at a rate that is determined by a surface integral,

$$\frac{d}{dt} \left(\int_D \vartheta^0 \, d^3x \right) + \oint \vartheta^k \, dS_k = 0, \tag{17.13}$$

always provided the field equations are satisfied. It remains to calculate the expressions ϑ^μ. Utilizing Eqs. (16.8), (17.7), and (17.8), we find:

$$\vartheta^\mu = \frac{1}{8\pi\varkappa} (\xi^{\mu\varrho}{}_{,\varrho} + 2 \mathfrak{G}^\mu_\varrho \xi^\varrho) + \left(\psi_{A,\varrho} \frac{\partial \mathfrak{M}}{\partial \psi_{A,\mu}} - \delta^\mu_\varrho \mathfrak{M} \right) \xi^\varrho - c_{A\varrho}{}^{\sigma} \xi^\varrho{}_{,\sigma} \frac{\partial \mathfrak{M}}{\partial \psi_{A,\mu}} . \tag{17.14}$$

The symbol $\xi^{\mu\varrho}$ stands for

$$\xi^{\mu\varrho} = \sqrt{-g}\, g^{\mu\alpha} g^{\varrho\beta} (\xi_{\alpha,\beta} - \xi_{\beta,\alpha}) \equiv \sqrt{-g}\, (g^{\varrho\beta} \xi^\mu{}_{;\beta} - g^{\mu\alpha} \xi^\varrho{}_{;\alpha}) . \tag{17.15}$$

The first parenthesis in Eq. (17.14) by itself, and the second and third terms together (if ψ_A is a tensor or a tensor density), are contravariant vector densities. Hence the (ordinary) divergence of ϑ^μ is manifestly a scalar density.

Eq. (17.14) may be further developed. By substituting from Eq. (16.10), we obtain the alternative expression for ϑ^μ:

$$\vartheta^\mu = \frac{1}{8\pi\varkappa}\,\xi^{\mu\varrho}{}_{,\varrho} + \left(\psi_{A,\varrho}\frac{\partial\mathfrak{M}}{\partial\psi_{A,\mu}} - \mathfrak{P}^\mu_\varrho - \delta^\mu_\varrho\,\mathfrak{M}\right)\xi^\varrho - c_{A\varrho}{}^\sigma\frac{\partial\mathfrak{M}}{\partial\psi_{A,\mu}}\,\xi^\varrho{}_{,\sigma}. \tag{17.16}$$

This expression has the interesting property that it does not vanish if the sources of the gravitational field are zero. We can assign to the gravitational field itself a non-vanishing energy-stress complex, the first term of the expression (17.16)[1].

All these expressions depend on the arbitrary displacement field ξ^μ. Given this field, the conserved integral

$$\Xi = \int \vartheta^\mu\, d\Sigma_\mu \tag{17.17}$$

is invariant for any fixed three-dimensional domain of integration. If the integral is extended over a domain at the boundary of which the field $\vartheta^\varkappa$ vanishes, or vanishes asymptotically sufficiently strongly, then Ξ is independent of the surface of integration as well, a constant of the motion.

In a certain sense, the vector field ϑ^μ, and the integral Ξ, may be considered the generating density and the generator, respectively, of the infinitesimal coordinate transformation described by the displacement vector field ξ^μ. If we consider in particular the case of a rigid displacement along one coordinate direction,

$$\xi^\mu = \alpha^\mu, \tag{17.18}$$

then the expression (17.16) reduces to the following:

$$\left.\begin{aligned}\vartheta^\mu &= \vartheta^\mu_\nu\,\alpha^\nu,\\ \vartheta^\mu_\nu &= \frac{1}{8\pi\varkappa}\left[\sqrt{-g}\,g^{\mu\varkappa}g^{\varrho\lambda}(g_{\nu\varkappa,\lambda} - g_{\nu\lambda,\varkappa})\right]_{,\varrho} + \left(\psi_{A,\nu}\frac{\partial\mathfrak{M}}{\partial\psi_{A,\mu}} - \delta^\mu_\nu\,\mathfrak{M} - \mathfrak{P}^\mu_\nu\right).\end{aligned}\right\} \tag{17.19}$$

The coefficients ϑ^ν_μ, whose divergence must also vanish,

$$\vartheta^\nu_{\mu,\nu} = 0, \tag{17.20}$$

may be considered as the components of the energy-stress complex. This expression contains second derivatives of the metric tensor. From the point of view of PALATINI's formalism, however, the energy-stress complex (17.19) may also be interpreted as containing no higher than first derivatives of the components of the affine connection. From that point of view, we may write, if we prefer,

$$\vartheta^\mu_\nu = \frac{1}{8\pi\varkappa}\left(\mathfrak{g}^{\sigma\varrho}\,\Gamma^\mu_{\nu\sigma} - \mathfrak{g}^{\sigma\mu}\,\Gamma^\varrho_{\nu\sigma}\right)_{,\varrho} + \left(\psi_{A,\nu}\frac{\partial\mathfrak{M}}{\partial\psi_{A,\mu}} - \delta^\mu_\nu\,\mathfrak{M} - \mathfrak{P}^\mu_\nu\right). \tag{17.21}$$

Even in the expression (17.19) the components ϑ^0_ν, which represent the energy and momentum densities, respectively, are free of second-order *time* derivatives[2], because of the antisymmetry of the square bracket with respect to the superscripts μ and ϱ.

In the next section we shall examine the degree of arbitrariness left in the construction of energy-stress complexes.

---

[1] A. KOMAR: Phys. Rev. **113**, 934 (1959).
[2] C. MØLLER: Ann. Physics **4**, 347 (1958).

18. Alternative energy-stress complexes. The derivation of a particular expression for the energy-stress complex as given in the preceding section is beset with a number of ambiguities. Even if we accept an identity of the form (17.11) as our point of departure, the choice of "independent" field variables is to some extent arbitrary. The replacement of field variables by a new set which are *algebraic* functions of the original set (such as the transition from covariant to contravariant components of the metric tensor, or vice versa) will have no effect on the product $\frac{\delta L}{\delta y_A}\bar{\delta} y_A$, $y_A = (\psi_A, g_{\mu\nu})$. However, we do get different expressions when we consider the components of the affine connection as a set of separate variables, and when we fix them initially to be the Christoffel symbols.

Aside from the choice of field variables and corresponding field equations, Eq. (17.11) does not give us ϑ^μ, but rather its divergence. Hence, given a solution of Eq. (17.11), we can construct an infinite variety of alternative solutions in accordance with the principle:

$$\overline{\vartheta^\mu} = \vartheta^\mu + \sigma^{[\mu\varrho]}{}_{,\varrho}, \tag{18.1}$$

where the field $\sigma^{[\mu\varrho]}$ is to be antisymmetric with respect to its two superscripts but otherwise arbitrary[1]. It can be shown that the arbitrariness expressed by Eq. (18.1) includes the lack of uniqueness implied by the freedom to choose one's set of field variables.

A number of choices have been made for the field ϑ^μ, and for the energy-stress complex ϑ^μ_ν; some lead to different values for the total energy and linear momentum of a system than others, whereas in many other cases the choice of expression affects only the local values of energy and linear momentum densities, not the integrals. From Eq. (18.1) we find that the integral $\bar{\Xi}$ takes the value

$$\bar{\Xi} - \Xi \equiv \int \sigma^{[\mu\varrho]}{}_{,\varrho}\, d\Sigma_\mu \equiv \oint \sigma^{[\mu\varrho]}\, d\Sigma_{\mu\varrho}, \tag{18.2}$$

where the last integral on the right is to be taken over the two-dimensional bounding surface of a three-dimensional hypersurface. When taken to spatial infinity, this surface integral may vanish, may be finite, or may even diverge, depending on the choice of the expression $\sigma^{[\mu\varrho]}$.

There is one expression for ϑ^μ that is free of all second derivatives (not only derivatives with respect to x^0) of the gravitational potentials $g_{\mu\nu}$ and which was the first energy-stress complex constructed by EINSTEIN. We find this "canonical energy-stress complex" by introducing a preliminary expression

$$T^\mu \equiv \frac{1}{8\pi\varkappa}\, U^{[\mu\varrho]}{}_{,\varrho}, \qquad U^{[\mu\varrho]} \equiv U^{[\mu\varrho]}_\sigma\, \xi^\sigma, \tag{18.3}$$

where[2]

$$U^{[\mu\varrho]}_\sigma \equiv \mathfrak{g}_{\sigma\tau}(\mathfrak{g}^{\mu\tau}\,\mathfrak{g}^{\varrho\alpha} - \mathfrak{g}^{\mu\alpha}\,\mathfrak{g}^{\varrho\tau})_{,\alpha}. \tag{18.4}$$

The $\mathfrak{g}^{\alpha\beta}$ are the expressions introduced in Eq. (7.3), the $\mathfrak{g}_{\alpha\beta}$ are the components of the reciprocal matrix, thus:

$$\mathfrak{g}_{\alpha\varrho}\,\mathfrak{g}^{\varrho\beta} = \delta^\beta_\alpha. \tag{18.5}$$

The divergence of the expression T^μ vanishes identically. We can now eliminate from T^μ all second derivatives of the metric by combining it with the field equations:

$$\tau^\mu \equiv T^\mu + \left(\frac{1}{4\pi\varkappa}\,\mathfrak{G}^\mu_\varrho + \mathfrak{P}^\mu_\varrho\right)\xi^\varrho. \tag{18.6}$$

[1] J.N. GOLDBERG: Phys. Rev. **111**, 315 (1958). — P.G. BERGMANN: Phys. Rev. **112**, 287 (1958).

[2] PH. V. FREUD: Ann. Math. Princeton **40**, 417 (1939). — L. LANDAU and E. LIFSHITZ: The Classical Theory of Fields (English translation), p. 317f. Reading (Mass, U.S.A.): Addison-Wesley 1951.

The divergence of τ^μ vanishes only if the field equations are satisfied. The combination

$$T^\mu + \frac{1}{4\pi\varkappa}\mathfrak{G}^\mu_\varrho \xi^\varrho \equiv \frac{1}{8\pi\varkappa} U^{[\mu\varrho]}_\sigma \xi^\varrho{}_{,\varrho} + \left(\frac{1}{8\pi\varkappa} U^{[\mu\varrho]}_{\sigma}{}_{,\varrho} + \frac{1}{4\pi\varkappa}\mathfrak{G}^\mu_\sigma\right)\xi^\sigma \quad (18.7)$$

is free of second derivatives of the metric tensor. This fact is obvious for the first term on the right. For the parenthesis in the second term we have:

$$\left.\begin{aligned} U^{[\mu\varrho]}_{\sigma}{}_{,\varrho} + 2\mathfrak{G}^\mu_\sigma \equiv \mathfrak{g}^{\iota\varkappa}(\Gamma^\mu_{\iota\varkappa}\Gamma^\varrho_{\sigma\varrho} + \Gamma^\mu_{\iota\sigma}\Gamma^\varrho_{\varkappa\varrho} - 2\Gamma^\mu_{\iota\lambda}\Gamma^\lambda_{\varkappa\sigma}) + {} \\ + \mathfrak{g}^{\mu\iota}(\Gamma^\varkappa_{\iota\sigma}\Gamma^\varrho_{\varkappa\varrho} - \Gamma^\varkappa_{\iota\varkappa}\Gamma^\varrho_{\sigma\varrho}) + \delta^\mu_\sigma \mathfrak{g}^{\iota\varkappa}(\Gamma^\tau_{\iota\varrho}\Gamma^\varrho_{\varkappa\tau} - \Gamma^\varrho_{\iota\varkappa}\Gamma^\tau_{\varrho\tau}). \end{aligned}\right\} \quad (18.8)$$

The lack of uniqueness of the energy expression is not confined entirely to general relativity. In classical mechanics the energy of a system of mass points is defined only up to a constant of integration. Similarly in ordinary non-relativistic or Lorentz-covariant field theories we may add to the energy-stress complex a "curl", that is to say an expression whose four-dimensional divergence vanishes identically. What is peculiar about the energy-stress complex of general relativity is that we may adopt an expression that vanishes if the field equations are satisfied, and that, accordingly, any non-vanishing expression equals a "curl", the last term on the right-hand side of Eq. (18.1). This peculiarity of general relativity is the direct result of the general covariance of the theory[1]—the action integral of the theory is invariant with respect to a transformation group (the "invariance group") that depends on a set of four arbitrary functions, ξ^ϱ.

For proof we shall consider a theory whose field consists of one or several geometric objects y_A whose infinitesimal transformation law is

$$\bar{\delta} y_A = c_{A\varrho}{}^\sigma \xi^\varrho{}_{,\sigma} - y_{A,\varrho}\xi^\varrho. \quad (18.9)$$

Let the Lagrangian density be some function of the field variables and their first derivatives,

$$S = \int L(y_A, y_{A,\varrho})\, d^4x, \quad (18.10)$$

constructed so that an infinitesimal coordinate transformation confined to the interior will not change the value of S if the form of L as a function of its arguments is assumed to be the same in all coordinate systems. We have, then,

$$\bar{\delta} S \equiv \int \bar{\delta} L\, d^4x \equiv \int \left[\frac{\delta L}{\delta y_A}\bar{\delta} y_A + \left(\frac{\partial L}{\partial y_{A,\varrho}}\bar{\delta} y_A\right)_{,\varrho}\right] d^4x \equiv 0, \quad (18.11)$$

as long as $\bar{\delta} y_A$ has the form (18.9) and the four functions ξ^ϱ, arbitrary in the interior of the domain of integration, vanish on the boundary, along with their derivatives. Because of the latter condition, the last term under the integral on the right-hand side of Eq. (18.11) may be disregarded.

Next we substitute the expression (18.9) into the reduced identity (18.11), and obtain the identity

$$\int \frac{\delta L}{\delta y_A}\bar{\delta} y_A\, d^4x \equiv \int \frac{\delta L}{\delta y_A}(c_{A\varrho}{}^\sigma \xi^\varrho{}_{,c} - y_{A,\varrho}\xi^\varrho)\, d^4x \equiv 0, \quad (18.12)$$

valid if the ξ^ϱ are different from zero in the interior. By a simple integration by parts, we may give the relationship (18.12) the form

$$\int \frac{\delta L}{\delta y_A}\bar{\delta} y_A\, d^4x \equiv \oint \frac{\delta L}{\delta y_A} c_{A\varrho}{}^\sigma \xi^\varrho\, d\Sigma_\sigma - \int \left[\left(\frac{\delta L}{\delta y_A} c_{A\varrho}{}^c\right)_{,\sigma} + \frac{\delta L}{\delta y_A} y_{A,\varrho}\right]\xi^\varrho\, d^4x \equiv 0. \quad (18.13)$$

[1] E. Noether: Nachr. Ges. Wiss. Göttingen **1918**, 235. — A. Trautman: Conservation Laws in General Relativity, in volume edit. by L. Witten. New York: Wiley (to be published) [*15*].

The identity between the first and the second expressions holds whether or not the boundary condition is satisfied; but both expressions vanish identically only on condition that on the boundary ξ^ϱ and its derivatives vanish. Because in the interior the ξ^ϱ are arbitrary, we conclude that in the last four-dimensional integral the square bracket must vanish identically;

$$\left(\frac{\delta L}{\delta y_A} c_{A\varrho}{}^{\sigma}\right)_{,\sigma} + \frac{\delta L}{\delta y_A} y_{A,\varrho} \equiv 0. \tag{18.14}$$

But then it follows, regardless of boundary conditions, that

$$\frac{\delta L}{\delta y_A} \bar{\delta} y_A \equiv \left(c_{A\varrho}{}^{\sigma} \frac{\delta L}{\delta y_A} \xi^\varrho\right)_{,\sigma} = 0. \tag{18.15}$$

That is to say, the divergence on the right vanishes if the field equations

$$\frac{\delta L}{\delta y_A} = 0 \tag{18.16}$$

are satisfied. However, if they are satisfied, then not only the divergence of the parenthesis vanishes, but the parenthesis itself. To this extent it is proper to claim that the stress-energy complex in a general-relativistic theory may be chosen so as to vanish wherever the field equations are satisfied.

19. Theory of motion. The conservation laws that we have discussed in the last two sections lead to a relationship between the field equations and the ponderomotive laws that is peculiar to general-relativistic theories. Assuming that there are in the field world lines, or world tubes, which represent the sources of the field, then it turns out that certain integrals over two-dimensional surfaces surrounding the singular region satisfy dynamical laws that are, in effect, ponderomotive laws.

Formally, the existence of the ponderomotive laws depends on the simultaneous presence of conservation laws that hold identically (i.e. irrespective of whether the field equations are satisfied) and those that hold "weakly" (only on condition that the field equations are obeyed). We shall first obtain a "strong" conservation law by means of STOKES' theorem in four dimensions.

Suppose that an m-dimensional subspace is imbedded in an n-dimensional manifold. The imbedment relationship may be stated by means of a parametric description: If we introduce on the m-dimensional hypersurface m parameters u^k, then the n functions $x^\alpha(u^1, \ldots, u^m)$, $\alpha = 1, \ldots, n$, will give us the coordinate values of each point of the hypersurface that is identified by the values of the parameters u^k, $k = 1, \ldots, m$. We introduce then, by means of LEVI-CIVITA's completely antisymmetric tensor density $\delta_{\varrho_1 \ldots \varrho_n}$, the $(n-m)$-indexed hypersurface element

$$d\Sigma_{\varrho_1 \ldots \varrho_{n-m}} = \delta_{\varrho_1 \ldots \varrho_{n-m}\sigma_1 \ldots \sigma_m} \frac{\partial x^{\sigma_1}}{\partial u^1} \cdots \frac{\partial x^{\sigma_m}}{\partial u^m} du^1 \ldots du^m. \tag{19.1}$$

Given a complex $A^{\sigma_1 \cdots \sigma_{n-m}\tau}$ which is skewsymmetric in all of its $(n-m+1)$ superscripts, STOKES' theorem (or rather a corollary of the theorem) assumes the form

$$\oint A^{\varrho_1 \cdots \varrho_{n-m}\tau}{}_{,\tau}\, d\Sigma_{\varrho_1 \cdots \varrho_{n-m}} \equiv 0, \tag{19.2}$$

provided our hypersurface is closed and orientable.

We introduce now, in addition to the skewsymmetric complex $U^{[\mu\nu]}$, Eqs. (18.3), (18.4), a vector field r^ϱ, which we shall leave unspecified, for the time being. With its help we form the skewsymmetric complex

$$V^{[\mu\nu\varrho]} \equiv U^{[\mu\nu]} r^\varrho + U^{[\nu\varrho]} r^\mu + U^{[\varrho\mu]} r^\nu, \tag{19.3}$$

which, in accordance with STOKES' theorem, satisfies the identity

$$\oint V^{[\mu\nu\varrho]}{}_{,\varrho}\, d\Sigma_{\mu\nu} \equiv 0, \tag{19.4}$$

the integral to be extended over any closed two-dimensional surface. When we carry out the differentiation indicated on the expression (19.3), we obtain the identical relationship

$$\left.\begin{aligned}&\oint [r^\varrho\, U^{[\mu\nu]}{}_{,\varrho} + r^\varrho{}_{,\varrho}\, U^{[\mu\nu]} + 8\pi\varkappa\, (r^\mu\, T^\nu - r^\nu\, T^\mu) + \\ &\qquad + (r^\mu{}_{,\varrho}\, U^{[\nu\varrho]} - r^\nu{}_{,\varrho}\, U^{[\mu\varrho]})]\, d\Sigma_{\mu\nu} \equiv 0.\end{aligned}\right\} \tag{19.5}$$

From this relationship we go over to a "weak" relationship, which will be valid only if on the surface of integration the field equations are satisfied. With the help of Eq. (18.6) we replace T^μ by τ^μ, thus:

$$\left.\begin{aligned}&\oint [r^\varrho\, U^{\mu\nu}{}_{,\varrho} + r^\varrho{}_{,\varrho}\, U^{\mu\nu} + 8\pi\varkappa\, (r^\mu\, \tau^\nu - r^\nu\, \tau^\mu) + (r^\mu{}_{,\varrho}\, U^{[\nu\varrho]} - r^\nu{}_{,\varrho}\, U^{[\mu\varrho]})]\, d\Sigma_{\mu\nu} \\ &\equiv 8\pi\varkappa \oint \left[r^\mu \left(\mathfrak{P}^\nu_\varrho + \frac{1}{4\pi\varkappa}\, \mathfrak{G}^\nu_\varrho\right) - r^\nu \left(\mathfrak{P}^\mu_\varrho + \frac{1}{4\pi\varkappa}\, \mathfrak{G}^\mu_\varrho\right)\right] \xi^\varrho\, d\Sigma_{\mu\nu} = 0.\end{aligned}\right\} \tag{19.6}$$

The validity of this integral relationship depends on the field equations holding on the surface of integration, but not on the choice of the two fields ξ^ϱ and r^ϱ.

In order to obtain the conservation laws proper, we shall specialize the field r^ϱ and assume that it is constant in the coordinate system chosen. With this choice we obtain the four laws:

$$\frac{1}{2}\, \frac{\partial}{\partial x^\varrho} \oint U^{[\mu\nu]}\, d\Sigma_{\mu\nu} + 8\pi\varkappa \oint \tau^\nu\, d\Sigma_{\varrho\nu} = 0. \tag{19.7}$$

The derivative in front of the first integral is to be understood in the sense that the whole surface of integration is displaced by a constant (coordinate) amount. The integrands of both the first and the second terms are free of any second-order or higher derivatives of the field variables, including the components of the metric tensor. The first integrand is linear in the first derivatives (they might be termed "gravitational field strengths"), the second quadratic.

Laws having the structure (19.7) are of particular interest if the surface of integration is chosen so that it surrounds a singular world tube, if, in other words, the region in which the (vacuum) field equations are satisfied is multiply connected. Depending on the choice of the field ξ^ϱ, the first integrand may then be thought of as the total energy (mass), linear momentum, or angular momentum enclosed in the singular region, whereas τ^ν represents the density or flux of these quantities. This interpretation is not exact, for the simple reason that relations of the form (19.7) do not have an invariant, or even simple covariant, character. The interpretation given is suggested by the weak-field approximation, which is a Lorentz-covariant approximation.

A completely covariant set of integral laws may be obtained as follows. One can show that if both the gravitational and the nongravitational field equations are satisfied, then the vector density (17.16) equals a divergence,

$$\vartheta^\mu = \overline{U}^{\mu\varrho}{}_{,\varrho} \equiv \overline{U}^{\mu\varrho}{}_{;\varrho}, \qquad \overline{U}^{\mu\varrho} \equiv \frac{1}{8\pi\varkappa}\, \xi^{\mu\varrho} + \frac{1}{2}\left(c_{A\sigma}{}^{\mu}\, \frac{\partial\mathfrak{M}}{\partial\psi_{A,\varrho}} - c_{A\sigma}{}^{\varrho}\, \frac{\partial\mathfrak{M}}{\partial\psi_{A,\mu}}\right) \xi^\sigma. \tag{19.8}$$

$\overline{U}^{\mu\varrho}$, in contrast to $U^{[\mu\varrho]}$, is a contravariant skewsymmetric tensor density of weight 1. If we form from it an expression $\overline{V}^{\mu\nu\varrho}$, in analogy to Eq. (19.3), that expression will also be a tensor density, and so will be the integrand of the integral corresponding to (19.4). Accordingly, we may write the invariant relationship

$$\oint [(r^\varrho\, \overline{U}^{\mu\nu})_{;\varrho} + 8\pi\varkappa\, (r^\mu\, \vartheta^\nu - r^\nu\, \vartheta^\mu) - (r^\mu{}_{;\varrho}\, \overline{U}^{\varrho\nu} + r^\nu{}_{;\varrho}\, \overline{U}^{\mu\varrho})]\, d\Sigma_{\mu\nu} = 0. \tag{19.9}$$

This equality may be simplified further. By a somewhat lengthy but straightforward calculation (which will not be reproduced here) it may be shown that the first and the last parenthesis together represent the change in the value of the invariant integral $\oint \overline{U}^{\mu\nu}\, d\Sigma_{\mu\nu}$ if the surface of integration is displaced at each point infinitesimally so that $\delta x^\varrho = \varepsilon\, r^\varrho$, ε being an infinitesimal constant. We shall designate this infinitesimal change as a result of displacement by the symbol δ_r. Eq. (19.9) then turns into the very intuitive equality:

$$\delta_r \oint \overline{U}^{\mu\nu}\, d\Sigma_{\mu\nu} = 16\pi\varkappa \oint r^\mu \vartheta^\nu\, d\Sigma_{\mu\nu}, \tag{19.10}$$

where $\overline{U}^{\mu\nu}$ is the expression (19.8) and ϑ^μ the expression (17.16). The integrand on the left depends on the choice of the vector field ξ^ϱ, the one on the right on both ξ^ϱ and r^ϱ.

We obtain conventional ponderomotive laws if in the relation (19.7) we set the free index ϱ equal to zero (in other words, if we examine the change of a surface integral in the course of "time" while the "spatial" coordinates remain unchanged), and if then we set up the four different relationships that are obtained if we let the vector ξ^ϱ be constant and parallel to a coordinate axis. If we set $\xi^\mu = \delta^\mu_0$, we get the law for the change of energy, whereas $\xi^\mu = \delta^\mu_k$ leads to the k-th component of the force equation. In the lowest non-trivial approximation stage of a weak-field expansion of the metric tensor, the integrals on the left represent energy and linear momentum, respectively, those on the right the Poynting vector of the gravitational field (energy flux) and the stress (flux of linear momentum).

The Einstein-Infeld-Hoffmann theory (often abbreviated to EIH theory) of the ponderomotive laws in general relativity[1] consists of a combination of weak-field expansion with an expansion with respect to the rapidity of motion of the mass points of the system: that is to say, the EIH method assumes that all material velocities are small compared to the speed of light. By this device all radiative effects are relegated to a very high stage of the approximation, a stage that has been explored within the framework of the EIH approach only recently[2].

The investigations carried out within the framework of the EIH method are to some extent supplemented by "fast" approximation methods. The terms "fast" and "slow" (which denotes the EIH approach) refer to assumptions concerning the state of motion of the mass points responsible for the gravitational field; in the fast approximations, there is also an expansion in terms of deviation of the metric from the Minkowski values, but v/c is permitted to be of the order of one[3].

"Fast" approximation methods are manifestly Lorentz-covariant. As a result, the field equations lead at every stage to inhomogeneous d'Alembert equations, rather than to three-dimensional POISSON's equations. To make these problems unambiguous we must adopt definite GREEN's functions, or equivalently radiation conditions, and thus solve the problems consistently with retarded potentials, or with half retarded, half advanced potentials, etc. The worst part of fast approaches to the problem of motion is, however, that we cannot begin to solve the field equations unless we already assume definite histories of the sources

[1] A. EINSTEIN, L. INFELD and B. HOFFMANN: Ann. Math. Princeton **39**, 65 (1938). A review and bibliography in: L. INFELD and J. PLEBANSKI: Motion and Relativity. Warszawa: Panstwowe Wydawnictwo Naukowe and London: Pergamon Press 1960 [*3*].

[2] A. TRAUTMAN: Bull. Acad. Polon. Sci., sér. sci. math. astr. et phys. **6**, 627 (1958).

[3] P. HAVAS: Phys. Rev. **108**, 1351 (1957). — B. BERTOTTI and J. PLEBAŃSKI: Ann. Phys. USA **11**, 169 (1960).

of the field; that is to say, we must begin with an assumed solution of the ponderomotive equations before we can construct the metric field out of which we can hope to gain the ponderomotive laws. These laws then will enable us to correct the assumed motions of the mass points, calculate corrected fields, etc.

Incidentally, a similar difficulty arises in the "slow" approximation, but at a relatively late stage. In general at some stage of the approximation the right-hand sides of the Poisson equations will decrease at spatial infinity so slowly that there is no solution with acceptable behavior at spatial infinity. What saves us is that all the observationally accessible Newtonian and relativistic effects are to be found at a much earlier stage of the approximation.

Aside from the distinctions between "slow" and "fast" approximation procedures, there are two alternative methods of dealing with the ponderable bodies themselves. In their early papers, EIH represented these bodies as mass points, i.e. as singular world lines; the objective of their calculations was to obtain the three spatial coordinates of each mass point as functions of the coordinate x^0. FOCK[1] described bodies in terms of extended mass-stress distributions. Originally it was thought that the mass-point description could avoid the ambiguities inherent in the adoption of form factors, equations of state, and other devices that would fix the histories of FOCK's extended bodies; subsequently it was discovered that the two approaches are fairly closely related, and that a description of a ponderable body in terms of a delta function does not remove ambiguities, either. In the mass-point approach this ambiguity resides in the need to specify, for the ten components of the usual energy-stress tensor, not only the amplitude of the delta function, but also that of the various derivatives of the delta function, which assume the role of the dipole, quadrupole, and higher moments of FOCK's extended distributions.

The original belief that the field equations of general relativity determine the equations of motion of the ponderable bodies has been borne out only in part; it has been confirmed to the extent that the development in time of either a point-like or an extended region with non-vanishing energy-stress is constrained by the requirement that the covariant divergences of this source tensor must vanish. The resulting conditions on the motion of such a structure can be given a very intuitive, but perhaps misleading interpretation within a given approximation procedure. For instance, if we expand the first non-trivial portions of the metric field in an EIH approximation about such a structure into a Laurent series with respect to the three-dimensional coordinate distance from the center r and interpret the terms with negative powers of r as resulting from the mass, the mass dipole moment, etc., whereas the positive powers of r represent the external field, then we find that the gravitational force acting on the structure is composed of the product of its mass by the gravitational field strength, the product of its dipole moment by the first gradient of that field strength, etc.

Unfortunately, this interpretation is open to the criticism that the whole power expansion is not covariant and that the splitting-up into self-field and external field suggested by the Laurent expansion might come out entirely differently in a different coordinate system. As a matter of fact, it has not yet been possible to characterize the properties of mass distributions invariantly[2]. Thus it would seem that the foundations of the theory are yet in need of further clarification.

---

[1] V.A. FOCK: J. Phys. Moscow **1**, 81 (1939). — Ž. eksp. teor. Fiz. **9**, 375 (1939). — Theory of Space, Time, and Gravitation. London: Pergamon Press 1960 [*2*].

[2] R. SACHS and P.G. BERGMANN: Phys. Rev. **112**, 674 (1958). This paper attempts to characterize singular regions within the framework of the linearized theory; whether its method can be extended to the full theory has not been established.

20. Critique of the energy concept. The energy concept penetrates almost all branches of physics and presents us with a powerful tool for obtaining results in situations where we cannot analyze all of the details. In mechanics any isolated system of mass points possesses a number of absolute conservation laws, which are the direct result of the invariance of mechanics with respect to the inhomogeneous Galilean group of transformations. In relativistic mechanics Lorentz covariance gives rise to the same number of laws, ten in number. In Lorentz-covariant field theories, energy and the other constants of the motion occur in the form of three-dimensional volume integrals, whose conservation corresponds to local laws concerning energy density, linear momentum density, etc.; these local laws always take the form of four-dimensional divergence relations, so-called equations of continuity. The integrals that represent the total energy, linear momentum, etc., will be conserved if the fields under consideration are confined to finite regions. Under normal conditions fields are, however, not confined; at least they do not remain confined in the course of time. For a field that extends throughout space we must require that two integral conditions are satisfied, (a) that the integral defining the quantity to be conserved should converge, and (b) that the surface integral over the flux (e.g. POYNTING's vector) will tend to zero if we move the surface of integration out to infinity.

It has always been considered desirable to extend the energy concept to the general theory of relativity, if only to maintain the "correspondence" of general relativity with the remainder of physics. But we have found that in general relativity the energy concept is beset with ambiguities. We can explain these ambiguities in terms of the relationship between the invariance properties of a theory and its constants of the motion. The constants are simply the generators of the invariant transformations. But it is a result of NOETHER's investigations[1] that the constants of the motion in a theory that is invariant with respect to a Lie group (whose transformations can be identified by a finite number of parameters, e.g. 10 in the case of the inhomogeneous Lorentz group) differ profoundly from those that are related to an invariance group with arbitrary functions (such as general relativity). In the latter type of theory the conserved quantities vanish, i.e. they are constraints.

Even in general relativity we may obtain non-vanishing quantities that obey equations of continuity if we add to the constraints expressions that are (three-dimensional) divergences. The time derivative of a divergence is, of course, a divergence itself. The resulting equations of continuity are not identities, however, because the original constraint is not an identity but an algebraic combination of field equations that is free of time derivatives in the Hamiltonian formulation, or free of second time derivatives in the Lagrangian formulation. Of course, the addition of a divergence to a constraint is a highly arbitrary procedure; besides, the resulting non-vanishing densities, and their integrals, usually do not obey any simple transformation law. Knowledge of a finite number of pieces of data in one coordinate system will not enable us to calculate the value of the "energy" in another coordinate system.

We think of a gravitational field as sufficiently "confined" in space if its potentials drop off at spatial infinity as $1/r$, and their derivatives as r^{-2}.[2] Under these circumstances the integrals for the total energy and the total linear momentum will converge and be constant in the course of time; the affine connection

---

[1] Cf. Ref. in Sect. 18.

[2] A. EINSTEIN: Ber. preuss. Akad. Wiss. **1918**, 448. — F. KLEIN: Nachr. Ges. Wiss. Göttingen **1918**, p. 394. — Critical review by A. TRAUTMAN: Conservation Laws in General Relativity, to be published in volume on gravitation, L. WITTEN, Ed. New York: Wiley [*15*].

will be "asymptotically integrable", that is to say, if we displace a vector parallel to itself along a closed curve whose total length is proportional to r, then, as we remove the curve to infinity, the change of the vector that results from the circuit about the curve will tend to zero.

In the presence of gravitational radiation the total energy will not be conserved, because the waves carry some energy with them; analogous statements apply to the linear momentum, etc. But that is not all; if there is no coordinate system in which the field strengths drop off as $1/r^2$, then there is no possibility to generate out of one vector "at infinity" a whole field of parallel vectors "at infinity". Thus we are unable in the presence of radiation to define, even at infinity, a "rigid displacement", the type of coordinate transformation that is presumably generated by the energy integral. Under these circumstances it is very difficult to see how one can define the "free vector" energy—linear momentum in a convincing manner.

These ambiguities of course do not imply that general relativity lacks quantities that obey equations of continuity; rather, general relativity suffers in this respect from an *embarras de richesse*. There is an infinity of such quantities, and our difficulty is to single out a subset and to present these as the "natural" expressions for energy, linear momentum, etc. BERGMANN[1] has published a prescription for the construction of generators of arbitrary infinitesimal coordinate transformations free of all second derivatives of the metric tensor, and KOMAR[2] a different expression that is manifestly covariant. MØLLER'S[3] expression and that published by LANDAU and LIFSHITZ[4] are special cases, as are those by GOLDBERG[5]. At this time, the problem of energy does not appear susceptible to further constructive clarification except in physical situations possessing special symmetries[6].

B. Invariant description of the field.

21. Introduction. Throughout the history of the general theory of relativity, the intuition of physicists has been taxed by the ambiguity of description of any physical situation that is caused by the general covariance of the theory. Given two formally different solutions of EINSTEIN'S field equations, it is not at all an easy matter to decide whether the two sets of functions $g_{\mu\nu}(x^\varrho)$ represent two intrinsically different physical situations, or whether they describe the same situation, in terms of two different curvilinear coordinate systems. Similar problems are met with elsewhere in physics, but nowhere are they as formidable, and as difficult to overcome, as in a general-relativistic theory. In fact, this so-called *problem of equivalence* (that is to say, equivalence of two manifestly different metric fields), though solved "in principle" some time ago, is only now being made manageable from the point of view of the physically oriented relativist. In this introductory section we shall indicate some of the problems that arise in the interpretation of research results in general relativity, and the similarities and dissimilarities with equivalence problems elsewhere in physics.

Even in Newtonian physics one and the same physical situation admits an infinity of different descriptions in terms of inertial frames of references. These

[1] P. G. BERGMANN: Phys. Rev. **112**, 287 (1958).

[2] A. KOMAR: Phys. Rev. **113**, 934 (1959).

[3] C. MØLLER: Ann. Phys. USA **4**, 347 (1958). — Mat. Fys. Medd. Dan. Vid. Selsk. **31**, No. 14.

[4] L. LANDAU and E. LIFSHITZ: The Classical Theory of Fields (English translation). Cambridge: Addison-Wesley 1951.

[5] J. N. GOLDBERG: Phys. Rev. **111**, 315 (1958).

[6] A. TRAUTMAN: Conservation Laws in General Relativity [*15*].

different descriptions may be converted into each other by Galilean transformations. There is, however, a sufficient number of transformation invariants available so that one may, without difficulty, decide whether or not two physical situations are equivalent: As for configuration, the distance between correspond. ing mass points do not change under Galilean transformations, nor do relative speeds. If we add angles between directed quantities, there is obviously available a sufficient quantity of invariant quantities to establish equivalence uniquely. This situation does not change materially if we go over to a special-relativistic theory, as the group of Lorentz transformations is a Lie group, just as the group of Galilean transformations.

The group of gauge transformations in electrodynamics is not a Lie group. Topologically the group manifold has not a finite dimensionality, but represents a function space. Suppose we are presented with two vector fields representing electromagnetic potentials. Aside from the possibility of Lorentz transformations, which must be handled in a manner similar to the treatment of the equivalence problem in classical mechanics, there is the possibility of a gauge transformation. To test this possibility, we may subtract one vector field from the other and ascertain whether the difference is a pure gradient field (i.e. its curl must vanish everywhere); if it is the two fields are equivalent. Alternatively, we may form from the potential field the field intensity tensor, as the field intensities are gauge-invariant. Two fields will be equivalent if, and only if, their field intensities are equal throughout the four-dimensional domain.

The situation in general relativity resembles the one in electrodynamics, with significant differences: First, the "gauge" transformations are identical with the changes of frame. Hence we cannot approach the equivalence problem in two stages, as we may in electrodynamics, first pass over to gauge-invariant quantities, then perform any necessary Lorentz transformations. Second, there are no simple coordinate invariants, comparable in structure to the simplicity of the electrodynamic field strengths. There is, then, no *simple* test for equivalence. What can be done in this respect will be presented in later sections. There are, however, numerous occasions when we must concern ourselves with equivalence problems in one form or another.

In physical applications and interpretations of general relativity, we deal with static solutions of the theory, such as the Schwarzschild solution, as representatives of (classical) ponderable particles at rest. That such solutions are static is established by the availability of a coordinate system in which the components of the metric tensor are all independent of the coordinate x^0 and in which the mixed metric components g_{0k} $(k=1, 2, 3)$ vanish. It is desirable to express the existence of a coordinate system with these properties in a coordinate-independent manner. As for the independence of the metric from x^0, in an arbitrary coordinate system there exists a vector field (a Killing field) whose symmetrized covariant derivatives vanish,

$$A_{\varrho;\sigma} + A_{\sigma;\varrho} = 0. \tag{21.1}$$

This vector field generates a family of curves (whose tangents are parallel to the Killing field everywhere), whose curvature k_μ is everywhere given by the expression

$$k_\mu = \alpha_{\mu;\varrho}\,\alpha^\varrho, \qquad \alpha^\nu = (g_{\varrho\sigma} A^\varrho A^\sigma)^{-\frac{1}{2}} A^\nu. \tag{21.2}$$

The requirement that there should exist a coordinate system in which $g_{0k}=0$ may then be expressed by the covariant condition on the unit vector field α_μ,

$$\alpha_{\mu,\nu} - \alpha_{\nu,\mu} = k_\mu \alpha_\nu - k_\nu \alpha_\mu. \tag{21.3}$$

The existence of a vector field with these properties is presumably an intrinsic property of the metric, to be cast into the form of an integrability condition. However, this integrability condition involves differential relations between the components of the curvature tensor of a relatively high order, which we shall not develop here.

A much more complicated problem than that of the intrinsic characterization of static solutions of EINSTEIN's field equations arises in the theory of motion of ponderable "particles". Suppose we consider the interaction of several "mass points"; the actual motions carried out by each member of the mechanical system depends on its mass polarizability, even in Newtonian mechanics. It is well known, for instance, that the Earth-Moon system is affected by the capability of the Earth to develop a quadrupole moment in an inhomogeneous gravitational field, the tides. In general relativity the motion of a singularity is determined by the surrounding field only if we know all of its mass multipole moments. In a weak-field approximation, and in the early stages of the expansion of the field, the meaning of a mass multipole moment is perfectly clear; most of the work on the ponderomotive laws in the literature has been carried out on the explicit assumption that all moments but the mass itself vanish. If the only permissible transformations were Lorentz transformations, the mass multipole moments would remain unique if we understood that these quantities are to be determined in the co-moving frame of reference. It has not yet been determined whether any invariant significance can be given to these moments if we consider arbitrary curvilinear coordinate transformations. It is in fact possible to construct invariant moments in the linearized weak-field approximation, but this construction does not lend itself to routine generalization, for the simple reason that the weak-field approximation will certainly break down in the immediate vicinity of each mass point. The structure of the so-called gauge transformations in the weak-field linearized theory is sufficiently different from the structure of the group of coordinate transformations that we may not take it for granted that to each covariant construction in the linearized theory there corresponds a similar possibility in the full theory; this whole question must still be considered open.

A third, and by far the most general problem of intrinsic characterization arises in the need to correlate results of the formal theory with observations that can be made in nature. The "classical" predictions of general relativity invariably refer to observations that can be made by an observer who is sufficiently far removed from the regions of appreciable deviations from flat space that he is able to construct quasi-Lorentzian coordinates or directions.

Take the famous three classical observational tests: In the gravitational red shift (Einstein effect) we compare with each other clocks that are falling freely but are instantaneously at rest in a static coordinate system and clocks that are located far away from the region of appreciable gravitational potential. In the modern modification of that experiment, which makes use of the Mössbauer effect, the two clocks are both in the gravitational field, at two locations with slightly different gravitational potentials, with known relative velocities. For the interpretation of the Einstein effect, the existence of a static coordinate system [i.e. Killing field as well as validity of Eq. (21.3)] is thus assumed. With respect to the group of transformations that leads from one to another static coordinate system the metric component g_{00}, which controls the Einstein effect, behaves like a scalar.

The precession of perihelion of elliptic orbits (such as that of Mercury) must be referred to an assumed inertial frame of directions defined at spatial infinity, which in practice is defined by the galaxies far from our own. Likewise, the bending of light rays passing near the limb of the Sun must be interpreted in terms of an asymptotically flat frame of directions. Especially the last effect is typically an S-matrix effect (in the terminology of quantum field theory): We predict the transition from one free-field state (propagation of plane electromagnetic waves in flat space-time in a particular null direction) into another free-field state (characterized by a different wave propagation null vector), the transition resulting from interaction with a localized gravitational field. The whole effect cannot be described without the assumption of initial and final free-field states.

Clearly, for the comprehensive interpretation of general relativity in terms of conceptually possible observations and experiments the reliance on static, and on asymptotically flat, frames of reference is a potential source of embarrassment. General relativity was conceived as a local theory, with locally well defined physical characteristics. For the clarification of the contents of the theory its interpretation in terms of locally observable quantities is eminently desirable. We shall call such quantities *observables*. The remainder of this article will be devoted to a report on the present status of the search for observables.

22. Observables. Let us consider a general-relativistic theory, with the gravitational field represented by means of the metric tensor, and other fields possibly by additional field variables which we shall collectively denote by the symbol ψ. If we knew the field variables $g_{\mu\nu}$, ψ as functions of some chosen four-dimensional coordinate system x^ϱ, assuming, of course, that these fields satisfy all the field equations, this knowledge would include all the physical characteristics of the situation to be described. However, our description should be redundant in two respects: (1) Because the field variables satisfy field equations, some of the information provided is redundant; if the field equations could be cast into the form of a Cauchy initial-value problem, then it should be possible to provide sufficient information on a three-dimensional space-like hypersurface alone to fix the physical situation uniquely. (2) Because in a general-relativistic theory the choice of coordinate system is arbitrary, some of the information that we have provided in our hypothetical case does not refer to the physical situation but to our chosen mode of description, i.e. to the choice of coordinate system. Accordingly, we could conceivably remove from our total description a very great portion without losing those pieces of information that refer to the particular physical situation to be described and which serve to distinguish it from other physical situations.

We shall call *observables* physical quantities that are free from the ephemeral aspects of choice of coordinate system and contain information relating exclusively to the physical situation itself. Any observation that we can make by means of physical instruments results in the determination of observables; because by assumption all coordinate systems are in principle equivalent, a physical measurement by its very nature is either empty (that is to say, we could have predicted the result of the observation without any specific knowledge of the physical situation), or it provides us with some physical information. It cannot provide us with information about the coordinate system, because the coordinate system is a matter of free choice, for us to make without any restriction caused by the characteristics of a particular physical situation.

We shall call a set of observables *complete* if knowledge of their numerical values determines the physical situation uniquely; complete sets may be *redundant*, whenever knowledge of a subset of observables determines the value of at least one additional member of the set. Ideally, we should like to possess a complete, non-redundant set of observables for general relativity. This aim has not yet been achieved.

The observables are presumably functions, or functionals, of the conventional field variables. Whereas the conventional field variables will change their values at a world point identified by the values of its coordinates whenever we pass from one coordinate system to another one, the observables must be constructed so that under a coordinate transformation they retain their individual numerical values. Hence observables are *invariants* under arbitrary curvilinear coordinate transformations. To avoid confusion, and because the terms "invariant" and "scalar" are used with different meanings by many authors, we shall make this distinction: A scalar is a local field variable which retains its numerical value at the same world point under coordinate transformations. Suppose a coordinate transformation is described by the functional dependence of the new coordinates $\bar{x}^\varrho$ of a fixed world point on the original coordinate values x^σ,

$$\left.\begin{aligned} \bar{x}^\varrho = f^\varrho(x^0, \ldots, x^3), \qquad x^\sigma = g^\sigma(\bar{x}^0, \ldots, \bar{x}^3), \\ g^\mu[f^0(x^0, \ldots, x^3), \ldots, f^3(x^0, \ldots, x^3)] \equiv x^\mu, \end{aligned}\right\} \tag{22.1}$$

then the transformation law of a scalar U is

$$\overline{U}(y^0, \ldots, y^3) = U[g^0(y^0, \ldots, y^3), \ldots, g^3(y^0, \ldots, y^3)]. \tag{22.2}$$

$y^0, \ldots, y^3$ are four freely chosen numbers, the values of the arguments of the function $\overline{U}$. Eq. (22.2) tells us how to construct the function $\overline{U}$ of four arguments from the function U, also of four arguments. Clearly, these two functions are not identical. By contrast, an invariant S is a functional of the given fields which has been constructed so that if we substitute the coordinate transforms of the field variables into the arguments of S instead of the originally given field variables, then the numerical value of S remains unchanged. In all that follows, the terms "observable" (which is physically motivated) and "invariant" (which is formally defined) are interchangeable.

The most successful method for the construction of observables that has been invented so far consists of the introduction of *intrinsic* coordinates. If it were possible to single out one coordinate system from all the others by means of specifications that are based on local characteristics of the four-dimensional manifold, then any field variables in that particular coordinate system will have the characteristics of observables. In order to exhibit these characteristics, we should have to cast the definition of the intrinsic coordinate system into a form in which they become functionals of the field as presented in conventional coordinates. The transformation equations leading to the field variables in terms of intrinsic coordinates then turn into an algorithm that will enable us to construct them from any field described in terms of arbitrary conventional coordinates. This algorithm must lead to identical results regardless of the original choice of coordinate system, as long as the physical situation is the same. Accordingly, the new field variables have become invariant functionals of the original field variables, and that is the requirement we have imposed on observables.

We shall describe one method for the definition of intrinsic variables later on. For the time being, we shall continue our discussion of the general characteristics of this approach. If we were to specify all of the field variables, $g_{\mu\nu}$ and ψ, in terms of the intrinsic coordinate system, we should certainly have a complete set of observables, for the amount of information thus provided is sufficient to determine all the field variables everywhere in terms of any coordinate system whatsoever. On the other hand, this set of observables is highly redundant. They are restricted by two types of relations, one being the field equations, the other the condition that the coordinate system chosen is the intrinsic coordinate system; just as in any other coordinate system we may proceed to construct the intrinsic coordinates, but this construction is required to lead us to the system already being employed. This requirement of circularity represents a set of coordinate conditions, sufficient to specify precisely the intrinsic coordinates. If we could solve the two sets of relations, the field equations and the coordinate conditions, with respect to a subset of observables, then the remainder should represent a complete and non-redundant set. This goal has not been achieved, and it cannot be achieved with the methods presently employed, at least not without destroying the local character of the observables.

A certain measure of reduction of the redundance of observables has, however, been achieved: It is now possible to restrict oneself to presenting a limited number of observables at every point of a space-like three-dimensional hypersurface, with the assurance that the continuation off the hypersurface is uniquely determined by the initial-value data. The remaining redundance within the chosen hypersurface may be expressed in terms of a set of partial differential equations to be satisfied by the initial-value data. This partial reduction is achieved by a Hamiltonian approach to the problem.

In the next few sections we shall first indicate methods for constructing intrinsic coordinate systems in four dimensions. Subsequently, we shall go over to a presentation of the Hamiltonian formulation of general-relativistic theories, to be followed by the amalgamation of the two approaches.

23. Intrinsic coordinates. As explained in Sect. 22, we call a coordinate system intrinsic if it can be defined uniquely on the basis of local characteristics of the physical situation. Of the various methods for accomplishing this purpose we shall discuss here one; the construction of four scalar fields which will serve as the coordinates. For intuitive reasons, it is desirable that these four scalar fields be constructed from the metric and its differential covariants. Such a construction does not force us to specify what other fields, if any, we wish to consider, and we are free to treat all non-gravitational fields in a conventional manner.

The lowest differential covariants of the metric are the components of the Riemann-Christoffel curvature tensor; all higher differential covariants may be expressed in terms of the curvature tensor and its derivatives. The Riemann-Christoffel tensor itself has 20 components in four dimensions. Of these, half are algebraically determined by the non-gravitational fields through the field equations. The remaining 10 components may be collected into WEYL's tensor, also known as the conformal curvature tensor, which was introduced through Eq. (9.1). From this tensor one can form algebraically exactly four scalars, as was pointed out by GÉHÉNIAU[1]. This number may be understood most simply in this manner: If we introduce a coordinate system so that at one world point the metric tensor assumes MINKOWSKI's values, then the components of the Weyl tensor may be grouped into two symmetric three-dimensional matrices, A and B (cf. Sect. 9), whose traces vanish individually. These two matrices behave as three-dimensional tensors with respect to local orthogonal transformations not involving x^0; with respect to Lorentz transformations they form linear combinations. We may now (except for special situations) first employ a Lorentz transformation so as to make the two forms A and B commute with each other. Then, by means of a spatial orthogonal transformation we may diagonalize A and B

[1] J. GÉHÉNIAU and R. DEBEVER: Bull. Acad. Roy. Belg. Cl. Sci. **42**, 114 (1956). — Helv. phys. Acta Suppl. **4**, 101 (1956) [*17*].

simultaneously. We have now reduced the two "tensors" A and B to three diagonal elements each, which are, moreover, restricted by the requirement that the sum of these three "eigenvalues" vanishes. Hence, effectively, there are four eigenvalues left, which can no longer be changed by any admissible transformation. The construction just described is not very practical for actually obtaining four scalar numerics, but we may instead introduce the coefficients of the secular equation associated with the eigenvalue problem

$$[C_{\iota\varkappa\lambda\mu} - \Lambda(g_{\iota\lambda} g_{\varkappa\mu} - g_{\iota\mu} g_{\varkappa\lambda})] V^{\lambda\mu} = 0. \tag{23.1}$$

In this equation the skewsymmetric tensor $V^{\lambda\mu}$ is the "eigenbivector" of the problem and Λ the eigenvalue. With the help of the three 6×6 matrices C (the Weyl tensor), G of Eq. (9.4), and ε of Eq. (9.5) we may form the following four scalars:

$$\left.\begin{aligned} A^1 &= \operatorname{tr}\{C\,G\,C\,G\}, & A^2 &= \operatorname{tr}\{C\,G\,C\,\varepsilon\},\\ A^3 &= \operatorname{tr}\{(C\,G)^3\}, & A^4 &= \operatorname{tr}\{(C\,G)^2\,C\,\varepsilon\}.\end{aligned}\right\} \tag{23.2}$$

The two scalars A^1 and A^2 are quadratic in the components of the Weyl tensor, whereas A^3 and A^4 are cubic. Any other formulation of the four scalars leads to expressions that are functions of the four A's.

For the four A's (or four independent functions of the A's) to serve as intrinsic coordinates it is necessary that they be algebraically independent of each other, that is to say, that the determinant

$$\Delta \equiv \det|A^{\varrho}{}_{,\sigma}| \tag{23.3}$$

be non-zero. It is well known that not all solutions of the field equations satisfy this requirement. First of all, most of the known closed-form solutions possess Killing vector fields, which is to say, there exist coordinate systems in which the components of the metric tensor depend on fewer than all four coordinates; the components of the Schwarzschild metric, for instance, depend on but one coordinate in a suitably chosen frame of reference, the "radius". It follows that the A's will not be universally suitable to serve as intrinsic coordinates. If there are one or more Killing vector fields, and if the remaining fields also are independent of the same coordinates, then there is no possibility at all of constructing intrinsic coordinates that will permit us to distinguish all world points from each other. In that case the statement of a lower-dimensional coordinate system, along with the group of "motions" (non-identical mappings of the manifold on itself that leaves all fields unchanged), is the best we can do in describing the intrinsic characteristics of the manifold. In the absence of Killing fields, the determinant Δ may still vanish. In that case we can construct an intrinsic coordinate system by replacing some of the A's of Eq. (23.2) by a higher-order scalar of the metric or, if necessary, of another field.

In what follows we shall suppose that we deal with a manifold in which the functions (23.2) are available as intrinsic coordinates. If so, any specified set of four algebraically independent functions of the A's will do as well. We may then assume that we have chosen such an intrinsic coordinate system, and proceed thence. Any field that we may describe in terms of a specified intrinsic coordinate system will now represent a set of scalars, in the sense that the values of the field variables in the intrinsic coordinate system may be calculated from a knowledge of the field in any conventional coordinate system, and the result is unique. Hence, information about the field variables as functions of the intrinsic coordinates represents knowledge of *coincidences:* We are told that specified values of five or more scalars occur at the same world point. Unless the chosen scalars are algebraically dependent on each other (in which case our information is either trivially correct or false), such coincidence information represents information concerning a particular manifold; if sufficient coincidence information is made available, it will characterize the manifold completely.

In a purely gravitational situation, i.e. in the absence of any fields distinct from the metric tensor, the 10 scalars

$$\bar{g}^{\varrho\sigma} = g^{\mu\nu} A^{\varrho}{}_{,\mu} A^{\sigma}{}_{,\nu} \tag{23.4}$$

represent the scalar products between the gradients of any two of the four scalars (23.2). At the same time, they represent contravariant components of the metric tensor in the intrinsic coordinate system of the A's. Given these 10 scalars as functions of the 4 A's, a given Riemannian manifold is uniquely identified. As discussed previously, these 10 scalars cannot be chosen freely at each world point, but they are related to each other both by the field equations and by the coordinate conditions, which specify that if we form the Weyl tensor from the $\bar{g}$'s, and the four scalars $A^1, \ldots, A^4$ in accordance with the instruction (23.2), then these scalars must equal at each world point the values of the coordinates. Various investigations, some of which we shall describe later, have led to the result that a Riemann-Einstein manifold (i.e. a Riemann manifold that obeys EINSTEIN's field equations for the vacuum) can be determined completely if at each point of a three-dimensional space-like hypersurface four pieces of numerical information are provided[1].

It should also be pointed out that the approach described here is a purely local approach. We do not ask whether the intrinsic coordinates over the manifold in its entirely, or whether they represent only a "coordinate patch".[2] The local character of our intrinsic coordinates carries with it certain advantages and certain weaknesses. Whether our intrinsic coordinates are appropriate to a given physical situation we can decide by examining the vicinity of one world point; we must verify whether at that point the Jacobian (23.3) is non-zero. We are not forced to consider the global aspects of the situation, whether the metric is free of singularities, whether it is asymptotically flat, and similar questions. The probable drawbacks are that local intrinsic coordinates may not be well adapted to the consideration of weak fields, to the examination of questions of correspondence, etc.

Typically, any attempt to construct an intrinsic coordinate system by non-local procedures involves the solution of partial differential equations with boundary conditions at spatial infinity. Generally, the existence and uniqueness of solutions of such problems requires an elaborate investigation of its own, which in most cases has not yet been completed. On the other hand, knowledge of the existence of solutions of the boundary value problems involved probably would carry with it some understanding of global properties of classes of Riemann-Einstein manifolds. In this sense the effort involved in the solution of boundary value problems might well carry its own reward.

Whenever the boundary conditions imposed amount to the asymptotic transition of the coordinate system looked for into one of the appropriate coordinate systems of specially symmetric manifolds (e.g. Lorentz frames in a Minkowski universe), it appears a reasonable conjecture that the coordinate systems looked for

[1] W. PAULI and M. FIERZ: Helv. phys. Acta **12**, 297 (1939). — A. KOMAR: Phys. Rev. **111**, 1182 (1958). — R. ARNOWITT, S. DESER and C. W. MISNER: Phys. Rev. **116**, 1322 (1959); **117**, 1595 (1960).

[2] The program by ARNOWITT and coworkers (cf. preceding footnote) aims at a construction of observables on a non-local basis. Another non-local approach is by E. NEWMAN and P. G. BERGMANN: Rev. Mod. Phys. **29**, 443 (1957).

are determined only up to a transformation group of at least the same dimensionality as the group of motions of the symmetric standard manifold, such as the Lorentz group for the Minkowski universe. To this extent, coordinate systems defined by non-local criteria are probably not intrinsic in the strict sense of our definition, but they may yet be useful for the definition of observables. To carry out the analogy to Lorentz-covariant theories, the value of a certain field variable at a world point identified only by its Lorentz coordinates is not strictly an observable; but a finite number of landmarks (not generally at the world point in which we are interested) will identify the coordinate system uniquely, and the value of any field variable in terms of a properly identified Lorentz coordinate system is, of course, an observable according to our definition. In actual laboratory experiments or astronomical observations, inertial frames of reference are always identified by a finite set of landmarks.

24. Observables and quantum theory. The search for observables is in part motivated by the program of quantizing the gravitational field. Though there is some difference of opinion concerning the need for interpreting the gravitational field as one of the several quantum fields of nature, it appears at least not unreasonable to explore the possibility of treating the gravitational field like all other physical fields. In the famous paper by BOHR and ROSENFELD[1] it was shown that the assumption of a classical electromagnetic field with sources subject to quantum mechanics leads to contradictions; briefly, if it were possible to give all components of the electromagnetic field classical meaning, not subject to uncertainty relations, then the locations and velocities of the sources could be given similar classical meaning. Similarly, it could be argued that with a classical gravitational field surrounding each particle, the classical determination of the full metric field would permit us, by mathematical exploitation of the field equations, to determine effective locations and motions of the sources of the gravitational field and thus to circumvent the uncertainty relations.

The execution of such a program would depend on our ability to measure with sufficient accuracy the metric field produced by one elementary particle, a feat far beyond our present experimental technique. As a result, it is not contradictory to any facts known at present to assume that a classical (c-number) metric field is produced by a c-number source field consisting of the expectation values of particle density, linear momentum density, and linear momentum flux. If we were to accept this assumption, we should also have to accept that in the event of observations concerning the sources that involve the process called "reduction of the wave packet", the effective sources of the field would change discontinuously, and hence the c-number metric field would undergo similar changes that would no longer be governed by the field equations along with the ponderomotive equations. This conclusion, unpalatable as it may be, is qualified by our present inability to measure the metric field with an accuracy remotely approaching the accuracy needed for a verification of such observation-induced jumps, and the consideration that any observation to be made on particles involves physical intervention on a scale that would have effects on the metric field in any case. On the other hand, according to EINSTEIN, PODOLSKY and ROSEN[2], the physical intervention need not take place in the immediate vicinity of the location where observation-induced jumps are to take place.

In this section we shall explore some of the conclusions flowing out of the opposite physical hypothesis, the hypothesis that the metric field is indeed

---

[1] N. BOHR and L. ROSENFELD: Dan. Mat.-Fys. Medd. **12**, No. 8 (1933).
[2] A. EINSTEIN, B. PODOLSKY and N. ROSEN: Phys. Rev. **47**, 777 (1935).

subject to field quantization. We shall argue that in this case not all components of the metric tensor may be regarded as operators capable of inducing mappings of the Hilbert space on itself, but that the role of Hilbert operators is reserved to observables.

Without going into the details of a quantum construction, we may argue, in any case, that the state vectors that can form the base of a Hilbert space for a quantum theory of the gravitational field must represent states that are physically possible. The physical requirements involve, of course, the field equations. It is characteristic of general-relativistic theories that the field equations are not all dynamical in the sense that they describe the unfolding of a physical situation in the course of time, but that four of the field equations impose conditions on the initial-value data on any chosen three-dimensional hypersurface.

The field equations of electrodynamics possess similar properties. For purposes of illustration we shall discuss the structure of the field equations of electrodynamics first. We have a formal choice of formulating Maxwell theory either in terms of the four electrodynamic potentials, or in terms of the gauge-invariant field intensities; in the former mode of description the field equations are of the second differential order, whereas the Maxwell equations in terms of field strengths are of the first differential order. In either mode of description there is one field equation,

$$\operatorname{div} \boldsymbol{D} = 4\pi\sigma = 0, \tag{24.1}$$

which does not involve time derivatives of the highest differential order. Among the first-order Maxwell equations there is also another similar equation,

$$\operatorname{div} \boldsymbol{B} = 0, \tag{24.2}$$

which in terms of electromagnetic potentials reduces to an empty relationship. Whatever our chosen mode of description, we cannot permit quantum states in which the field strengths violate either of these two Eqs. (24.1), (24.2) to have a non-zero probability of existence. Hence, if we wish the state vector of the quantum field to be some functional of the potentials or the field strengths at a particular time, we must construct this functional so that the expectation values of $(\operatorname{div} \boldsymbol{D} - 4\pi\sigma)$ and of $\operatorname{div} \boldsymbol{B}$ vanish. Accordingly, a Hilbert space constructed from mutually orthogonal state vectors that are physically admissible must assign to these two operators only zero matrix elements.

Let us now turn our attention to the requirement of Lorentz covariance. Both Eqs. (24.1) and (24.2) are single components of world vector equations. That is to say, even if we were to assume that the expectation values of their left-hand sides vanish throughout the four-dimensional universe (and not merely at some particular time), these equations will not reproduce themselves under Lorentz transformations involving transition to a relatively moving frame of reference unless the expectation values of the left-hand sides of the remaining Maxwell equations are zero as well. In order to obtain a manifestly Lorentz-covariant quantum electrodynamics, we must build our Hilbert space of permissible state vectors not from instantaneous states but from histories, that is to say from fields that satisfy the whole set of Maxwell equations.

If we adopt this point of view, we must ask what sorts of operators will map such a Hilbert space on itself, that is to say, what properties operators must possess if they are to map permissible histories (fields defined on a four-dimensional domain and satisfying all field equations) on permissible histories. The answer is well known, both in classical mechanics and in quantum mechanics: Generators of canonical (or unitary) transformations that map solutions of the dynamical laws on other solutions are constants of the motion. A constant of the motion in electrodynamics is, incidentally, gauge-invariant, because any quantity that is not may have its value changed through a gauge transformation differently at different times. This is because the group of gauge transformations involves a function with arbitrary time dependence. Our conclusion is, then, that the Lorentz-covariant formulation of quantum electrodynamics involves as possible Hilbert space operators only gauge-invariant variables, which, moreover, must be constants of the motion. After this excursion we shall return to the quantization program in general relativity.

In general relativity there are, among the ten field equations, also four, for any choice of the initial three-dimensional hypersurface, that are free of highest surface-normal derivatives. Given some such surface, and a normal vector n_ϱ, the four field equations (or combinations of field equations)

$$L^\varrho \equiv n_\sigma (G^{\sigma\varrho} + 4\pi\varkappa P^{\sigma\varrho}) = 0 \tag{24.3}$$

have this property. This property of the field equations is not accidental; it can be shown that any set of field equations that can be obtained from a general-relativistic action principle must include four such field equations. This is because the field equations cannot exclude the possibility of transformation of any solution into a formally different solution through coordinate transformations that may be chosen so as to leave any set of initial conditions on a three-dimensional hypersurface unchanged. Accordingly, it must be impossible to solve the field equations with respect to the highest-order derivatives of the field variables, even though the number of field equations equals the number of field variables whenever the field equations are Euler-Lagrange equations of a variational principle. Hence, there must exist algebraic combinations of the field equations free of these highest normal derivatives. Eqs. (24.3) represent these combinations for Einstein's theory of gravitation, on the assumption that the field variables chosen are the contravariant components of the metric tensor. For any other choice (such as metric tensor plus affine connection) the appropriate set is obtained as the coefficient of that combination of field variables whose transformation law under infinitesimal coordinate transformations involves the highest-order normal derivatives of the variations of the coordinates, δx^ϱ.

We may now proceed with an argument that is almost the exact parallel of the one presented in electrodynamics. If we are looking for state vectors that are functionals of initial-value data on a three-dimensional hypersurface, we must assign zero probabilities to all states that would violate those field equations representing restrictions on the initial data. Accordingly, the Hilbert space must be constructed only from these state vectors for which the expectation values of the four expressions L^ϱ, Eq. (24.3), vanish. For the next step in our proceedings, we should distinguish between coordinate transformations and substitutions of new initial-value hypersurfaces. These two operations coincide, of course, if the hypersurface is always chosen to be a coordinate hypersurface (e.g. $x^0 = 0$); otherwise, the set of four special field equations goes over into itself under coordinate transformations but combines with the remaining field equations whenever the substitution of new hypersurfaces leads to a change in direction of the hypersurface normal n_ϱ. Unless we propose to assign a particular set of hypersurfaces permanent significance, the natural requirement that our representation should be independent of the choice of initial-value hypersurface leads to the condition that the left-hand sides of all field equations, not merely those of the type (24.3) should have zero expectation values. With this requirement, the

Hilbert space again must be built up from a base consisting of permissible histories; and the Hilbert space operators become constants of the motion, hence invariants.

The relationship between generators and infinitesimal canonical transformations is best known, at least in classical physics, in the Hamiltonian formulation of the equations of mechanics. The connection may, however, also be formulated without recourse to the Hamiltonian formalism. In general, if we change our field variables infinitesimally, and add, at the same time, a divergence $Q^{\varrho}{}_{,\varrho}$, to the Lagrangian density (which, as we know, has no effect on the Euler-Lagrange equations), the change in the functional dependence of the Lagrangian density L on its arguments, the field variables y_A and their partial derivatives $y_{A,\varrho}$ is given by the expression

$$\left.\begin{aligned} \delta' L &= - C^{\varrho}{}_{,\varrho} - L^A\,\delta y_A, \\ C^{\varrho} &\equiv - Q^{\varrho} + \delta y_A \frac{\partial L}{\partial y_{A,\varrho}}, \\ L^A &\equiv \frac{\partial L}{\partial y_A} - \left(\frac{\partial L}{\partial y_{A,\varrho}}\right)_{,\sigma}. \end{aligned}\right\} \tag{24.4}$$

If the expression $\delta' L$ is free of second-order normal derivatives of the field variables, $y_{A,\varrho\sigma}\,n^{\varrho}\,n^{\sigma}$, then we may say that the transformation δy_A is a canonical transformation that is generated by the generator Γ,

$$\Gamma = \int C^{\varrho}\,d\Sigma_{\varrho}, \tag{24.5}$$

a functional defined on a three-dimensional hypersurface, with the surface element

$$d\Sigma_{\varrho} = \delta_{\varrho\alpha\beta\gamma}\,\frac{\partial x^{\alpha}}{\partial u^1}\,\frac{\partial x^{\beta}}{\partial u^2}\,\frac{\partial x^{\gamma}}{\partial u^3}\,du^1\,du^2\,du^3 \tag{24.6}$$

in a parametric representation of the hypersurface

$$x^{\varrho} = f^{\varrho}(u^1, u^2, u^3). \tag{24.7}$$

The expression (24.6) is manifestly invariant with respect to a transformation of the three parameters u^k among themselves. With respect to coordinate transformations the hypersurface element $d\Sigma_{\varrho}$ transforms as a covariant vector density of weight -1; hence Γ will be invariant with respect to coordinate transformations if C^{ϱ} is a contravariant vector density of weight $+1$. We shall call C^{ϱ} the generating density. It is worthy of note that the surface element (24.6) is defined without reference to a metric. Accordingly our formalism does not require the existence of a Riemannian structure of the manifold. In other words, if we should consider the metric as a q-number field, the possibility of constructing c-number surface elements is thereby not affected, whereas the generating density and the generator would presumably be q-numbers in any quantum field theory.

Given an infinitesimal variation δy_A, we may succeed in eliminating second normal derivatives from $\delta' L$ only by the judicious choice of Q^{ϱ}. Given the possibility of such a choice (which is not always available), it is possible to add to the generating density arbitrary functions of the undifferentiated y_A without affecting the δy_A. Such an addition will, however, affect the canonical momentum densities,

$$\pi^A = \frac{\partial L}{\partial y_{A,\varrho}}\,n_{\varrho}, \tag{24.8}$$

in a manner that is consistent with the customary relationship between generators and infinitesimal canonical transformations.

For a generator to be a constant of the motion, the integral (24.5) should be independent of the choice of hypersurface of integration. Because of Eq. (24.4), Γ will not change as the result of an infinitesimal displacement of the hypersurface of integration (we either must confine this displacement to a limited domain of the hypersurface, or we must supplement our theory by boundary conditions imposed on all field variables at spatial infinity) if we consider *invariant transformations* of the field variables, i.e. transformations which leave the form of the action functional S unchanged, and if we assume that the field equations, $L^A=0$, are satisfied throughout the four-dimensional domain of definition of the action integral, $S=\int L\, d^4x$. In our quest for a manifestly covariant formulation of a quantum theory of the gravitational field, we are then led to a consideration of constants of the motion as the candidates for Hilbert space operators; constants of the motion are *ipso facto* observables, that is to say invariants under curvilinear coordinate transformations, because any non-invariant functional on a sequence of three-dimensional hypersurfaces can have its value changed by different amounts on different hypersurfaces by means of curvilinear coordinate transformations.

Once we have discovered the appropriate group of transformations of solutions of the field equations, those that carry solutions over into solutions, and their generators, the constants of the motion the group character of the transformations defines on the generators group-theoretical commutators, Poisson brackets in the case of a c-number theory, quantum commutators in a q-number theory. In what follows, we shall describe methods for obtaining both permissible generators and commutator brackets between them.

25. The function group of observables. Among all the invariant canonical transformations there is one set, in fact a subgroup, which is immediately known, the group of (infinitesimal) curvilinear coordinate transformations. Their generators must also be constants of the motion, on the strength of the general theorem that we have already quoted in the previous section. As before we shall designate the infinitesimal change in the field variables y_A (that is to say, the change in their functional dependence on the coordinates x^ϱ) by the symbol $\bar{\delta} y_A$. Then we have for the generating density the identity

$$C^\varrho{}_{,\varrho} + L^A \bar{\delta} y_A \equiv 0. \tag{25.1}$$

Of course, an identical relationship holds for the generating density of, say, Lorentz transformations in a field theory that is merely Lorentz-invariant. What distinguishes Lorentz-covariance from general covariance is that the field $\bar{\delta} y_A$, besides depending on the field variables themselves and their derivatives, in the case of Lorentz covariance depends on a finite number of parameters that may be freely chosen, whereas under general covariance the $\bar{\delta} y_A$ depend on four arbitrary functions of the four coordinates, the functions

$$\delta x^\varrho(x^\varkappa) \equiv \xi^\varrho(x^\varkappa). \tag{25.2}$$

In other words, whereas the Lorentz group is a Lie group, whose members form a topological manifold of ten dimensions, the group of general coordinate transformations has the dimensionality of four functions of four arguments, i.e. of a particular function space. We shall find that as a result we can construct a generating density field that vanishes.

Let us substitute into Eq. (25.1) the actual expressions corresponding to a purely gravitational field. We have, then:

$$L^A \bar{\delta} y_A \equiv \frac{\sqrt{-g}}{4\pi\varkappa} G^\varrho_\mu \xi^\mu{}_{;\varrho}. \tag{25.3}$$

This expression, in view of the fact that the semicolon stands for covariant differentiation, contains the "descriptors" ξ^ϱ, Eq. (25.2), both undifferentiated and their first partial derivatives. However, because of the contracted Bianchi identities the whole right-hand side of Eq. (25.3) equals an ordinary divergence,

$$\sqrt{-g}\, G^\varrho_\mu \xi^\mu{}_{;\varrho} \equiv \left(\sqrt{-g}\, G^\varrho_\mu \xi^\mu\right)_{;\varrho} \equiv \left(\sqrt{-g}\, G^\varrho_\mu \xi^\mu\right)_{,\varrho}. \tag{25.4}$$

Hence, we may solve Eq. (25.1) by the expression

$$C^\varrho = -\frac{\sqrt{-g}}{4\pi\varkappa} G^\varrho_\mu \xi^\mu, \tag{25.5}$$

which vanishes if the field equations hold.

The solution (25.5) of Eq. (25.1) is not unique. We may add to one such solution terms of the structure

$$C^{\varrho *} = C^\varrho + V^{[\varrho\sigma]}{}_{,\sigma}, \tag{25.6}$$

where $V^{[\varrho\sigma]}$ is any field whose components are antisymmetric in the two superscripts. Its transformation law need not be specified, but the new generating density will be a vector density, and hence the generator invariant, if V is a tensor density of weight 1. The generator Γ^* will differ from Γ, Eq. (24.5), which vanishes, by a two-dimensional surface integral, because of GAUSS's theorem,

$$\int V^{[\varrho\sigma]}{}_{,\sigma}\, d\Sigma_\varrho = \oint V^{[\varrho\sigma]}\, d\Sigma_{\varrho\sigma}. \tag{25.7}$$

The integral on the right, to be taken over the closed two-dimensional boundary of the three-dimensional domain of integration that defines the generator, will vanish if the structure $V^{[\varrho\sigma]}$ satisfies conventional boundary conditions on the boundary. Its values in the interior of the three-dimensional domain of integration contribute nothing to the generator. The transformation (25.6) is the field analog to the constants of integration in classical mechanics that always enter into the defining equation of the energy of a system.

Aside from this ambiguity, the generating density of a coordinate transformation is a combination of field equations, and hence will vanish. This result holds necessarily for the generators of the members of any invariance group that possess the dimensionality of a four-dimensional function space. Clearly, the generators of the infinitesimal members of the invariance group will depend on the "descriptors" and their derivatives linearly. If they are to be constants of the motion, in spite of the arbitrary time-dependence of the descriptors, the coefficients of these linear functionals of the descriptors must vanish, and hence the generators themselves will be zero as well.

Of course, these vanishing generators generate canonical transformations that map every solution of the field equations on another solution that represents the same physical situation, only in terms of a different coordinate frame. To this extent it is, perhaps, not too disturbing that they vanish. With the help of the coordinate transformations, we can in fact divide the totality of all solutions of the field equations into equivalence classes. All members of one equivalence class represent the same physical situation; coordinate transformations map each equivalence class on itself; and under the group of curvilinear coordinate transformations there exist no invariant subdivisions within one equivalence class. Equivalence classes are, of course, non-overlapping, and they cover the totality of all solutions.

We shall now show that the coordinate transformations form an invariant subgroup of all the invariant canonical transformations, and furthermore that the factor group formed with respect to this invariant subgroup consists of all those canonical transformations that map equivalence classes of solutions of the field equations on each other intact. First, to show that the coordinate transformations form an invariant subgroup we must demonstrate that the commutator of an infinitesimal coordinate transformation with any infinitesimal invariant transformation is again a coordinate transformation. But the commutator of two infinitesimal canonical transformations may be interpreted also as the transformation law of one generator under the canonical transformation generated by the other. In view of the fact that the generator of any invariant transformation is itself invariant under coordinate transformations, the commutator must have the value zero, hence it generates a coordinate transformation, if not the identity transformation.

What about the factor group? We obtain the factor group by identifying all those invariant canonical transformations with each other which differ only by a coordinate transformation. Thus any one coset of the invariant subgroup of coordinate transformations maps an equivalence class of solutions on another equivalence class. Moreover, the generators of such a coset differ from each other only by linear combinations of the field equations, that is to say, they are numerically equal. Hence we have a reversibly unique relationship between the members of the factor group of infinitesimal invariant transformations and the set of invariants of different values. The structure of the factor group imposes on the constants of the motion commutators, which we shall call the Poisson brackets of the theory or, if necessary, following the terminology of DIRAC[1], "modified" Poisson brackets. These commutators have the property that if either of the two invariants involved vanishes, then their commutator will vanish as well. We shall now consider how to construct actual Poisson brackets.

One method which is probably valid but which has not been carried out because of the immense labor involved, consists of constructing integrals that are constants of the motion and which contain a certain number of arbitrary coefficients on an initial three-dimensional hypersurface. Once such an integral has been obtained, we may construct the transformation generated by it, in accordance with Eq. (25.1). This transformation is, of course, not entirely unique but determined only up to an infinitesimal coordinate transformation. But if we study the infinitesimal change of a second constant of the motion under a transformation generated by the first, this change will represent the Poisson bracket between these two constants of the motion, and it will itself be a constant of the motion. If we should succeed in determining in this manner the Poisson brackets between a complete set of observables (complete in the sense defined in Sect. 22), we should have accomplished our aim.

In the next sections we shall follow a somewhat different approach, by utilizing a Hamiltonian formalism. This approach is based completely on the conceptual framework laid up to now. The purpose of the Hamiltonian, or canonical, formulation is only to systematize, and to facilitate, the computational work.

26. Hamiltonian theories with constraints. The development of the Hamiltonian, or canonical, formulation of general relativity is characterized by the occurrence of *constraints*, that is to say relations imposed on the canonical field variables in addition to the canonical field equations (which may also be referred to as the

[1] P. A. DIRAC: Canad. J. Math. **2**, 129 (1950); **3**, 1 (1951). — P. G. BERGMANN and I. GOLDBERG: Phys. Rev. **98**, 531 (1955). — P. G. BERGMANN, I. GOLDBERG, A. JANIS and E. NEWMAN: Phys. Rev. **103**, 807 (1956). — P. A. M. DIRAC: Proc. Roy. Soc. Lond. A **246**, 333 (1958). — Phys. Rev. **114**, 924 (1959).

canonical equations of motion). These constraints will occur necessarily as long as not all of the canonical field variables are observables. They arise as follows.

Consider a field theory that is invariant with respect to curvilinear coordinate transformations, whose field variables y_A obey some transformation laws. Given the form of the Lagrangian density, which for the sake of definiteness we assume is a function of the field variables themselves, and their first partial derivatives, we should normally define the canonical momentum densities as the expressions

$$\pi^A = \frac{\partial L}{\partial y_{A,0}} \equiv \frac{\partial L}{\partial \dot{y}_A}. \tag{26.1}$$

It is not difficult to show that these momentum densities cannot all be independent of each other, but that in view of their defining equations (26.1) they will satisfy relations that do not involve any time derivatives of the canonical variables y_A, π^A.

Let us consider the field equations. They may be cast into the form

$$\dot{\pi}^A = \frac{\partial L}{\partial y_A} - \left(\frac{\partial L}{\partial y_{A,k}}\right)_{,k} \qquad k = 1, 2, 3. \tag{26.2}$$

Here, and in what follows, we shall use the dot to denote partial differentiation with respect to x^0. In terms of the field variables y_A, the one term on the left is the only one containing second-order derivatives with respect to x^0. Specifically, we may write the term on the left in the form

$$\dot{\pi}^A = L^{AB}\ddot{y}_B + \cdots, \qquad L^{AB} \equiv \frac{\partial^2 L}{\partial \dot{y}_A\, \partial \dot{y}_B}, \qquad L^{AB} \equiv L^{BA}. \tag{26.3}$$

That is to say, the field equations are linear in $\ddot{y}_A$, and the symmetric matrix L^{AB} represents the coefficients. The same matrix also has the significance

$$L^{AB} = \frac{\partial \pi^A}{\partial \dot{y}_B}. \tag{26.4}$$

We shall now prove that this matrix is singular. If we formulate in any general-relativistic theory the transformation properties of the Lagrangian density under coordinate transformations, we shall find that the left-hand sides of the field equations satisfy a set of four differential identities, similar to those obtained in Eq. (18.14). The precise form of these identities will depend on the transformation law of the field variables, but they always exist, and they will always be linear in the L^A, the left-hand sides of the field equations. In these identities there will be some terms containing the highest time derivatives of the L^A; these terms in Eq. (18.14) are

$$c_{A\varrho}{}^{0}\,\dot{L}^A + \cdots \equiv 0. \tag{26.5}$$

More particularly, the terms linear in the third time derivatives of the y_A will be of the form:

$$c_{A\varrho}{}^{0}\,L^{AB}\dddot{y}_B \equiv 0, \tag{26.6}$$

and hence:

$$c_{A\varrho}{}^{0}\,L^{AB} \equiv 0. \tag{26.7}$$

In other words, we have constructed four eigen "vectors" for the matrix L^{AB} which have the eigenvalue 0. Our derivation will hold precisely for theories in which the infinitesimal transformation law for the field variables involves no higher than first-order partial derivatives for the field variables; this type includes EINSTEIN's theory of gravitation. For theories with more involved transformation laws the four null-eigen "vectors" of L^{AB} will be constructed differently but they always exist. It follows, then, that the Jacobian between the $\dot{y}_A$ and the π^A always vanishes, according to Eq. (26.4), and that the rank of that matrix does not exceed $(n-4)$, n being the number of field variables.

Given the canonical momentum densities π^A as functions of the "velocities" $\dot{y}_A$, and of the y_A, $y_{A,s}$ $(s=1, 2, 3)$, there are four different infinitesimal changes of the $\dot{y}_A$ that leave the π^A unchanged. Thus any set of possible values of π^A corresponds to a 4-parametric set of possible values of $\dot{y}_A$. Conversely, by spanning the n-dimensional manifold of possible values for the $\dot{y}_A$, we span only an $(n-4)$-dimensional manifold of the π^A; hence not all values of π^A are possible, and the π^A (together with y_A, $y_{A,s}$) must satisfy at least 4 relationships. These are *constraints.*

In EINSTEIN's theory of the gravitational field we may derive these constraints by starting from the action density $(8\pi\varkappa)^{-1}\sqrt{-g}\,R$, but by modifying this action principle in such a manner that it will contain no more second-order derivatives of the metric tensor. By adding to the original Lagrangian density a suitable divergence, we get the equivalent principle

$$\left.\begin{aligned} L &= \frac{1}{8\pi\varkappa}\,\mathfrak{g}^{\varkappa\lambda}\left(\left\{{\sigma \atop \varrho\sigma}\right\}\left\{{\varrho \atop \varkappa\lambda}\right\} - \left\{{\sigma \atop \varkappa\varrho}\right\}\left\{{\varrho \atop \lambda\sigma}\right\}\right) \\ &= \frac{\sqrt{-g}}{16\pi\varkappa}\,g^{\varkappa\lambda}g^{\mu\nu}g^{\varrho\sigma}\left(g_{\varkappa\mu,\nu}\,g_{\varrho\sigma,\lambda} - g_{\varkappa\mu,\varrho}\,g_{\lambda\sigma,\nu} + \frac{1}{2}\,g_{\mu\varrho,\varkappa}\,g_{\nu\sigma,\lambda} - \frac{1}{2}\,g_{\mu\nu,\varkappa}\,g_{\varrho\sigma,\lambda}\right). \end{aligned}\right\} \tag{26.8}$$

If we differentiate this expression with respect to $g_{\alpha\beta,\iota}$, we get the three-indexed set of quantities usually denoted by the symbol $\pi^{\alpha\beta,\iota}$,

$$\frac{\partial L}{\partial g_{\alpha\beta,\iota}} \equiv \pi^{\alpha\beta,\iota} = \frac{1}{16\pi\varkappa}\left[g^{\alpha\varrho}\,\mathfrak{g}^{\beta\iota}{}_{,\varrho} + g^{\beta\varrho}\,\mathfrak{g}^{\alpha\iota}{}_{,\varrho} - g^{\alpha\beta}\,\mathfrak{g}^{\iota\varrho}{}_{,\varrho} - g^{\iota\varrho}\,\mathfrak{g}^{\alpha\beta}{}_{,\varrho}\right]. \tag{26.9}$$

By setting $\iota=0$, we obtain the canonical momentum densities $\pi^{\alpha\beta}$,

$$\pi^{\alpha\beta} \equiv \pi^{\alpha\beta,0} = \frac{1}{16\pi\varkappa}\left(g^{\alpha\varrho}\,\mathfrak{g}^{0\beta}{}_{,\varrho} + g^{\beta\varrho}\,\mathfrak{g}^{0\alpha}{}_{,\varrho} - g^{\alpha\beta}\,\mathfrak{g}^{0\varrho}{}_{,\varrho} - g^{0\varrho}\,\mathfrak{g}^{\alpha\beta}{}_{,\varrho}\right). \tag{26.10}$$

These canonical momentum densities satisfy the four relations

$$\left.\begin{aligned} g_{n\tau}\,\pi^{\tau 0} &= \frac{1}{16\pi\varkappa}\,\mathfrak{g}^{00}{}_{,n}, \qquad & n &= 1, 2, 3, \\ g_{0\tau}\,\pi^{\tau 0} &= -\frac{1}{16\pi\varkappa}\,\mathfrak{g}^{0s}{}_{,s}, \qquad & s &= 1, 2, 3, \end{aligned}\right\} \tag{26.11}$$

which are free of any time derivatives of the canonical field variables. These expressions are formed directly in accordance with the prescription implied by Eq. (26.7). As pointed out before, these constraints are the consequence of the defining equations for the canonical momentum densities, (26.10) and hold regardless of any dynamical laws.

DE WITT[1], DIRAC[2], and ANDERSON[3] discovered, independently of each other, that the constraints (26.11) can be greatly simplified if we perform a canonical transformation that changes the definition of the canonical momentum densities but not of the "configuration" field variables $g_{\mu\nu}$. We can perform such a canonical transformation most readily by adding to the Lagrangian density a divergence; such a divergence term adds nothing to variations of the action

[1] B. S. DE WITT: Unpublished.
[2] P. A. M. DIRAC: Proc. Roy. Soc. Lond. A **246**, 333 (1958).
[3] J. L. ANDERSON: Phys. Rev. **111**, 965 (1958).

confined to the interior of the four-dimensional domain of integration. The appropriate divergence to be added to the expression (26.8) is:

$$\Delta L = \frac{1}{8\pi\varkappa}\left[\left(\frac{g^{0s}}{g^{00}}\,\mathfrak{g}^{00}{}_{,s}\right)_{,0} - \left(\frac{g^{0s}}{g^{00}}\,\mathfrak{g}^{00}{}_{,0}\right)_{,s}\right]. \tag{26.12}$$

With this addition to the Lagrangian density, the canonical momentum densities (26.10) change by the amounts:

$$\left.\begin{aligned}\Delta\pi^{\alpha\beta} = \frac{1}{16\pi\varkappa}\,\frac{1}{g^{00}}\,&[(2g^{0\alpha}g^{0\beta} - g^{00}g^{\alpha\beta})\,\mathfrak{g}^{0s}{}_{,s} + \\ &+ (g^{0s}g^{\alpha\beta} - g^{\alpha s}g^{0\beta} - g^{\beta s}g^{0\alpha})\,\mathfrak{g}^{00}{}_{,s}].\end{aligned}\right\} \tag{26.13}$$

The constraints (26.11) then reduce to the four conditions:

$$\pi^{0\alpha} = 0. \tag{26.14}$$

These constraints, which we shall call *primary constraints*, must be satisfied at every world point of the four-dimensional manifold. It follows that the variational derivatives of the Hamiltonian density with respect to the canonical conjugates of $\pi^{0\alpha}$, the field variables $g_{0\alpha}$, must vanish. However, these derivatives do not vanish automatically: They have to be set equal to zero, and this requirement leads to a second set of four constraints, the *secondary constraints* of the theory.

27. DIRAC's invariance requirements. In addition to simplifying the primary constraints, DIRAC improved the canonical formulation of general relativity considerably by setting up a new invariance requirement as a principle of selecting dynamical field variables. In view of the fact that the Hamiltonian formalism is primarily adapted to the discussion of initial-value problems, DIRAC considered the degree to which among all the field variables a smaller set may be selected that depends only on the choice of coordinates on the initial three-dimensional hypersurface and which, taken by itself, already obeys conditions of continuation off the hypersurface. To understand this point of view, we consider a general infinitesimal coordinate transformation, which we describe, as in Eq. (25.2), by means of the four functions ξ^ϱ. Let us now examine the transformation laws of various geometric objects, such as the ones given in Eqs. (17.3) through (17.8). From the point of view of a fixed initial hypersurface, we may classify the transformation terms into those that depend only on the undifferentiated ξ^ϱ and on their *spatial* derivatives (i.e. derivatives within the three-dimensional hypersurface), $\xi^\varrho{}_{,m}$, $\xi^\varrho{}_{,mn}$, ..., $(m, n = 1, 2, 3)$, and terms that depend on off-surface derivatives as well (such as $\xi^\varrho{}_{,0}$, $\xi^\varrho{}_{,m0}$, $\xi^\varrho{}_{,00}$, ...). Quantities whose infinitesimal transformation laws contain only terms of the first kind we shall call *D*-invariant. Clearly, their values depend only on the choice of coordinates on the three-dimensional hypersurface, and not on their continuation into the four-dimensional manifold.

Among the covariant components of the metric tensor, g_{mn} are *D*-invariant, whereas $g_{0\mu}$ are not. We shall also introduce the *D*-invariant quantities e^{mn},

$$e^{mn} = g^{mn} - \frac{g^{m0}g^{n0}}{g^{00}}, \tag{27.1}$$

which obey the algebraic relations

$$e^{ms}g_{ns} = \delta^m_n. \tag{27.2}$$

The variables g_{mn}, or alternatively the reciprocal matrix e^{mn}, completely describe the intrinsic geometry of the initial three-dimensional hypersurface.

Given a D-invariant field variable, its partial derivatives within the three-dimensional surface are also D-invariant, as a matter of course. In general, there exists no D-invariant partial derivative in a direction out of the three-dimensional hypersurface, but in special cases such derivatives can be constructed. There exists a unit normal vector at each point of the three-surface, whose components are:

$$l_\varrho = (g^{00})^{-\frac{1}{2}} \delta^0_\varrho, \quad l^\varrho = (g^{00})^{-\frac{1}{2}} g^{0\varrho}. \tag{27.3}$$

With its help we may construct from any covariant vector four D-invariant formations, A_s and A_L,

$$A_L = l^\varrho A_\varrho = (g^{00})^{-\frac{1}{2}} g^{0\varrho} A_\varrho = (g^{00})^{-\frac{1}{2}} A^0. \tag{27.4}$$

In the same manner we may construct D-invariant components of covariant tensors of arbitrary rank. Suppose, now, that we consider the covariant normal derivatives of the spatial components of a covariant tensor (which are D-invariant, to begin with),

$$A_{ab\ldots;L} \equiv A_{ab\ldots;\varrho} l^\varrho, \tag{27.5}$$

then these quantities are also D-invariant.

From the remarks above, given a D-invariant vector component A_a, both its partial and its covariant derivatives with respect to x^b are D-invariant, and hence also their difference,

$$A_{a,b} - A_{a;b} \equiv \left\{ {\varrho \atop ab} \right\} A_\varrho. \tag{27.6}$$

Though l_ϱ is not, strictly speaking, a vector, a short calculation shows that the expression

$$l_\varrho \left\{ {\varrho \atop ab} \right\} \equiv (g^{00})^{-\frac{1}{2}} \left\{ {0 \atop ab} \right\} \equiv v_{ab} \tag{27.7}$$

is also D-invariant. This expression is connected with the canonical momentum densities π^{mn} as follows:

$$\pi^{mn} = \frac{K}{8\pi\varkappa} (e^{mn} e^{ab} - e^{ma} e^{nb}) v_{ab}, \tag{27.8}$$

where K denotes the absolute value of the square root of the determinant of the three-dimensional metric on the hypersurface,

$$K = \sqrt{-{}^3g} = \sqrt{-|g|\, g^{00}}. \tag{27.9}$$

In summary, then, the 12 field variables g_{mn}, π^{mn} are all D-invariant. They satisfy among themselves standard Poisson bracket relations.

Because these quantities are D-invariant, it must be possible to express their infinitesimal transformation laws as canonical transformations generated by functionals that involve only the ξ^ϱ on the three-surface, not the $\dot{\xi}^\varrho \equiv \xi^\varrho{}_{,0}$. As indicated previously, these generators must vanish. In this manner we can establish a relationship between the transformation laws of the D-invariant canonical field variables and the remaining constraints of the theory. The question arises whether the constraints themselves are also D-invariant; this question is equivalent to the one whether the infinitesimal transformation laws of D-invariant quantities are themselves D-invariant.

Put in this form, the answer is trivially negative. For the transformation law of any quantity, except of invariants, involves the arbitrary four-dimensional functions ξ^ϱ; in any further transformation the arbitrary dependence of these descriptors on the fourth coordinate will appear. Hence, the only transformations of D-invariant quantities that are themselves D-invariant will be those that do not involve the x^0-dependence of the ξ^ϱ. Obviously, infinitesimal transformations of the three-surface coordinates among themselves are well defined without reference to the continuation off the three-surface. Hence the transformation law of any D-invariant quantity under infinitesimal coordinate transformations that map the three-surface on itself,

$$\xi^0 = 0, \tag{27.10}$$

will be D-invariant. We shall now show that the transformation of a D-invariant quantity under an infinitesimal transformation for which ξ^ϱ is proportional to the unit normal l^ϱ is also D-invariant. Such a transformation is formally described by the condition

$$\xi^\varrho = \varepsilon\, l^\varrho, \tag{27.11}$$

where ε is constant throughout the three-surface. To carry out this proof we require a lemma on the transformation properties of transformation laws.

Let us consider some mathematical quantity G satisfying a transformation law U under the group of transformations $U, V, \ldots$. We shall designate the transform of G by $U(G)$, without implying that the transformation law is necessarily linear. By the transformation law of $U(G)$ under a new transformation V we mean a new transformation $U'(G')$ that is obtained if both G and $U(G)$ are subjected to the transformation V. This new transformation U' is then

$$U'(G') = V[U(G)] = V\{U[V^{-1}(G')]\}, \tag{27.12}$$

or, if we omit the reference to the object of the transformation and also all parentheses,

$$U' = V\,U\,V^{-1}. \tag{27.13}$$

If the transformations to be considered are infinitesimal, then we find that the infinitesimal transformation law u' is the commutator

$$u' = v\,u - u\,v. \tag{27.14}$$

This is the lemma we need.

Let us return to the issue of D-invariance. We call D-invariant a transformation law that contains no reference to $\dot{\xi}^\varrho$. If we consider two infinitesimal coordinate transformations $\xi^\varrho(x_0, \ldots, x^3)$, $\eta^\varrho(x_0, \ldots, x^3)$, the commutator ξ'^ϱ is given by the expression

$$\xi^{\varrho\prime} = \eta^\varrho{}_{,\sigma}\,\xi^\sigma - \xi^\varrho{}_{,\sigma}\,\eta^\sigma = \cdots - \dot{\xi}^\varrho\,\eta^0. \tag{27.15}$$

Hence, in general the descriptors of the commutator will depend on the x^0-derivatives of the descriptors of the commutants; thus the transformation law of a D-invariant quantity is in general not itself D-invariant. The exception we have already noted previously is borne out by Eq. (27.15): If the coordinate transformation η^ϱ is three-dimensional, if, in other words, $\eta^0 = 0$, then the $\xi^{\varrho\prime}$ will be independent of $\dot{\xi}^\varrho$, and D-invariance is preserved.

If η^ϱ obeys the requirement (27.11), the simple formula (27.15) is insufficient to obtain the commutator, because the descriptors do not only depend on the coordinates but also on certain of the field variables themselves, the $g^{0\mu}$. The more elaborate expression for the commutator of two infinitesimal coordinate transformations whose descriptors are functions of the field variables as well is:

$$\left.\begin{aligned} \bar{\delta}_{\xi'}\, y_A &= \frac{\partial\,\bar{\delta}_\eta\, y_A}{\partial y_B}\,\bar{\delta}_\xi\, y_B + \frac{\partial\,\bar{\delta}_\eta\, y_A}{\partial y_{B,\varrho}}\,\bar{\delta}_\xi\, y_{B,\varrho} + \cdots \\ &\quad - \left(\frac{\partial\,\bar{\delta}_\xi\, y_A}{\partial y_B}\,\bar{\delta}_\eta\, y_B + \frac{\partial\,\bar{\delta}_\xi\, y_A}{\partial y_{B,\varrho}}\,\bar{\delta}_\eta\, y_{B,\varrho} + \cdots\right). \end{aligned}\right\} \tag{27.16}$$

If we apply this formula to the special case in which η^ϱ is of the form (27.11), the descriptor of the commutator turns out to be

$$\xi^{r\prime} = l_0 \, e^{rs} \xi^0{}_{,s}, \qquad r, s = 1, 2, 3, \qquad \xi^{0\prime} = 0. \tag{27.17}$$

Hence, in this case also D-invariance is preserved. We may note, as a special example, that the transformation of g_{mn} under the transformation (27.11) leads to the D-invariant result:

$$\bar{\delta} g_{mn} = 2 v_{mn}. \tag{27.18}$$

The result (27.17) may be generalized still further. We may construct the most general infinitesimal coordinate transformation as follows:

$$\xi^\varrho = \delta^\varrho_s \xi^s + l^\varrho \xi^L, \tag{27.19}$$

where ξ^s, ξ^L are four arbitrary functions of the coordinates $x^0, \ldots, x^3$ only. If we form the commutator of two such coordinate transformations, the result is:

$$\xi^{\varrho\prime} = \delta^\varrho_r [(\eta^r{}_{,s} \xi^s - \xi^r{}_{,s} \eta^s) + e^{rs} (\eta^L \xi^L{}_{,s} - \xi^L \eta^L{}_{,s})] + l^\varrho (\xi^s \eta^L{}_{,s} - \eta^s \xi^L{}_{,s}). \tag{27.20}$$

This result is remarkable insofar as the difference between the choice of infinitesimal coordinate transformations with descriptors $\xi^\varrho(x^0, \ldots, x^3)$ and those with descriptors (27.19) appears at first sight trivial, both containing the same number of arbitrary functions of all four coordinates. Nevertheless, these two sets of infinitesimal coordinate transformations represent different mappings of the function space of metric manifolds on itself; and whereas the former set forms an infinitesimal subgroup of all coordinate transformations (the full group including those whose descriptors are functionals of the metric as well), the set (27.19) does not have group property, as evidenced by the fact that the quantities

$$\xi^{r\prime} = \eta^r{}_{,s} \xi^s - \xi^r{}_{,s} \eta^s + e^{rs} (\eta^L \xi^L{}_{,s} - \xi^L \eta^L{}_{,s}), \tag{27.21}$$

obtained from (27.20) are not pure functions of the coordinates but depend on the metric as well. But, and that is more important for our purposes than the group property, the commutator (27.20) is independent of the x^0-derivatives of the arbitrary functions describing the commutants ($\xi^s, \xi^L; \eta^s, \eta^L$), whereas the commutator (27.15) is not. In this respect, the Poisson brackets of the generators of the transformations (27.19) will faithfully represent the generators of the corresponding infinitesimal coordinate transformations. In earlier versions of Hamiltonian formulations of general relativity, this point had been troublesome.

We are now in a position to introduce D-invariant constraints, the generators of infinitesimal coordinate transformations of the form (27.19). For any D-invariant functional Q, defined on our initial three-dimensional hypersurface, we shall define:

$$\left.\begin{aligned} \bar{\delta} Q &= [Q, \Gamma] \\ \Gamma &= -\int (H_s \xi^s + H_L \xi^L) \, d^3x, \end{aligned}\right\} \tag{27.22}$$

where H_s and H_L are the D-invariant constraints,

$$H_s = 0, \qquad H_L = 0. \tag{27.23}$$

The Hamiltonian of the theory is obtained if we set $\xi^\varrho = -\delta^\varrho_0$. From Eq. (27.19) we find that this requirement is equivalent to setting

$$\xi^L = -l_0, \qquad \xi^s = +\frac{g^{0s}}{g^{00}} = -e^{sn} g_{n0}. \tag{27.24}$$

Hence the Hamiltonian is:

$$H = \int [l_0 H_L + g_{0n} e^{ns} H_s] \, d^3x . \tag{27.25}$$

It remains to determine the constraints H_s, H_L as functions of DIRAC's D-invariant variables, g_{mn} and π^{mn}.

28. The secondary constraints. The constraints H_s, H_L may be obtained by several different methods. We may form those four field equations that in the ordinary formalism are free of second-order time derivatives. For the choice of the $g_{\mu\nu}$, these field equations are $\mathfrak{G}^{\mu 0} = 0$. The constraints must be linear combinations of these expressions, which are required to vanish but which do not vanish identically. As a matter of fact, the constraints are the following combinations:

$$H_s = \frac{1}{4\pi\varkappa} \mathfrak{G}^0_s , \qquad H_L = \frac{1}{4\pi\varkappa} \mathfrak{G}^0_\varrho l^\varrho = \frac{1}{4\pi\varkappa} \frac{1}{\sqrt{g^{00}}} \mathfrak{G}^{00} , \tag{28.1}$$

in accordance with Eqs. (27.19) and (25.5). The computational task then consists merely of expressing all the Christoffel symbols occurring in these field equations in terms of manifestly D-invariant variables, i.e. of the g_{mn}, their spatial derivatives, and the π^{mn}.

Another method of approach would consist of calculating the Hamiltonian density from the standard expression

$$l_0 H_L + e^{rs} g_{0r} H_s = -L + \pi^{\mu\nu} \dot{g}_{\mu\nu} . \tag{28.2}$$

This standard expression leads to a momentary difficulty. In the last term on the right, there occur two kinds of terms,

$$\pi^{\mu\nu} \dot{g}_{\mu\nu} \equiv \pi^{mn} \dot{g}_{mn} + 2\pi^{0n} \dot{g}_{0n} + \pi^{00} \dot{g}_{00} . \tag{28.3}$$

In each of these terms we are enjoined to express the "velocities" $\dot{g}_{\mu\nu}$ in terms of the canonical momentum densities, as no velocities are permitted to occur in the Hamiltonian as such. But whereas the elimination of the $\dot{g}_{mn}$ is straightforward, we have found that none of the canonical momentum densities contains any reference to any of the four $\dot{g}_{0\mu}$. Fortunately, precisely those velocities that cannot be expressed in terms of canonical variables are multiplied by vanishing factors. Thus only the first term on the right acts on D-invariant variables in any Poisson brackets (which in turn we need in order to formulate dynamical laws), and we may omit the remaining terms as long as we apply our Hamiltonian only to D-invariant quantities. Because of DIRAC's choice of Lagrangian, (26.8) plus (26.12), the troublesome velocities do not occur in L, either, and the computation of the constraints out of Eq. (28.2) is thus straightforward. The results of such computations are the following:

$$\left. \begin{aligned} H_s &= g_{mn,s} \pi^{mn} - 2 (g_{ms} \pi^{mn})_{,n} , \\ H_L &= \frac{8\pi\varkappa}{K} \left(g_{mr} g_{ns} - \frac{1}{2} g_{mn} g_{rs} \right) \pi^{mn} \pi^{rs} + \frac{K}{8\pi\varkappa} {}^3R . \end{aligned} \right\} \tag{28.4}$$

The symbol ${}^3R$ in the last term of H_L denotes the three-dimensional curvature scalar calculated from g_{mn}, e^{mn}. K is, as before, the square root of the absolute value of the determinant of that same three-dimensional metric[1].

As required by the commutators of the infinitesimal transformations generated by the "secondary" constraints H_s, H_L, the Poisson brackets between them are

[1] P. A. M. DIRAC: Phys. Rev. **114**, 924 (1959).

linear in the secondary constraints themselves, hence they vanish if the constraints are satsfied. According to DIRAC's classification of constraints, they are first-class constraints, second-class constraints being those whose Poisson brackets do not all vanish.

According to our definition of observables, an observable in DIRAC's canonical formulation of general relativity is a functional (defined on a three-dimensional hypersurface) of the g_{mn}, π^{mn}, whose Poisson brackets with all four secondary constraints vanish. Such a functional will be invariant with respect to arbitrary infinitesimal coordinate transformations. So far, such observables have been constructed only with the help of coordinate conditions. In fact, the construction to be detailed in the next section is equivalent to the one described in Sect. 23, the only difference being that the construction is performed completely within a Hamiltonian formalism. Practically, this means that commutator brackets between the observables thus obtained are simply Poisson brackets, and thus obtainable by routine computations.

29. Commutators in the Hamiltonian formalism. In Sect. 23 we introduced the four Géhéniau-Komar scalars A^ϱ as intrinsic coordinates. We shall express these scalars in terms of DIRAC's canonical field variables; but before doing this we shall prove that not only these but all D-invariant expressions formed from the metric tensor and its derivatives can be expressed in terms of the g_{mn}, π^{mn}, provided the field equations are satisfied. Such a quantity Q (which may be a scalar, or the component of some geometric object) will be a function of the components of the metric, and of a finite number of derivatives of several orders; let the highest-order derivative with respect to x^0 be of the n-th order in this regard (we shall not count "spatial" derivatives in all that follows). Clearly, the infinitesimal transformation law of Q will be of the form

$$\bar{\delta} Q = \frac{\partial Q}{\partial g_{\mu\nu}} \bar{\delta} g_{\mu\nu} + \frac{\partial Q}{\partial g_{\mu\nu,\varrho}} \bar{\delta} g_{\mu\nu,\varrho} + \frac{\partial Q}{\partial g_{\mu\nu,\varrho\sigma}} \bar{\delta} g_{\mu\nu,\varrho\sigma} + \cdots. \tag{29.1}$$

By assumption, all references to $\dot{\xi}^\mu$, $\ddot{\xi}^\mu$, and to higher x^0-derivatives on the right must cancel out. But the transformation law of the n-th x^0-derivative of g_{mn} will contain terms with the n-th x^0-derivative of ξ^μ, whereas the transformation law of the n-th x^0-derivative of $g_{\mu 0}$ will contain $(n+1)$-th x^0-derivatives of the ξ^μ. It follows, if we write out all the possible combinations that might arise, that Q cannot contain any n-th x^0-derivatives of $g_{\mu 0}$, but that the highest x^0-derivatives that occur must be those of g_{mn}.

Provided $n \geq 2$, we may substitute for these highest x^0-derivatives from the 6 field equations that express $\ddot{g}_{mn}$ in terms of lower x^0-derivatives. Such a substitution will not change the transformation law if we continue to assume that the field equations are satisfied. By this substitution, we have converted Q into a new expression for which n is smaller by at least 1 than it was to begin with. We now repeat our procedure for the new expression for Q, continuing until n has been reduced below 2, that is to say to 1 or to 0. In either case, the argument that the highest x^0-derivative must be of g_{mn} still holds. If $n=0$, then there is nothing further to say. If $n=1$, then we can show that the only D-invariant combination containing $\dot{g}_{mn}$ is v_{mn}, and algebraic combinations of v_{mn} and g_{mn}, such as π^{mn}. Hence our proof is complete.

Armed with this result, we now proceed to show that the scalars A^ϱ, Eq. (23.2), consist of D-invariant constituents. We notice first that of any covariant vector b_ϱ we may form the four D-invariant quantities b_r, $b_L = b_\varrho l^\varrho$; analogous statements hold for covariant tensors. If any tensor has contravariant indices, we shall lower these indices first with the help of the metric tensor and then proceed as before. If we form the scalar product of two vectors b_ϱ and c_ϱ, this product is, of course, a scalar and hence D-invariant; however, we may form the same product of D-invariant constituents according to the equivalence:

$$\left.\begin{aligned} g^{\varrho\sigma} b_\varrho c_\sigma &\equiv (e^{\varrho\sigma} + l^\varrho l^\sigma)\, b_\varrho c_\sigma \equiv e^{rs} b_r c_s + b_L c_L, \\ e^{\varrho\sigma} &\equiv g^{\varrho\sigma} - l^\varrho l^\sigma, \quad e^{\varrho\sigma} l^\sigma \equiv 0, \quad e^{\varrho 0} \equiv 0. \end{aligned}\right\} \tag{29.2}$$

Given these facts, the D-invariant constituents of the expression (23.2) are then the components of the Weyl tensor, C_{klmn}, C_{klmL}, C_{kLmL}, and the non-vanishing components of the Levi-Civita tensor, ε^{Lkmn}. These expressions are, in terms of D-invariant variables,

$$\left.\begin{aligned} C_{klmn} &= {}^3R_{klmn} + v_{lm} v_{kn} - v_{km} v_{ln}, \\ C_{klmL} &= v_{mk|l} - v_{ml|k}, \\ C_{kLmL} &= e^{rs} C_{rkms}, \end{aligned}\right\} \tag{29.3}$$

and

$$\varepsilon^{Lkmn} = \varepsilon^{kmn}. \tag{29.4}$$

The vertical bars in the expression for C_{klmL} indicate covariant differentiation with the help of Christoffel symbols formed with the 3-metric tensor g_{mn}, e^{mn}, and the symbol ε^{kmn} denotes the Levi-Civita tensor in the 3-surface formed with the help of its metric. The four scalars A^ϱ are algebraically composed of the expressions (29.3), (29.4). We shall not write them down, as they are not very instructive.

Consider now that we have constructed four scalars, functions of DIRAC's canonical field variables, which we shall denote by the symbol f^ϱ. If we adopt them as intrinsic coordinates, the new coordinates will be identified by the equations

$$f^\varrho - x^\varrho = 0, \qquad \varrho = 0, \ldots, 3. \tag{29.5}$$

We consider now only infinitesimal canonical transformations, and their generators, which preserve the form of the original constraints $H_s = 0$, $H_L = 0$, and of the coordinate condition (29.5), which are four additional constraints. The Poisson brackets between these eight constraints at each point no longer vanish. In view of the fact that H_s, H_L generate coordinate transformations, their Poisson brackets with the expressions (29.5) represent the infinitesimal change of coordinates generated by each of the H-type constraints. According to DIRAC's terminology, through the addition of the constraints (29.5) all eight have become second-class constraints.

We now contend that for any functional A of DIRAC's canonical field variables defined in the presence of an intrinsic coordinate system it is possible to find a new functional A^*, equal in value to A, which has vanishing Poisson brackets with all constraints[1]. A^* will then be the generator of an infinitesimal canonical transformation that preserves the form of all constraints but maps a given physical situation (as described by the canonical field variables) on a different physical situation. We construct A^* by adding to A a linear combination of all eight constraints,

$$A^* = A + \int [\gamma_\varrho (f^\varrho - x^\varrho) + \varepsilon^s H_s + \varepsilon^L H_L]\, d^3x. \tag{29.6}$$

The coefficients γ_ϱ, ε^s, and ε^L are determined by the requirement that the Poisson brackets of A^* with every one of the constraints must vanish.

[1] P. G. BERGMANN and A. B. KOMAR: Phys. Rev. Letters 4, 432 (1960).

Because the four functions f^ϱ are scalars, it follows that

$$[f^\varrho(x), \int (\xi^{s\prime} H_s' + \xi^{L\prime} H_L')\, d^3x'] = f^\varrho{}_{,\sigma} (\delta^\sigma_s \xi^s + l^\sigma \xi^L) = \delta^\varrho_s \xi^s + l^\varrho \xi^L. \tag{29.7}$$

Substituting this result into the equations expressing the requirement that the Poisson brackets of A^* with any linear combination of the H_s, H_L must vanish, we obtain immediately the following expressions for half of the coefficients:

$$\left.\begin{aligned} \gamma_k &= [H_k, A], \\ \gamma_0 &= l_0 [(H_L - l^s H_s), A]. \end{aligned}\right\} \tag{29.8}$$

Whereas the first three of these coefficients are already in final form, and independent of the intrinsic coordinates f^ϱ that we may have adopted, the last coefficient, γ_0, depends, in addition to the canonical variables g_{mn}, π^{mn}, on the metric components $g_{0\mu}$. The latter, however, must also be expressible in terms of the canonical variables, in view of the fact that we have adopted a uniquely determined coordinate system, in which no part of the metric can remain undetermined. As a consequence of the coordinate conditions (29.5) we must have:

$$f^\mu{}_{,\varrho} (\delta^\varrho_s \xi^s + l^\varrho \xi^L) \equiv \delta^\mu_s \xi^s + l^\mu \xi^L = [f^\mu, \int (\xi^s H_s + \xi^L H_L)\, d^3x]. \tag{29.9}$$

Hence the last four components of the metric are determined from the conditions:

$$\left.\begin{aligned} \sqrt{g^{00}}\, \delta(x, x') &= [f^0, H_L'], \\ g^{0s}\, \delta(x, x') &= \sqrt{g^{00}}\, [f^s, H_L']. \end{aligned}\right\} \tag{29.10}$$

Because all four f's are by assumption scalars, the Poisson brackets on the right will contain undifferentiated delta functions only; hence these equations will not give rise to any inconsistencies.

We can now proceed to determine the remaining four coefficients in the expression for A^*. We require that the Poisson bracket of A^* with every one of the coordinate constraints vanish. We find then:

$$\left.\begin{aligned} \varepsilon^L &= [[J, l_0 (f^0 - x^0)], A], \\ \varepsilon^s &= [[J, (\delta^s_\varrho - l^s l_\varrho) (f^\varrho - x^\varrho)], A], \\ J &\equiv \int [l_0 (f^0 - x^0) H_L + (\delta^s_\varrho - l_\varrho l^s) (f^\varrho - x^\varrho) H_s]\, d^3x. \end{aligned}\right\} \tag{29.11}$$

J is quadratic in the constraints; its Poisson bracket with an expression linear in the constraints will itself be linear in the constraints. Finally, the Poisson bracket of that latter expression with A will not vanish at all.

Herewith the construction of A^* is completed. Given two functionals of the g_{mn}, π^{mn}, say A and B, it can be shown that

$$\begin{aligned} [A^*, B^*] &\equiv [A^*, B] \equiv [A, B^*] \\ &\equiv [A, B] + [[A, J], [B, J]]. \end{aligned} \tag{29.12}$$

This is because the Poisson bracket may be interpreted as the transformation law of one of the two factors under an infinitesimal canonical transformation generated by the other. Inasmuch as $A^* = A$, $B^* = B$, the transformation laws of the unstarred and the starred quantities under a transformation that preserves the choice of coordinate system must be the same.

Bibliography.

This list consists of books and of major sources of reference that might serve as starting points for a survey of both the special and the general theories of relativity.

Books.

[1] BERGMANN, P.: Introduction to the Theory of Relativity. New York: Prentice-Hall 1942.
[2] FOCK, V. A.: Theory of Space, Time, and Gravitation. London and New York: Pergamon Press 1960.
[3] INFELD, L., and J. PLEBANSKI: Motion and Relativity. Warszawa: Państwowe Wydawnictwo Naukowe, and London and New York: Pergamon Press 1960.
[4] JORDAN, P.: Schwerkraft und Weltall, 2. Aufl. Braunschweig: Vieweg 1955.
[5] LAUE, M. v.: Die Relativitätstheorie. 2 Bde. Braunschweig: Vieweg 1921.
[6] LICHNEROWICZ, A.: Théories Relativistes de la gravitation et de l'électromagnétisme. Paris: Masson & Cie. 1955.
[7] LORENTZ, H. A., A. EINSTEIN, H. MINKOWSKI, H. WEYL, with notes by A. SOMMERFELD: Das Relativitätsprinzip, 4. Aufl. Leipzig: Teubner 1922. English translation: The Principle of Relativity. New York: Dover 1951.
[8] MØLLER, C.: The Theory of Relativity. London: Oxford Univ. Press 1952.
[9] PAULI, W.: Theory of Relativity. London and New York: Pergamon Press 1958. Translation (with recent supplementary notes by the author) from "Relativitätstheorie". Enzyklopädie der mathematischen Wissenschaften. Leipzig: Teubner 1921.
[10] RZEWUSKI, J.: Field Theory, Part I, Classical Theory. Warszawa: Państwowe Wydawnictwo Naukowe 1958.
[11] SCHILPP, P. A. (Ed.): ALBERT EINSTEIN, Philosopher-Scientist. Evanston: Library of Living Philosophers 1949.
[12] SYNGE, J. L.: Relativity; The Special Theory (1956); The General Theory (1960). Amsterdam: North Holland Publishing Co., and New York: Interscience Publ. 1956 and 1960.
[13] TOLMAN, R. C.: Relativity, Thermodynamics, and Cosmology. London: Oxford University Press 1934.
[14] WHITTAKER, E.: History of the Theories of Aether and Electricity. New York: Philosophical Library 1951.
[15] WITTEN, L.: Handbook of Relativity. New York: Wiley (in process of publication).

Others.

[16] Reviews of Modern Physics **21**, No. 3 (1949). Collection of papers in honor of 60th birthday of A. EINSTEIN.
[17] Helvetica Physica Acta Suppl. **4** (1956). Proceedings of first international conference on relativity („Fünfzig Jahre Relativitätstheorie") at Bern, 1955.
[18] Reviews of Modern Physics **29**, No. 3 (1957). Collection of papers from the conference on relativity at Chapel Hill, 1957.
[19] Proceedings of the Colloque International at Royaumont, 1959. (In process of publication.)
[20] Collection of papers in honor of L. INFELD. (In process of publication.)
[21] JORDAN, P., J. EHLERS u. W. KUNDT: Akad. Wiss. Lit. Mainz, math.-nat. Kl. **1960**, No. 2. First of a series of articles on closed-form solutions, aiming at a comprehensive review of the whole field.
[22] TRAUTMAN, A.: Lectures on General Relativity. King's College, London, 1958.
[23] Centro Internazionale Matematico Estivo, 1° Ciclo, Sestriere, 1958. Vedute e problemi attuali in relatività generale. Collection of lectures on relativity.

Sachverzeichnis.

(Deutsch-Englisch.)

Bei gleicher Schreibweise in beiden Sprachen sind die Stichwörter nur einmal aufgeführt.

Subject Index.

(English-German.)

Where English and German spelling of a word is identical the German version is omitted.